The Construction of Cognitive Maps

The GeoJournal Library

Volume 32

The titles published in this series are listed at the end of this volume.

The Construction of Cognitive Maps

edited by

JUVAL PORTUGALI

Department of Geography,
Tel Aviv University,
Israel

KLUWER ACADEMIC PUBLISHERS
DORDRECHT / BOSTON / LONDON

A C.I.P. Catalogue record for this book is available from the Library of Congress.

ISBN 0-7923-3949-5

Published by Kluwer Academic Publishers,
P.O. Box 17, 3300 AA Dordrecht, The Netherlands.

Kluwer Academic Publishers incorporates
the publishing programmes of
D. Reidel, Martinus Nijhoff, Dr W. Junk and MTP Press.

Sold and distributed in the U.S.A. and Canada
by Kluwer Academic Publishers,
101 Philip Drive, Norwell, MA 02061, U.S.A.

In all other countries, sold and distributed
by Kluwer Academic Publishers Group,
P.O. Box 322, 3300 AH Dordrecht, The Netherlands.

Printed on acid-free paper

Printed in the Netherlands

For Bili, Mamushka and Badi

CONTENTS

Part Two: Transformations

Part Three: Specific Themes

CONTRIBUTORS

Mark Blades, Department of Psychology, University of Sheffield.
James M. Blaut, Department of Geography, University of Illinois, Chicago.
Rex Cammack, Department of Geography, University of South Carolina.
Luc Carité, Groupe Cognition Humaine, LIMSI-CNRS Université de Paris-Sud, Orsay.
Helen Couclelis, Department of Geography, University of California Santa Barbara.
Thea Ghiselli-Crippa, Department of Information Science, University of Pittsburgh.
Marie-Paule Daniel, Groupe Cognition Humaine, LIMSI-CNRS Université de Paris-Sud, Orsay.
Michel Denis, Groupe Cognition Humaine, LIMSI-CNRS Université de Paris-Sud, Orsay.
Nancy Franklin, Department of Psychology, State University of New York, Stony Brook.
Reginald G. Golledge, Department of Geography and Research Unit in Spatial Cognition and Choice, University of California Santa Barbara.
Sucharita Gopal, Department of Geography, Boston University.
Hermann Haken, Institute for Theoretical Physics and Synergetics, Stuttgart University.
Harry Heft, Department of Psychology, Denison University.
Stephen C. Hirtle, Department of Information Science, University of Pittsburgh.
Roberta L. Klatzky, Department of Psychology, Carnegie-Mellon University, Pittsburgh.
Robert Lloyd, Department of Geography, University of South Carolina.
Jack M. Loomis, Department of Psychology, University of California Santa Barbara.
Paul Munro, Department of Information Science, University of Pittsburgh.
Juval Portugali, Department of Geography, Tel Aviv University.
M. Jeanne Sholl, Department of Psychology, Boston College, Chestnut Hill.
Christopher Spencer, Department of Psychology, University of Sheffield.
David Stea, Universidad Internacional de Mexico, Centro Internacional para la Cultura y el Ambiente and Mount Holyoke College.
Jennifer Stephens, Department of Geography, University of Illinois, Chicago.
Simon Ungar, Department of Psychology, University of Sheffield.

ACKNOWLEDGEMENTS

I am very grateful to the contributors to this book for the chapters they have produced as well as for their good advice and stimulating suggestions. I am also grateful to Hedva Erlich for text-editing some of the chapters, to Izhak Omer for helping me in preparing the index, and in particular to Ester Keret for her devoted work in preparing the text for publication.

THE CONSTRUCTION OF COGNITIVE MAPS: AN INTRODUCTION

Juval Portugali

> The aim of this book is to shed light on processes associated with the construction of cognitive maps, that is to say, with the construction of internal representations of very large spatial entities such as towns, cities, neighborhoods, landscapes, metropolitan areas, environments and the like. Because of their size, such entities can never be seen in their entirety, and consequently one constructs their internal representation by means of visual, as well as non-visual, modes of sensation and information: text; auditory, haptic and olfactory means for example, or by inference. Intersensory coordination and information-transfer thus play a crucial role in the construction of cognitive maps.
>
> Because it involves a multiplicity of sensational and informational modes, the issue of cognitive maps does not fall into any single traditional cognitive field, but rather into, and often in between, several of them. Thus, although one is not dealing here with some remote issue, but with processes associated with almost every aspect of our daily life, the subject has received relatively marginal scientific attention. The relative lack of scientific discourse in this domain, on the one hand, and its importance, on the other, gave rise to a growing demand for research on this subject. This state of affairs was the stimulus for the recent *Geoforum* theme issue on *Geography Environment and Cognition*, where I tried, as editor, to direct attention to the general process of environmental and geographical cognition (Portugali, 1992). In the present book my aim is to be more specific and focus directly on the cognitive processes by which one form of information, say haptic, is being transformed into another, say a visual image, and by which multiple forms of information participate in constructing cognitive maps.

The above quotation is taken from the abstract attached to an invitation sent to a group of students of cognitive mapping to participate in the writing of a new book on *The Construction of Cognitive Maps*. The invitation entailed a long process, the end products of which are the 14 chapters composing this book. The chapters grouped in Part One present four *Theoretical Frameworks*, those in Part Two outline three forms of inter- and intra-modal *Cognitive Transformations*, and the ones in Part Three discuss two *Specific Themes*.

J. Portugali (ed.), The Construction of Cognitive Maps, 1–7.

Part One opens with a discussion on IRN (*Inter-Representation Networks*). This new notion, introduced for the first time in this book, proposes to view the cognitive system as a whole, and the one associated with the construction of cognitive map – as a network composed of internal elements and representations constructed in the brain, and external elements and representations constructed in the environment. The system as a whole is a synergistic self-organizing system, the dynamics of which is an on-going interaction between its internal and external elements. The notion of IRN is discussed, first, in a chapter by Portugali, who introduces its basic principles and elaborates its various properties and potentialities in relation to the discourse in cognitive science, and second, in a chapter by Haken and Portugali, who show how the approach of IRN can be cast into the mathematical formalism of Haken's (1987) theory of synergetics, and as such be operationally applied to various aspects of cognition and cognitive mapping.

The theoretical framework of IRN attempts, in fact, to integrate the three theoretical approaches which are currently prevalent in the study of cognitive mapping: *the information processing approach* with its metaphoric computational mind, Gibson's (1979) *ecological approach* with its notions of *direct perception* and *affordance*, and *experiential realism* as elaborated by George Lakoff (1987) in the domain of language, and by Gerald Edelman (1992) in the domain of neurobiology. All three theoretical perspectives are represented in Part One.

The information processing approach, which is currently also the "classical" main-stream approach, perceives the mind and the brain as information processing devices and thus considers cognition as an essentially internal process, taking place in people's brains, very much like the processes which occur in computers' "brains". This view is represented in Part One by two chapters on the applicability of *neural networks* and *(neo)connectionism* to cognitive mapping – two notions which can be regarded as the latest statement of the view that the computer is a good metaphor for the study of cognitive processes in the mind/brain, and the mind/brain – a metaphor and a source of inspiration for the development of computers and models of AI. The first chapter, by Sucharita Gopal, focuses mainly on the fundamental level of cognition, which examines the very basic mechanisms of cognition as they appear in living organisms, including humans. In particular, O'Keefe and Nadel's (1978) *Hippocampus as a cognitive map* is considered here as a port of departure. Gopal introduces the principles of neural networks and their basic terminology and relates them to the various properties of cognitive maps and their construction. This is done by presenting and discussing five models which have been presented in the literature and the principles of a sixth model currently developed by the author. The second chapter, by Thea Ghiselli-Crippa, Stephen C. Hirtle and Paul Munro, complements Gopal's chapter by examining the usefulness of connectionist models to "higher level domains" of spatial cognition, that is to say, to representations and processes which are exclusive to humans in their encoding, storing, decoding and retrieving spatial knowledge for various tasks. The authors present and discuss connectionist models of cognitive maps which are based on local representation, versus models which are based on distributed representation, as well as connectionist models concerning language and spatial relations.

As is well known, Gibson's (1979) ecological approach suggests a view on cognition which is diametrically different from the classical main stream view: perception (and thus cognition) is direct, immediate and needs no internal information processing, and is thus essentially an external process of interaction between an organism and its external environment. The chapter by Harry Heft introduces J.J. Gibson's ecological approach and its implication to the construction of cognitive maps in general and to the issue of wayfinding in particular. According to Heft, main stream cognitive sciences are essentially Cartesian in nature and have not as yet internalized the implications of Darwin's theory of evolution. Gibson, in his ecological approach, has tried to do exactly this. The author introduces the basic terminology of the ecological approach and relates its various notions, in particular *optic flow*, *nested hierarchy* and *affordances*, to navigation and the way routes and places in the environment are learned. An interesting view, which is implied by this theoretical approach, is that one's cognitive map of an environment is not seen as a constructed bird-eye representation of it (or its configurational survey knowledge), but as one's orientation to the environment by "being everywhere at once" (Gibson 1979, 198).

Experiential realism is a theoretical perspective elaborated independently by George Lakoff (ibid) in the domain of language, and by Gerald Edelman (ibid) in the domain of neurobiology. This new approach, while accepting the view that cognitive processes involve information processing, suggests that the models and patterns for this internal information processing are derived from the very basic experiential relations of organisms in the environment. This view is presented in Part One in a chapter by Helen Couclelis. Her approach combines Lakoff's experiential realism with Johnson-Laird's (1983) notion of *mental models*. According to this view, linguistic and visual representations in the mind (e.g. imagery) are seen as two representations of a deeper cognitive level in which they are integrated. (This suggestion is similar to Langacker's notion of "cognitive grammar" and to Edelman's view on cognition – see further discussion and bibliography in the following chapter by Portugali.)

As noted above, cognitive maps refer to very large spatial entities such as landscapes, neighborhoods, cities and the like. Unlike small objects, such large entities usually cannot be seen and perceived in their entirety. Consequently, their construction as cognitive maps involves inter- and intra-modal transformations between various sensory modalities, as well as within single sensory modes. Such forms of transformation, which are central to the construction of cognitive maps, are discussed in Part Two. The first form is *from visual information to cognitive maps.* Vision is probably the most natural and immediate source of information for the construction of cognitive maps and thus a case of intra-modal transformation. One can consider here two forms of vision: *primary* or *direct visual information* as acquired, for example, through navigation in an environment, and *secondary*, or *indirect visual information* gained from maps, for example. The first chapter, by Jeane Sholl, considers the role of *direct visual information,* in particular the partial visual information acquired by navigating humans, in constructing various forms of cognitive maps. For this purpose she first surveys the literature on animals' cognitive mapping and considers the implications to humans, then turns to Gibson's ecology to learn about the optical flow and the kind of information afforded by the environment to navigating individuals. On the basis of these sources and the literature on cognitive mapping by visually impaired people, and on children and the developmental aspects of navigation, she puts forward a set of hypotheses, the essence of which is that the construction of cognitive maps develops in concert with the development of the act of walking and the consequent ability to differentiate perspective and invariant structure in visual flow.

The chapter that follows, by Robert Lloyd and Rex Cammack, presents a research concerning the relation of various forms of indirect visual information to the construction of cognitive maps. They report on five experiments which examine how maps, two- and three-dimensional drawings of an environment, in various orientations, affect subjects' cognitive maps of these areas and their responses concerning the spatial relations presented in these figures. In particular they investigate the circumstances under which various properties of the secondary knowledge create orientation-bias and orientation-free cognitive maps.

The second form of transformations discussed in Part Two concerns the construction of *cognitive maps by visually impaired people* – people who lack the most powerful and direct information source for spatial cognition and behavior. This research domain exposes the issue of cognitive transformations in all its complexity and it is also a domain where understanding of such transformations is of an immediate relevance and importance. Two chapters are devoted to this issue. The first, by Reginald G. Golledge, Roberta L. Klatzky and Jack M. Loomis, focuses mainly on adults without vision, with particular attention to the processes of cognitive mapping and the use of cognitive maps in the context of navigation. The authors introduce the general characteristics of cognitive mapping and wayfinding, and compare them to the specific problematics of wayfinding to people without vision. This provides the ground for a subsequent discussion on the skills, abilities and provisions required to successfully complete such tasks by blind people. This is illustrated by a survey of the literature and interpretations of case studies and experimental data.

The following chapter, by Simon Ungar, Mark Blades and Christopher P. Spencer, concerns "the construction of cognitive maps by children with visual impairments". This discussion complements Golledge et al.'s chapter in three respects. First, it examines both small-scale and large-scale spatial cognition (i.e. cognitive maps). Second, it concentrates mainly on children, and third, and as a consequence of the above, it deals also with developmental aspects of spatial cognition of people with visual impairments. The authors review and discuss the theoretical background and literature on the above issues, as well as some of their own experimental results. They argue that the finding that children with visual impairments perform less well on a variety of spatial tasks results not from lack of direct visual information, but from its effects on coding strategies. The implication is that in order to improve their learning and spatial representation of the environment, children with visual impairments should be encouraged, from early age, to adopt appropriate strategies, including external reference systems, electronic aids and tactile maps.

It would be difficult to exaggerate the importance of language in cognition in general and in cognitive mapping in particular. Language is one of the cognitive capacities which differentiate humans from non-humans. It is intimately related to thinking and to the way we categorize the world around us, including the visual information we perceive and process. This is specifically so with respect to cognitive maps – the large spatial entities which we cannot see in their entirety. In the absence of complete visual information, most of the complementary spatial and non-spatial information needed for the construction of cognitive maps is supplemented by language. Two chapters consider transformations *from language to cognitive maps*. The first, by Nancy Franklin, gives – as the title suggests – an overview on "language as means to construct cognitive maps". Starting with *spatial mental models* – the spatial representations constructed through language – she describes cognitive maps as large-scale mental models which enable the efficient integration of the spatial properties of linguistic information. She then discuss the spatial characteristics of cognitive maps as mental models (distance, direction, perspective), how they enable to integrate perceptual and linguistic information, the similarities and differences between representations acquired from language and direct and indirect vision, the effect of time and the implications to mental models in relation to long term memory, situations where spatial linguistic descriptions do not lead to mental models as cognitive maps, and what verbal or textual descriptions teach us about the way people conceptualize space.

The second chapter on language, by Marie-Paule Daniel, Lue Carite and Michel Denis, considers a more specialized issue: unconstrained "modes of linearization in the description of spatial configurations". That is to say, the procedures people use to describe the two- and three-dimensional configurations of an environment, by means of the essentially one-dimensional verbal or textual description, when they are not constrained by a specific starting point. After placing their specific issue in the context of studies on linearization in the description of small- and large-scale spaces, the authors report on their own research. In the latter, subjects were presented with a map of a fictitious island and asked to produce written descriptions which will enable an addressee to construct a spatial mental representation of it. Analyzing and interpreting the results of their research the authors illustrate various strategies by which subjects were able to solve the linearization problem and thus show one of the ways by which linguistic and visual cognition can interact efficiently.

Two specific themes regarding inter- and intra-modal cognitive transformations in the construction of cognitive maps are discussed in Part Three. One concerns the spatial knowledge and reasoning people employ for this purpose, and the other concerns the universality of the human capacity to construct external as well as internal maps. The theme of spatial reasoning is taken up in a chapter by Daniel R. Montello and Andrew U. Frank. Reviewing the literature on spatial reasoning and the more specialized topic of qualitative metrics, they indicate that modelers' assumption in these domains, that their models are in line with human behavior, have no empirical support. The aim of the chapter is thus to empirically evaluate existing qualitative metric models of human directional knowledge and to make a first attempt toward their improvement. This is implemented by a two-stage experiment. In the first they compare simulations of existing metric models to empirical data. They then improve the models in light of the results, and in stage two simulate them once again against the empirical data. By these two sets of experiments they confirm the assumption that human spatial knowledge is metric (though in a rather imprecise way), they expose the empirical weakness of current metric models and suggest an improvement. The latter is close in principle to the notion of spatial framework, that is to say, it is a Cartesian framework which functions as an organizer of egocentric spatial knowledge. In concluding their chapter the authors discuss the difference between linguistic and nonlinguistic models of spatial reasoning as well as directions for future research.

Is the capacity to transform vision, language, experience in the environment and the like, into internally represented cognitive maps, and the latter into externally represented "ordinary" maps, a characteristic specific to certain age-groups and cultures, or is it a capacity of all age-groups and all cultures? This question, which forms the second theme of Part Three, is discussed in a chapter by David Stea, Jim Blaut and Jennifer Stephens. The authors first pose the above dilemma, then survey the empirical data as it appears in several bodies of literature, ranging from developmental and learning psychology to anthropology and archaeology. Based on this empirical data the authors challenge Piagetian developmental perspective in two important respects. First, they show that mapping activities appear in young children much earlier than predicted by Piaget's theory of stages, and second, they refute Piaget's contention that "in many societies adult thought does not ... reach that [level] of propositional operations which develop between the ages of twelve and fifteen in our milieus (Piaget 1971, 61). The alternative view proposed by the authors accepts the basic concept of stages, but adds that structural development and environmental experience as well as learning interact, with the consequence that the ability to internally and externally map "is indeed a cultural universal".

References

Edelman, G.M. (1992). *Bright Air Briliant Fire: On the Matter of the Mind*, London: Penguin Books.

Gibson, J.J. (1979). *The Ecological Approach to Visual Perception*, Boston, MA: Houghton-Mifflin.

Haken, H. (1987). *Advanced Synergetics. An Introduction*. 2nd. print., Berlin, New York: Springer.

Johnson-Laird, P. N. (1983). *Mental Models: Towards a Cognitive Science of Language, Inference, and Consciousness*, Cambridge, MA: Harvard University Press.

Lakoff, G. (1987). *Women Fire and Dangerous Things: What categories reveal about the mind*, The Chicago, London: University of Chicago Press.

O'Keefe, J., and Nadel, L. (1978). *The Hippocampus as a Cognitive Map*. Oxford: Clarendon Press.

Piaget, J. (1971). *Psychology and Epistemology*. New York: Grossman Publishers.

Portugali, J. ed.(1992). *Geography, Environment and Cognition* – a special theme issue *Geoforum* 23, 2.

Juval Portugali
Department of Geography
Tel Aviv University
Tel Aviv 69978, Israel

Part One: Theoretical Frameworks

INTER-REPRESENTATION NETWORKS AND COGNITIVE MAPS

Juval Portugali

Abstract: The notion of Inter-Representation Networks (IRN) suggests that the cognitive system in general, and the one associated with cognitive maps in particular, extend beyond the individual's mind/brain into the external environment. Accordingly, the cognitive system is perceived as composed of elements in the mind/brain, internally representing the external environment, and elements in the environment, externally representing the mind. The dynamics of cognitive processes and the construction of cognitive maps is interpreted as a complex interaction between these internal and external representations. While this view somewhat departs from main-stream cognitive science, it was always present in its discourse in the writings of scholars such as Vygotsky, Gibson, Bartlett, and more recently Rumelhart Smolansky and Hinton, Alexander, Lakoff, and Edelman. The main body of the paper discusses the IRN element in the writings of these authors and by doing so develops and elaborates the various facets and potentialities of IRN and their role in the construction of cognitive maps.

Introduction

Early studies on cognitive maps, or mental maps studies as they were called (Golledge, 1993), were characteristically based on behaviorism – an approach whose main concern was the relations between stimuli and responses (S-R) as they can be observed in the external environment. The mind/brain itself was considered a black-box, the internal processes of which cannot be directly observed, and therefore cannot be a subject for a genuine scientific enquiry (Figure 1a).

More recently, following main stream or "classical cognitive science", studies on cognitive maps turned their attention to this black-box (B-B), in search of the internal processes of the mind, that is to say, to the way the mind/brain encodes, stores and decodes information from the environment (Gärling and Golledge, 1993; Portugali, 1992). S-R relations have now become means to reveal what's going on inside the head. This is presented graphically in Figure 1b. Comparing figures 1a and 1b one can see that beyond the differences just mentioned, the two approaches share a common property: in both mind and environment are perceived as two essentially independent, and causally related, entities. In behaviorism the relevant system of interest lies outside the B-B, in the

J. Portugali (ed.), The Construction of Cognitive Maps, 11–43.

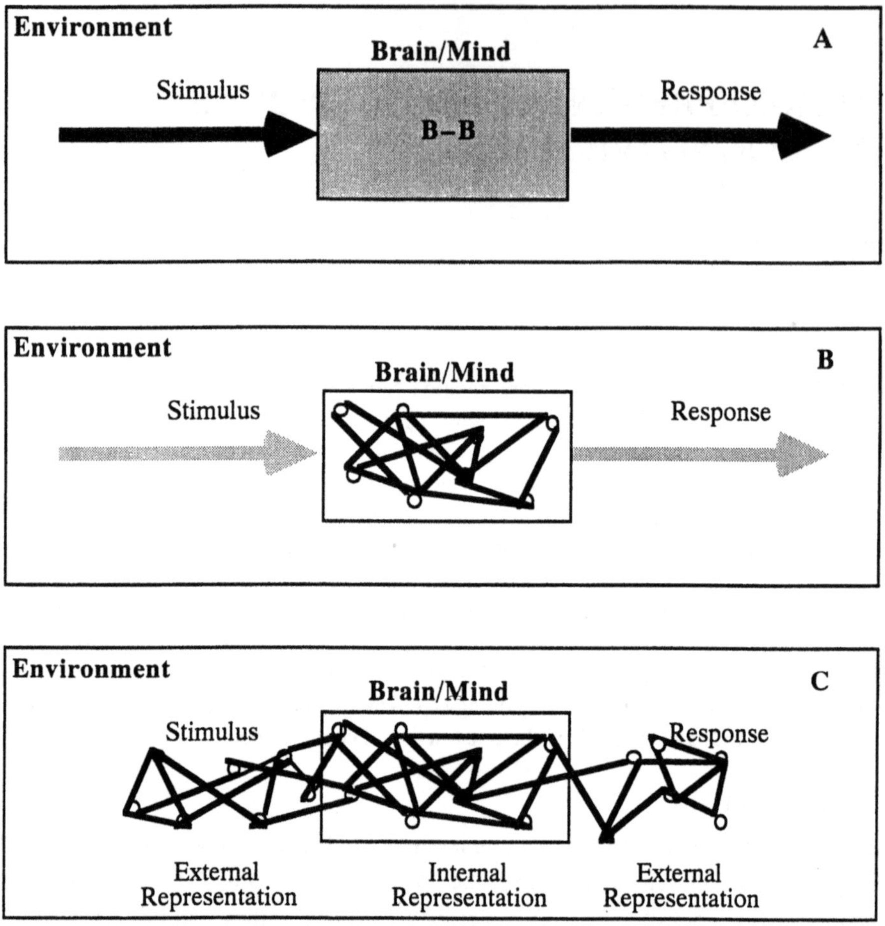

Figure 1: Three approaches to cognition and cognitive maps: (a) behaviorism, (b) the classical view, (c) the IRN approach.

environment; in classical cognitive science – inside the B-B; yet each of the two entities has its own independent existence.

The suggestion in this paper is to perceive the relations between mind and environment as in Figure 1c: indeed, as can be seen, much of the relevant network of interest lies inside the brain, and its elements thus form what is often termed "internal representations" and internal cognitive processes. However, some of the elements of the network lie outside the brain, in the environment, and thus form the content of what might be termed "external representations and processes". The internal part of the network corresponds to cognition as defined by classical cognitive science, that is, to processes by which the external environment, or elements of it, are encoded, stored and retrieved by the brain The external part of the network refers to the way in which the "internal environment" is externalized, stored, represented and retrieved in the outside environment. My suggestion is that the cognitive network as a whole is composed of internal and external elements and can thus be termed *Inter-Representation Network* (IRN).

This perception on cognition and cognitive maps follows directly from two previous studies (Portugali, 1990; Portugali and Haken, 1992) where an attempt was made to outline some general theoretical principles for the study of cognitive maps. The two papers suggested, that cognition in general and cognitive maps in particular should be considered as self-organizing systems, and that Bohm's notion of holomovement and Haken's notion of synergetics provide insight into their dynamics. These theoretical principles provide also the port of departure for the present study and they are summarized in the second section of the paper.

The view on cognition and cognitive maps as presented in Figure 1c and the notion of IRN are not in line with classical cognitive theory which, as shown in figure 1b, tends to concentrate on internal processes and representations (Gardner, 1984). Though this is the classical main-stream, there have been and still are exceptions. In the third and main section of this paper I present and discuss some of these exceptions. My aim here is not to provide an exhaustive survey, but to elaborate the theoretical context to the notion of IRN and its various facets by means of the literature surveyed, and relate them to the discourse on cognitive maps and their construction. The concluding section summarizes the discussion in section three and relates it to our theoretical port of departure as described in section two.

Implicate order, synergetics and cognitive maps

The notion of *implicate order* stands at the heart of David Bohm's (1980) theory and cosmology. It implies a new notion of order in which everything – an entity or a configuration of entities – is enfolded into everything else (Bohm, 1980, 177). As an analogy to this new form of order Bohm suggested the holographic record: In the latter you can observe a picture composed of various entities, and yet every point and entity in that picture enfolds all other entities and the whole configuration. Consequently, if you cut the picture into two, you'll see in each half the whole picture once again. In the holographic record we thus have two forms of order: the implicate order, which describes the above subtle property of enfoldment, and the order among the various entities that we see in the picture. The latter is termed by Bohm *explicate order* – it looks as if every entity in the picture is fully independent, but we know, and can prove by cutting the picture, that this explicate order is in fact a thought abstraction; the more subtle order is the implicate order.

Bohm's notions of order were related to the study of the mind by himself (Bohm, 1980), by others (Hiley and Peat, 1987), and in particular by Pribram's (1990) holographic or holonomic approach to brain structure and function. Pribram's assertion is that while the appropriateness of the holographic model of the brain is still problematic, it was nevertheless "perceived by many scientists as a starting point to what has become the 'connectionist' parallel distribution processing approach" (ibid, 166).

From the perspective of Bohm's orders, mind and environment are seen as two entities existing one inside the other or enfolding each other in implicate relations. In terms of cognitive maps this implies (i) that the environment is enfolded in the mind in the form of internal representation; (ii) that the minds of individuals are enfolded in the environment in the form of a multiplicity of external representations, and (iii) that mind and environment are only relatively independent and thus form a single interactive network which has implicate and explicate properties.

Synergetics is the name given by Haken (1983, 1987) to his theory of self-organization. Originating in physics, in the domains of laser and pattern formation in liquid, it was extended by Haken (1979, 1990, 1991) to the domain of pattern recognition and cognition in general. The formalism of synergetics of cognition and its relation to cognitive mapping are given below (Haken and Portugali, this volume). Here, on a more intuitive level it can be said, that synergetics is a theory of open, complex and thus self-organizing systems. It suggests that during their steady state such systems are governed by one or a few *order parameters*. The latter might refer to the macroscopic structure of the system or to the microscopic behavior of its parts. The theory which focuses on the way in which an order parameter comes to govern the system can be described by the following scenario: assume that as a consequence of some external or internal disturbance a previously stable system enters a state of instability. The latter is characterized by the co-existence of several competing order configurations or order parameters. This "competition" ends when a certain order parameter "wins" by *enslaving* the various elements of the system to its rhythm, structure or behavior. In a typical synergetics application to the cognitive process of pattern recognition, the system is given a few features of a certain pattern (i.e. face) out of a repertoire of patterns which are stored in the brain/computer as internal representations. This triggers a self-organization process, i.e. a competition among several order states until a certain order parameter "wins", enslaves the various features by means of associative memory, and a recognition is established. A similar process takes place in the construction of cognitive maps (Portugali, 1990): the cognitive system constructs a whole pattern/map out of a partial set of features, and this is achieved when a certain mapping principle, or mapping order parameter, enslaves the various features. There are, however, important differences between the two processes. Compared to ordinary pattern recognition, cognitive maps formation concerns very large patterns (e.g., cities), and this entails several qualitative implications discussed in Portugali (1990), Portugali and Haken (1992), and in Haken and Portugali (this volume).

The Holomovie metaphor was suggested to illustrate the possibility of integrating Bohm's theory of order with Haken's synergetics so that it may serve as a general framework for the study of society, environment and cognitive maps (Portugali, 1993). The metaphor starts with Bohm's analogy of the holographic record, in which each point enfolds all other points and the entire space. The holographic record can thus be described as a *spatial whole.* Another analogy is music (Bohm and Hilly, 1993), in which the various notes come in a sequential order, while for the listener, each sequential note enfolds all previous ones (as well as future notes in the form of expectations). This can be termed a *temporal whole.* Combining the properties of the two we arrive at an imaginary device which can be termed *holomovie.* An ordinary movie is made of still images; a holomovie is made of holographic images. In such an imaginary device every feature of an image enfolds not only its environment, as in a holographic record, but also its entire history and planned future.

Given this holomovie, the question is how can forms of relative space-time stability arise and be maintained? The answer, according to Bohm, lies in the notion of the *generative order*: Every order, implicate or explicate, is also generative, since it generates other orders. However, Bohm did not specify how is a generative order created in the first place. This, I submit, can be learned from Haken's synergetics: Imagine a holomovie as above; assume it enters a state of instability. At this stage several order configurations are in motion until a given configuration predominates the movement. The amplitude of this configuration is an *order parameter* in Haken's terminology, and it is a special case of what Bohm terms generative order. Once the motion of the order parameter is established, it enslaves the movie and generates a new reality and "a new movie".

The suggestion is (Portugali, 1990, 1993) to see social and cognitive processes in terms of the holomovie metaphor, that is to say, as events in space-time, which enfold their past and future, and their explicate space-time boundaries, shape and structure are generated by a specific generative order, or sets of order parameters, created in a process of self-organization. As open, complex and self-organizing systems both the cognitive system of the individual and the socio-spatial environment are seen as being always in movement, characterized by relatively long periods of structural stability, during which the evolution of the system is governed by one or a few order parameters, followed by short chaotic periods which can generate bifurcation, phase transition and structural change.

While the general structure of the above individual and social systems is similar, their motion and evolutionary tempo are different: the evolution of a social or an environmental system is much slower than that of the individual's cognitive system – the individual is usually born into, or comes to, an environment which is already self-organized and enslaved by a complex of interrelated order parameters. And thus, when the individual adapts to the environment, he or she internally represents, or enfolds, the order parameters of that environment. By acting in line with that order parameter, the individual is on the one hand enslaved by the environmental order parameter, while on the other participates in perpetuating it. From that perspective the environment externally represents, or enfolds, the action of the individual. The individual and the environment thus co-exist in *implicate relations.*

In terms of the holomovie, cognitive maps are seen as multi-stable/implicate patterns. Their various representational forms are not stored in any static way, neither with respect to geographical areas, nor with respect to modes of representation. They are dynamically created anew, each time, as ad hoc entities: The brain is capable of creating a multiplicity of cognitive maps with specific perspectives, scales and modes, by means of learned synaptic connection strengths that govern the cooperation between the neurons. These synaptic cooperations can be seen as order parameters generated by the continuous interplay between the person's internally represented environment and the physically and socially constructed external environment, through a process of reproductive recognition and learning. This is illustrated graphically in figure 2a.

Given a cognitive map, each new piece of internal or external information regarding that specific area stimulates the cognitive system and creates a local, small-scale perturbation, which if enslaved by the pre-existing order parameter, perpetuates and reproduces the prevailing order parameter. This has been termed *reproductive cognition* (or recognition). If, however, the new piece of information cannot be enslaved by the pre-existing order parameter, the result might be a cognitive dissonance, bifurcation, phase transition and a new order parameter which actually means a structurally new cognitive map. This has been termed *bifurcative cognition.* It has also been demonstrated, that with time reproductive cognition often leads to *systemic rigidity*, which implies a gradually increasing disparity between a reproduced internally represented cognitive map (or environmental behavior) and an externally changing environment. As can be seen in Figure 2b, systemic rigidity might lead to a bifurcation and phase transition and thus to a new order parameter and a new cognitive map. It might also lead to what in social theory is called "ideological false consciousness". Such a situation is closely related to the various cases of systematic distortions in cognitive mapping.

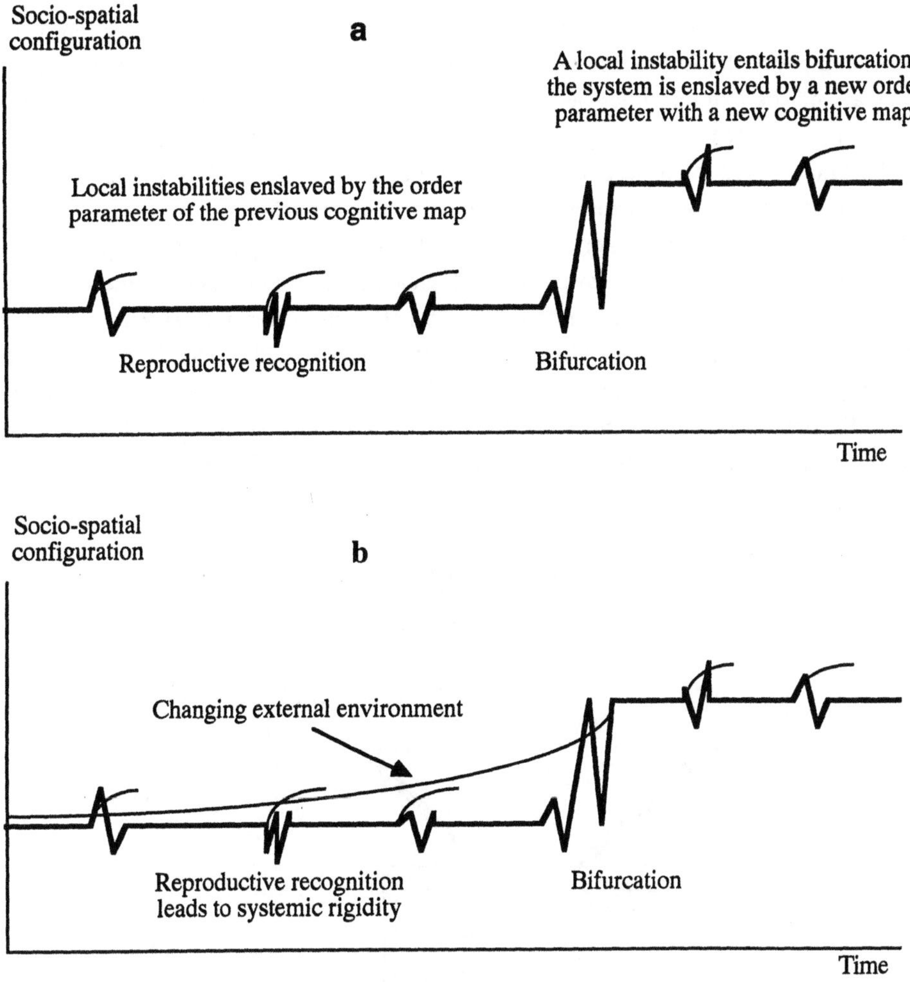

Figure 2: a) Reproductive and bifurcative cognition in cognitive mapping. b) The evolution of systemic rigidity. (Adapted from Portugali 1993.)

IRN: Theoretical Context and Properties

The notion of IRN implies that the external environment may be an integrative element in the process of cognition. This view, which goes counter to the classical view on cognition, is not new and has accompanied discourse in cognitive science from the start. In the following I shall discuss some of this discourse and through it the notion of IRN, define its scientific context and elaborate its various facets and its connection to cognitive maps.

Vygotsky

The project of the Russian and Marxist psychologist Vygotsky was an attempt to construct a psychological theory which (a) places equal emphasis on the internal intra-personal and the external inter-personal (i.e. social, cultural) dimension of a person's psychology and (b) put strong emphasis on change, evolution and development: psychological phenomena are seen as changing, developing and evolving out of a dialectical interplay between external and internal psychological elements.

Vygotsky has developed a series of interrelated conceptual pairs for this purpose: a distinction between *higher* and *lower* psychological processes; between *natural memory,* which is biological, personal and internal, and what might be termed *artificial memory*, which is inter-personal, social, cultural and external, and as such "extends the operation of memory beyond the biological dimensions of the human nervous system" (Vygotsky, 1978, 39); between *direct* stimulus–response (S–R) relations, as in Fig 3a, to *mediated* S–R relations, as in Figure 3b, where the mediator is an external object, a tool or a sign, for example.

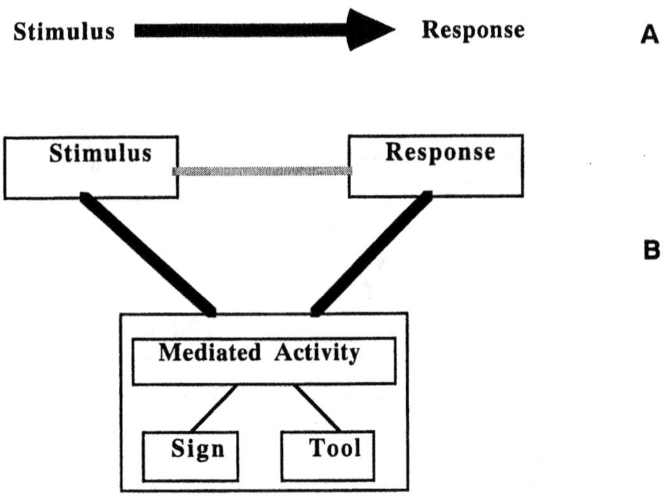

Figure 3: Direct (a) and mediated (b) S–R relations (according to Vygotsky 1978).

According to Vygotsky, the external mediators are not only full partners in the operation of higher psychological processes such as thinking and remembering, but the very elements and properties which make them *higher*:

> when a human being ... ties a knot .. as a reminder, ... she transforms remembering into an external activity. This fact alone is ... the fundamental characteristic of the higher forms of

> behavior. In the elementary forms *something is remembered*; in the higher form [of behavior] *humans remember something* .... humans personally create ... an artificial combination ... The very essence of human memory consists of the fact that human beings *actively remembe*r .. [they] personally influence their relations with the environment [and through it] their behavior ... civilization consists of purposely building monuments so as not to forget. In both the knot and the monument we have manifestations of the most fundamental and characteristic feature distinguishing human from animal memory (Vygotsky, 1978, 51, italics edded).

Vygotsky distinguishes between two types of mediators in the higher psychological processes: tools and signs. The tool is a mediating object, the function of which is "externally oriented ... it must lead to changes in objects ... is aimed at mastering ... nature. The sign, on the other hand, is internally oriented" (ibid, 55) and its aim is to act upon and master one's own and others' behavior. However, according to Vygotsky, "the mastering of nature and the mastering of behavior are mutually linked" phylo- and ontogenetically (ibid).

Mediating elements, such as tools and signs, as well as mediated activities, are constructed in the external environment, yet in the process of development they re-enter the individual's mind through a process of *internalization*. Internalization refers to "the internal reconstruction of an external operation" (ibid, 56). As an illustration Vygotsky gives the beautiful example of the child's attempt to grasp an object beyond his/her reach. The child's grasping is seen by the mother, who realizes its aim and mediates between the child and the object. By internalizing this triple interaction between the child, the object and the mother, the child's unsuccessful act of grasping an object is transformed into an act of pointing (to an object) directed towards the mother.

Vygotsky did not mention cognitive maps, but the weight he gave to the external environment makes his work directly relevant to the study of cognitive maps. The main points to emphasise in the present connection are the following:

1. The notion of internalization implies that a cognitive map, as the internal representation of the external environment, is not just a set of objects, their pattern and spatial relations, but their interactive, mediated and mediating nature. This view comes close to what has been said in the previous section – that the individual enters an environment which is already self-organized and enslaved by a set of order parameters, and constructing a cognitive map implies the internalization of the ordering principles of the environment.
2. Vygotsky distinguishes between something remembered and remembering something, that is to say, between passive and active remembering. As we shall see below, the two forms of remembering are relevant to cognitive mapping; active remembering is specifically relevant when we learn an environment or describe it to someone else for the purpose of navigation.
3. The external environment is full of socially and culturally constructed signs.

Gibson

Gibson's (1979) ecological approach is one of the more controversial conceptual frameworks suggested in cognitive sciences. The approach is "ecological" in its claim that the properties of the external environment structure our perception of it. This view is in contrast to classical cognitive sciences, according to which the environment is raw data and perception is a result of internal cognitive processes. An extensive exposition of Gibson's ecology and its relation to cognitive maps is given by Heft (this volume). Here my aim is to draw attention to some selected properties of Gibson's approach which are related to the conceptual notions suggested above and can contribute to the elaboration of IRN.

Like Vygotsky, Gibson directed attention to the external environment. However, unlike Vygotsky, who focused on "higher" psychological processes such as thinking, Gibson focused on "lower" psychological processes, notably on perception. Furthermore, in contrast to Vygotsky's interest in mediated external relations and their internalization, Gibson suggested the possibility of *direct* perception with no internal processing. Both, however, agreed that external relations are of primary importance, and that whether direct (Gibson) or mediated (Vygotsky), they first take place in the external environment and only then are internalized. Gibsonians are not so interested in internalization, whereas Vygotsky and his followers see it as an integral extension – a completion of the thought process.

Gibson's central and most provocative notion is *affordance*. Affordances are potentialities or "relations of possibility between animals and their environment. A particular environment has a given affordance if and only if it makes a given kind of action possible, whether that action is actually executed or not" (Neisser, 1987, 21). Thus people throw and grab things that are grabable, or sit on objects that are sitable. Note that the notion of affordance has the properties of Bohm's implicate order. From this perspective one would say that affordance is the implicate order of a certain animal–environment network which enfolds a multiplicity of actions. In Haken's terminology affordance would be an animal-environment system in a multi-stable state; in such systems initial conditions (e.g. an external stimulus) might determine which order of actions, that is, which order parameter, will eventually be realized and enslave the elements of the system. And when this has taken place, when a certain possibility is actually realized, the result is an explicate order of animal–environment relation.

An affordance is not the property of the individual taken alone, nor of the object or the environment taken alone, but a property of the *relation* between the organism and its object/environment (Neisser, ibid). This property helps to elucidate the notion of cognitive maps as IRN, which stands at the center of this chapter: the relational property of affordance can also be described as an organism–environment network. Gibsonians would probably claim that this network alone should be studied; the view suggested here is that the Gibsonian net is just a section of a larger network which extends inward in the form of internal representations, and outward in the form of external representations. And as the structure and architecture of the brain/mind afford certain cognitive potentialities, so do the structure and architecture of the environment. Both, the structure and architecture of the mind and the environment, must be studied if we are to understand cognitive mapping. From this perspective it is not surprising that Lynch's (1960) pioneering study on cognitive mapping – *The Image of a City* – starts by studying the architecture of the urban environment.

The externally represented part of the IRN includes not only the "lower" and relatively passive domain of Gibson's affordances, but also the external cultural and social activities which form Vygotsky's "higher" processes (thinking, concept formation, etc.). This latter extension is implied by Neisser's (1987) paper "From direct perception to conceptual structure", in which he writes that "human beings are thinkers as well as doers, and every environment offers intellectual opportunities as well as affordances for action" (Neisser, 1987, 21-2). With this statement Neisser takes us back to Vygotsky's example of the child's unsuccessful attempt at grasping. This environmental event offered a potential (implicated order) which was first realized (i.e. explicated) by the mother: she gave the object to the child. This emerging order generated a new implicate order, which was now explicated by the child in a specific way: He or she has internalized the event and transformed it into an internal representation of pointing. Note that the process is sequential, that it involves an interplay between internal and external representations, and that the external side of the process is at least partly collective, inter-personal and associated with cultural and social processes. These three elements are discussed in the next sections.

Rumelhart, Smolensky, McClelland and Hinton

In McClelland and Rumelhart's (1986) Volume 2 of *Parallel Distributed Processing* (PDP), Rumelhart et al. (1986) show how the PDP model can be related to complex cognitive notions such as logical thinking. Their port of departure is the following question: if PDP networks are conceptualized as *constraint networks,* and "if the human information-processing system carries out its computations by 'settling' into a solution rather than applying logical operations ... how can we do logic if our basic operations are not logical at all?" (Rumelhart et al., 1986, 44). Their answer: by our ability to create artifacts, that is, to create external representations and manipulate them sequentially. This answer is explicitly and directly inspired by Vygotsky's theory of thinking and it goes like this:

a. Humans (and animals) are good at pattern matching and can quickly "settle" on an interpretation of an input pattern. This property, which is basic to perception, for example, is illustrated in Figure 4a. As can be seen, this is an internally represented interpretation network which interacts with the external environment, and is composed of input, internal and output units.

b. Humans (and animals) are good at modeling the world, that is, at creating internal representations of it by means of Vygotsky's type internalization processes. The latter imply a more complex internal network, as described in Figure 4b: two relatively independent modules – an interpretation module, as above, and a new "model of the world", or an internal representation module. This complex internal network interacts with the environment.

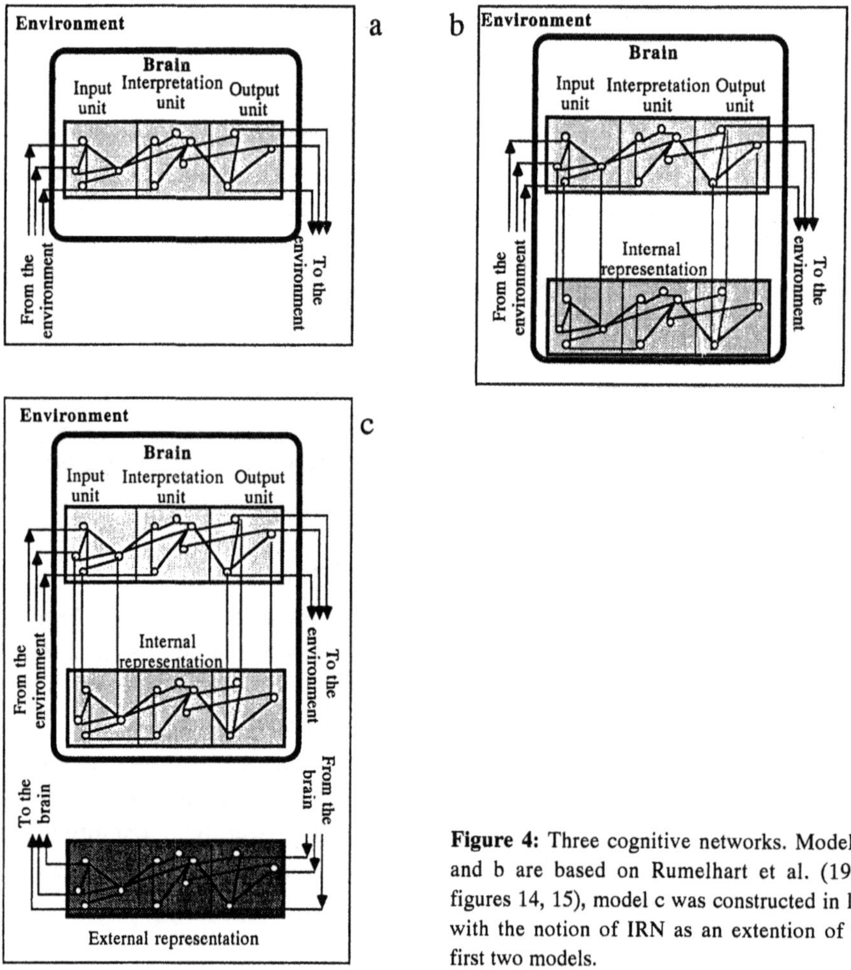

Figure 4: Three cognitive networks. Models a and b are based on Rumelhart et al. (1986, figures 14, 15), model c was constructed in line with the notion of IRN as an extention of the first two models.

c. Humans (but not animals) are good at manipulating the environment so that it comes to externally represent our internal world. External representation is the key to formal reasoning. This is illustrated in Figure 4c, where one can observe a network composed of two internal modules interacting with an external representation module, located in the external environment.

To illustrate the importance of external representation the authors give the example of multiplication. Due to the first property, humans are good at perceiving; due to the second property, that is, to their internally represented model, they can "see" or perceive the result of a simple multiplication such as 2x3. However, a complex multiplication such as 343x822 cannot be seen/perceived by our internal model. To solve this difficulty we break the problem to many small and simple multiplication steps, solve one by our internal model, externalize the result by writing it down ("so that we do not need to remember it"), we then connect this external representation to another internally represented activity, and so on in a sequence until the whole operation is completed. According to Rumelhart et al. (1986), "this dual skill of manipulating the environment and processing the environment ... [which] allows us to reduce very complex problems to a series of very simple ones", applies to logic in general – to mathematics, engineering etc. In this process "*the external environment becomes a key extension to our mind*" (ibid, 46, italics added).

Rumelhart et al. follow Vygotsky by emphasizing that our ability to internalize and create internal representations of the external environment includes the internalization of external representations we have created, and the creation of internal models of them. One can then imagine doing external multiplication and so on. "This ability to do the problem in our imagination is derivative from our ability to do it physically ... " (ibid). They further add that creating external representations is not a simple task – it is "the highest human intellectual ability [and] usually they are provided by our culture" (ibid, 47).

The above view is of immediate relevance to cognitive maps. Cognitive maps are internal representations of the external environment. This is a standard definition. They internally represent the external *natural environment* and the external *artificial environment* of artifacts such as buildings, roads, parks, towns, cities, neighborhoods, created through long and complex cultural, social and spatio-historical processes. This latter environment is artificial in the sense that it was created by humans as a result of their ability to manipulate the environment and to externally represent in it their internal model of the world – their thoughts, values, predictions, plans and aspirations.

Buildings, monuments, roads, parks ... which make neighborhoods, towns and cities are all external representations and at the same time they all are, to use Vygotsky's terminology, "tools" and "signs" . They are "tools" by virtue of their function as churches, palaces, townhalls, bell-towers .... and they are "signs" by virtue of the fact that they symbolize holiness, power, administration or money. Cognitive maps are thus complex networks, which are the product of a process of internalization of externally constructed environments, themselves external representations of our internally represented models of the world – of our cognitive maps. The picture which emerges out of Rumelhart's et al. PDP network of thought is that of a continuous interplay between internally represented cognitive maps and externally constructed artificial environments.

Bartlett

One of the main experimental methods used by Bartlett (1932/1961) in his book *Remembering* is that of serial reproduction: a person is shown a text or a figure and asked to memorize it. The object is then taken away and the person is asked to reproduce it from memory. The resultant reproduction is shown to the next person (or to the same person), the process repeats itself, and so on. Most of Bartlett's serial reproductions were made with texts, but he also reported on a few experiments with visual pattern like the one in Figure 5. It is typical of Bartlett's serial reproductions that at the beginning there are major changes from one reproduction to another, then the reproductions become more conventional and schematic, the texts/pictures stabilize and each sequential reproduction shows only minor changes.

The above Bartlett's scenarios from 1932 were recently taken up by Stadler and co-workers as the main experimental device in their attempt to construct a neo-Gestaltian cognitive theory on the basis of Haken's theory of synergetics (Stadler and Kruse, 1990; Kruse and Stadler 1993). Their basic suggestion is that Gestalt theory, in particular Kohler's, "has identified a number of fundamental principles of order formation in cognitive systems which anticipate in detail some of the concepts developed in the theory of complex nonlinear systems" (Stadler and Kruse, 1990, 33). They further suggest that Kohler's efforts "were limited by the physical thinking of his time, and that [Haken's theory of] synergetics allows a new approach to the old problem of cognitive research" (ibid, 32). In line with Haken's approach to pattern recognition, they suggest to see the cognitive system as a self-organizing system the dynamics of which is governed by the various conceptual ingredients of synergetics: order-parameters, the slaving principle, multi-stability and the like. The attraction of Bartlett's scenarios to this research program stems from the fact that cognitive self-organization processes are usually internal and happen very fast, nearly instantaneously, and therefore cannot be observed. Bartlett's method of serial reproduction seems to overcome both problems: it prolongs and externalizes, and thus makes observable the process of self-organization in cognition. Indeed, by conducting their own serial reproductions Studler and Kruse were able to

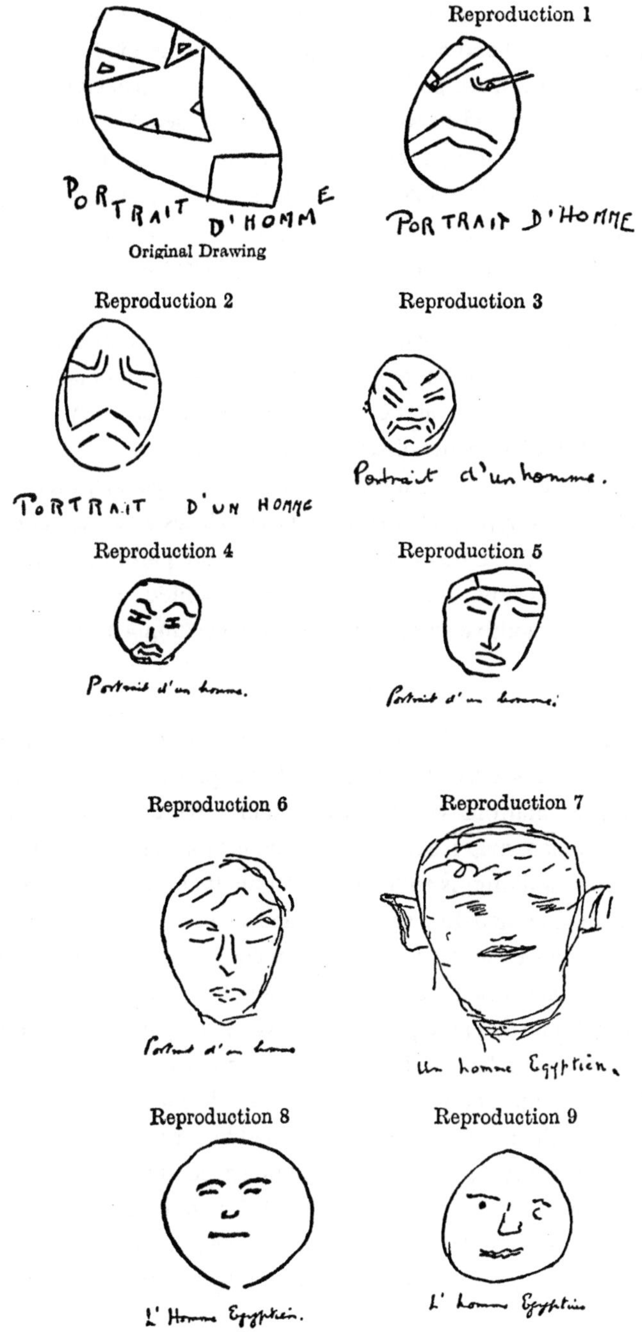

Figure 5: A figurative serial reproduction conducted by Bartlett (1961, 178-9).

show, first, how processes of order formation and phase transition in cognition are developed in relation to phenomena of multi-stability, emerging chaos and symmetry breaking; and second, how these processes are related to classical Gestaltian notions.

Both Bartlett and Stadler and co-workers are using the method of serial reproduction as means to find out what's going on inside the brain/mind – as means to externalize and thus expose the otherwise hidden internal processes. In this respect they follow classical cognitive science, which tends to concentrate on internal cognitive processes and representations. My interpretation of the various processes of serial reproduction is different: I submit the following:

a. The reproductions in the various scenarios are external representations in a sequential, self-organization, cognitive process, which proceeds as a continuous interplay between internal and external representations, very much in line with Rumelhart's et al. PDP model of sequential thinking.
b. The Bartlett scenarios, which involve more than one person, illustrate, first, how the internal-external-representation dynamics produces a certain collective decision, without the participants being aware that they take part in a collective cognitive process. Second, how the externally represented elements in the inter-representational cognitive system are gradually becoming a collective (i.e. cultural or social) entity.

The suggestion is that cognitive maps are constructed in a way similar to Bartlett's serial reproductions, that is to say, by a sequential interplay between internal and external elements and representations of the environment, which may come in different modalities, such as vision, map reading, stories and the like. A piece of input coming in a certain modality, say walking, is followed by an internal representation, which is then encountered with another modality, say a map, and so on.

To see how this is related to the construction of cognitive maps consider the following experiment. It is a game which may be seen as a public-collective Bartlett scenario of serial reproduction: Some 60 participants were asked to prepare models of buildings (at a 1:50 scale). They all assembled in a hall and sat around its central area (the floor), which was the "building site" for a new town in the game; they all observed the process as it developed. Each participant, in his/her turn, located his or her building on the site. The game started with an empty site and a single reference point (a railway, bus-station etc.) which connected the site to the rest of the world. The only rule of the game was that players could not block the entrances of already located buildings. Note that unlike the ordinary Bartlett scenarios, here the participants are fully aware that they are taking part in a collective process.

Figure 6: Stages in a public, collective, serial reproduction game. Top: first stage. Bottom: final stage.

The results of a typical game are illustrated in Figure 6. As can be seen, after several turns a certain spatial order was spontaneously created in the"town". The more this spatial order was observable, the more players tended to locate their buildings in line with this emerging order. By so doing they supported and strengthened the emerging order and at the very same time followed its ordering principles. In the language of synergetics, they were enslaved by the newly emerging order parameter. In a few cases a player decided to depart from the collective behavior and place his/her building in an "unexpected" location. This act often had the effect of a bifurcation in the evolving structure of the town – a new order parameter was created, with the effect that subsequent location decisions were enslaved by its ordering principles.

From the perspective of IRN what one sees here is an interplay between external and internal representations. Every individual act of location is an interaction between an internal representation created in the mind (say, a cognitive map of my ideal home in the town), and an external representation created in the environment. Every act of location not only externally constructs my internal cognitive map, but at the same time participates in constructing the external environment as a collective entity and memory.

Cognitive map construction in the above experiment is thus the construction of an IRN, the elements of which are partly internal and partly external, some are individual and some are collective; this IRN is an open and complex system, and as such evolves as a self-organizing system. The suggestion is that a structurally similar process takes place is other aspects of cognitive mapping, such as navigation, transformation between different modalities and so on.

Donald

In the *Origin of the Modern Mind*, Donald (1991) has taken the notions of external representation and external memory and made them the climax of his theory on the origin and evolution of the modern mind. *The externalization of memory* is the third and most recent cognitive transition in the evolutionary path from the *episodic mind* of the apes, through the *mimetic mind* of the Homo Erectus (first transition) and the *lexical mind* of the Homo Sapiens (second transition). Following Gould and Eldredge's (1977) model of "punctuated equilibria", which sees evolution as a series of radical changes (rather than a unitary gradual process), Donald's "central hypothesis is that there were three major cognitive transformations by which the modern mind emerged over several million years" (Donald, 1993, 737).

The episodic mind refers to the memory of the apes which are "brilliant perceivers" with "great sensitivity to ... environmental events, ... but with very poor episodic recall, ... they cannot self-trigger their memories ... independent of environmental cues. Thus, they are largely environmentally driven ... [with] ... very little independent thought" (ibid, 739). (Note the similarity to Figure 4a and to Gibson's "direct perception").

The mimetic mind was "the first truly human cognitive breakthrough". It "was a revolution in motor skill – mimetic skill – which enables hominids to use the whole body as [an external] representational device". ... Mimesis is based in a memory system that can rehearse and refine movement voluntarily ..., guided by a perceptual model of the body in its surrounding environment. ... a 'model of models' that allows ... a voluntary access route to memory and ... autocueing ... perhaps the most unifying feature of mimetic skill" (ibid, 739-40).

The mimetic model described by Donald (1991, 190, Figure 5) is not very different from the model in Figure 4c. It also has more than one internally represented module and several subsystems which form what in Figure 4c is termed an external representation network (facial, vocomotor, manual, whole body). Such a mimetic network has immediate socio-cultural implications in games, pedagogy, and toolmaking – "the most notable achievement of Homo Erectus..., [and] some degree of quasi-symbolic communication...[as well as the beginning of] a very simple shared semantic environment ... , communal sets of representations ... , first social costumes and the basis for the first truly distinctive hominid culture" (Donald, 1993, 741).

The *lexical mind*, with its capacity for lexical invention and innovation, was the key step towards the development of language. Once this capacity was developed, phonological evolution was accelerated, language system could evolve, together with its collective product – the narrative thought and its other corollaries.

The mechanism of lexical invention is not yet clear, according to Donald. Essentially it is a process of mapping meaning onto the form of a usually phonological symbol. It is "a complex process that involves labelling and differentiating our perceptions and conceptions of the world" (ibid, 743). What is specifically significant in this reciprocal process of form-meaning mapping is the tension created between form and meaning. This tension is "the driving force behind lexical invention – the need to define and redefine our maps of meaning onto word forms ..." (ibid).

"It is important to note", writes Donald (ibid, 744), "that these new representational acts – speech and mimesis – can be performed covertly as well as overtly. Covert speech has been called inner speech ... ", whereas covert mimesis – "imagination". In both internal representations are activated without actual motor execution. The human mind thus "became able to self-trigger recall from memory ... by means of mimetic imagination and by the use of word symbols, either of which could be overt or covert" (ibid).

Spoken language has not replaced mimetic external representation (in dance, athletics, craft, ritual and theater) but rather assumed "a dominant and governing role in human culture. ... The natural product of language is narrative thought ... [and] the normal use of language is storytelling about other people – gossip .. [this in turn] eventually produces collective, standardized narrative version of reality ... what we call the dominant 'myths' of society" (ibid, 745). Note that Bartlett's textual serial reproductions provide beautiful experimental illustrations of this process.

The externalization of memory is the third, and most recent, cognitive transition in Donald's evolutionary scheme, and no doubt his most provocative and radical suggestion. The starting point is a distinction between *biological memory,* which resides in the brain and within the body, and *external memory,* which resides in "a number of different external stores, including visual and electronic storage systems, as well as culturally transmitted memories that reside in other individuals. The key feature is that it is external to the biological memory of a given person" (Donald, 1991, 308-9). This new situation entailed major changes in the role of biological memory, resulting from the new architecture by which individual biological memories (which Donald terms *monades*) are linked to a network, in which part of the elements are externally represented memory units. This is illustrated in Figure 7. The essential property of external memory, or rather the basic metaphor for its operation, is that of a network:

> In a network, memory can reside anywhere in the system. ... Given a compatible network, the power of an element may become that of the entire network. ... Individual humans, utilizing their biological memories, may interact with their collective [external] memory apparatus in approximately similar ways. ... The major locus of stored knowledge is *out there*, not within the bounds of biological memory. Biological memories carry around the code, rather than a great deal of specific information (Donald, 1991, 312-4).

Donald's theory creates a rather interesting framework for the study of cognitive maps. Using his terminology one can speak of *episodic cognitive maps* typifying apes and

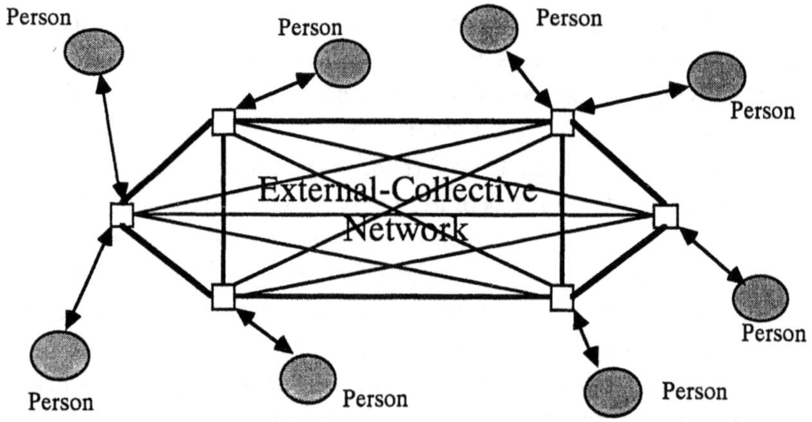

Figure 7: A network of biological internal memories connected to external, non-biological collective memory (adapted from Donald 1991, Figure 8.5).

other animals. The central characteristic of these maps is that they can be triggered only externally by stimuli which come from the environment. *Mimetic cognitive maps*, characterize humans from the age of the Homo Erectus onwards, and their central feature is that they can be self-triggered inside the mind in the form of images, and can also be externally triggered by will, using human mimetic skills. *Lexical cognitive maps* imply, first, a new internally represented capacity (module, network, etc.) to self-trigger a verbal map, in addition to, or in connection with, a mimetic imagined map. Second, they imply a capacity to produce an externally represented mimetic map, linguistic map or a combination thereof, describing people's relations with their extended environment. In terms of cognitive maps Donald's third transition would imply the emergence of cognitive *maps as external memories*. Various textual and graphic descriptions of the environment, maps of various forms, plans and so on, would all be typical examples here (Donald, 1991, 337), as well as computerized geographical information systems and remote sensing devices. According to Donald, each sequential evolutionary stage is built on top of its previous one and consequently the cognitive map (and the mind/brain) of the modern human can be seen as a composition of episodic, mimetic and lexical cognitive maps plugged into a variety of external memory maps and GIS storages. Given that cognitive maps reside in or are created by memory, the notion of external memory would imply that a person's cognitive map is distributed in the memory network, partly in the internal biological memory and partly in the external collective memory. The key feature of such a cognitive map would be the code or language that enables a person to plug in and thus gain access to the external memory elements of the network. In the next section I suggest to see these codes as order parameters of the cognitive system, that provide the interface between its internal and external elements.

Donald's primary concern is the distinction between biological and external memory, while the relations between internal-external representations (which formed the focus of the above discussions on Bartlett and Rumelhardt et al.) are of secondary interest in his project. Donald refers to these relations when he discusses the external motor skill associated with the internally represented cognitive units, and also when evaluating the social and cultural consequences of the various cognitive capacities. If, however, we consider Donald's scheme in terms of the relations between internal and external representations, then it appears that each of his evolutionary stages implies not only a more complex internal representation network, but also a parallel, more complex, externally represented network. And thus mimetic and lexical skills imply the ability to externally represent, by means of the whole body or parts of it, constructs created internally in the mind. In this respect some degree of external memory already existed before the third transition. Donald is aware of this, but claims that the existence of such an external memory (i.e. oral tradition) is constrained by the frames of biological memories, while his external memory is not. This view is in line with the distinction he made between external memory and culture or civilization and their products: "Culture and civilization are broader concepts, including material products, such as technologies and cities, and many aspects of human life that are not cognitive" (ibid, 309).

In what follows I try to suggest a different view: that artifacts such as tools, buildings or cities are cognitive entities by virtue of the property that they enfold information and as such are no less externally represented memories than books or GIS programs. They all exist in the outside and in order to gain access to their memory storage, humans need to know the code or the language by which to plug into them.

Alexander

According to Christopher Alexander (1979) artifacts such as stone tools, carpets, buildings, neighborhoods, cities and whole regions, are products of a language of patterns. His book, *A Pattern Language* (Alexander et al., 1977), can be regarded as a lexicon of patterns, starting with very large patterns of regions and metropolitan areas, through patterns of cities and neighborhoods, ending with patterns of very small details of alcoves, windows and door-handles. The book not only presents these patterns, but also shows how they are related to each other and form a whole language, structurally not very different from spoken or written languages.

Alexander's patterns are architectural entities of various sizes and scales. They can be compared to concepts, pictures, images (e.g. "an armchair by the fireside"), or schemata. Schemata, according to Rumelhart et al. (1986, 18), are "models of the outside world", "data structures for representing the generic concepts stored in memory", "conceptual structures", "a kind of generative thing, which is flexible but which can produce highly structured interpretations of events and situations". These schemata/patterns form the building blocks for the design and construction of space. The patterns of doors, windows, buildings, squares, neighborhoods and cities, are interrelated in a way similar to words, concepts, sentences, paragraphs, chapters and stories. The patterns are natural entities in the sense that they exist not only in physical structures in the environment, but also in people's minds. The languages which hold together the many patterns are "very complex sets of interacting rules [which] ... are actually there, in peoples' heads and are responsible for the way the environment gets its structure" (Alexander, 1979, 49-50). These rules form part of the human mind and are therefore timeless; they are *The Timeless Way of Building* (Alexander, 1979). In this respect Alexander's rules are similar to Chomsky's generative grammar which is innate to the human mind and as such prior to, or beyond, specific languages and cultures. But here the similarity ends. First, unlike Chomsky, Alexander's notion of language goes beyond the spoken/written ordinary language. Second, Alexander is more interested in semantics:

> Chomsky's work on generative grammar will soon be considered very limited. It happened to be brilliant in the sense that it was the first part of linguistics to receive this attention. But in fact, it does not deal with the interesting structure of language because the real structure of language lies in the relationships between words – the semantic connections. ... In that sense pattern languages are not like generative grammars. What they are like is the semantic structure, the really interesting part of language ... The structure which connects words together – such as "fire" being connected to "burn", "red" and "passion" – is much more like the structure which connects patterns together in a pattern language (interview in Grabow, 1983, 50).

Like schemata, Alexander's patterns are

> "something in the world" – a unitary pattern of activity and space, which repeats itself over and over again ... each time in a slightly different manifestation. ... these patterns are created by us ... in our minds ... [as] mental images of the patterns in the world: they are abstract representations of the very morphological rules which define the patterns in the world. However, unlike the patterns in the world ... the same patterns in our minds are ... generative. They tell us what to do ... (Alexander, 1979, 181-2).

The pattern language is more complex than simple mathematical languages or natural spoken languages. "From a mathematical point of view the simplest kind of language is a system which contains .. (1) a set of elements or symbols. (2) A set of rules". A natural spoken language is more complex: it has a set of elements (words), a set of rules which define the possible arrangements of words and in addition to these – "the complex network of semantic connections, which defines each word in terms of other words" (ibid, 184). A pattern language is still more complex in the sense that, like words, the patterns are elements and symbols, but unlike words, "each pattern is also a rule, which describes the possible arrangements of the elements – themselves again other patterns " (ibid, 185).

As in spoken languages, every person has his/her own personal pattern language which forms a personal variant of the language of a larger social and cultural collectivity. And the artificially built environment is the product of a conversation between a large number of individual pattern languages, which are the means with which people act on the environment. "And the enormous repetition of patterns, which makes up the world, come about because the languages which people use to make the world are widely shared" (ibid, 209-10).

Alexander's theory is a *par exellence* case of an IRN and a ready-made framework for the study of cognitive maps. Cognitive maps can thus be seen in terms of maps/patterns which form languages similar to Alexander's pattern languages. They exist in individuals' minds as internal representations which are both images of the environment and generative rules to act upon it. They exist also in the world as external representations and include Donald's external memories (maps, stories, texts, pictures) and Alexander's externally represented patterns – the artificial environment of buildings, neighborhoods, cities and so on. And as the written words, concepts and linguistic categories form stories, myths and histories, so do the physical shapes and forms which make the natural as well as the built environment: they afford or transmit power, poverty or wealth, they tell stories, form myths and describe histories, and by so doing form our external, collective, non-biological memory – that part of memory into which individuals with their biological memories can plug and thus be part of a rather complex cognitive network. We thus perceive the environment, cognize and act upon it, by means of a language the elements of which are maps and patterns of different modalities, sizes and scales. These maps and patterns are both the elements of the language and its generative rules, and they come into existence by interacting with other patterns. The construction of cognitive maps is thus a process which invokes this complex interacting network.

The generative property of patterns makes them in fact very similar to order parameters as described by Haken's synergetics. In particular, the process of self-organization by means of the slaving principle gives further insight into the process by which the various pattern languages of individuals are being enslaved by a larger and collective order parameter of, say, a building, and the process by which various patterns of buildings are being enslaved by the order parameter of a neighborhood, and so on.

In a way cognitive maps can be regarded as patterns and the pattern language as a language of cognitive maps. This is indeed so, but up to a point. Alexander confines his language to spatial or morphological forms: cities, buildings, artifacts, rags and so on. His language is similar to, but also distinguishable from, spoken and written language. Yet the phenomenon of cognitive maps is more general and complex than that. It refers to, or requires, not only the morphological representations of the environment, but also to its lexical and mimetic representations. As noted above, since we cannot see large scale environments in their entirety, we have to construct cognitive maps out of figurative as well as non-figurative lexical, conceptual patterns. Alexander's suggestion to see the world of architecture and artifacts by means of a pattern language is beautiful. But to do the same with cognitive maps would require a general purpose language, in which "ordinary" spoken and written language as well as pattern languages are but special and specific cases. The possibility of such a general cognitive language is discussed below in conjunction with Lakoff's and Gerald Edelman's studies.

Lakoff

Lakoff's (1987) *Women, Fire and Dangerous Things* can be seen as the psycho-linguistic parallel to Alexander's cognitive-architectural theory. Similarely to Alexander, whose data is the already constructed artificial environment of buildings and cities, Lakoff's data is the already constructed linguistic environment of concepts and categories as they appear in the world. He too sees cognitive entities as reflecting the interaction between the body-brain and the environment.

Lakoff's theoretical approach, "experiential realism", is based on two pillars. One is Johnson's (1987) insight, in his *The Body in the Mind: The bodily basis of meaning, imagination and reason,* "that experience is structured in a significant way prior to, and independent of, any concepts" (Lakoff, 1987, 271). Following Johnson, Lakoff suggests that our cognitive categories reflect conceptual embodiments, that is, concepts concerned with the interaction between the body-mind and the environment (an idea not very distant from Vygotsky's).

The second pillar is the discussion in cognitive sciences on the formation of concepts and categories. It includes the classical view that concepts and categories are characterized by some necessary and sufficient properties, Wittgenstein's (1953) criticism, which suggests that concepts and categories must be seen as networks of similarities, connected by means of what he has termed *family resemblance,* and Rosch and co-workers' (1976) addition that Wittgenstein's networks are nevertheless characterized by *prototypicality*: some instances of a concept or a category are more typical than others and thus form its *basic level.* According to Lakoff, the main merit of prototype theories is that they have

> isolated a significant level of human interaction with the external environment (the basic level). .... At this level, people function most efficiently and successfully in dealing with discontinuities ... our experience is preconceptually structured at that level. .... [This is] neither the highest nor the lowest level of conceptual organization (Lakoff, 1987,269-70).

Lakoff added to Rosch's prototypicality the notion that many categories have a *radial structure* with central category members related to other non-central members by various means: classical models, family resemblances and most importantly – by means of cognitive models and image schemas derived from the basic experiential level noted above. Lakoff discusses image schemas such as *container, source-part-goal, link, part-whole, center-periphery, up-down, front-back.* These schemas, he suggests, structure our experience of space, as well as our concepts; in fact they define most of what we commonly mean by the term "structure". This general view is termed by Lakoff *The Spatialization of Form Hypothesis,* implying a "metaphorical mapping from physical space into a conceptual space, ... and, metaphorical mappings themselves can also be understood in terms of image schemas" (ibid, 283).

The connection and relevance of Lakoff's view to cognitive mapping is clear and direct and indeed has already been acknowledged (Couclelis, 1988, and in this book; Mark, 1993; Mark and Frank, 1989; Portugali, forthcoming). If Lakoff's embodiment refers to the relations between body-brain and environment in general, then cognitive maps refer to a special type of this embodiment – between the body-brain and the large-scale extended environment by which it is surrounded. All of Lakoff's experiential schemas thus bear directly on this scale, and so does his notion of radial category structure: the relations between the individual and the radial horizon in the world around is a basic environmental experience (see, for example, the ancient Chinese and Roman maps of the world). In fact, many of the properties of cognitive maps as revealed by systematic distortion studies (Tversky, 1992; McNamara, 1992) and encoding-decoding studies in general (see Lloyd and Cammack, this volume), are formulated, or can be reformulated, in terms of Lakoff's image schemas.

From Lakoff follows that one can speak of cognitive maps on several interrelated levels: First, on the basic, person-environment, experiential level – some cognitive maps are constructed on this experiential, pre-conceptual level. Second, on the level of schemata and cognitive models. In this connection Portugali (1993a) has suggested that the schemata and models which form Lakoff's *Spatialization of Form Hypothesis* can be interpreted as experientially constructed order parameters with which internal and external information is self-organized. Some cognitive maps can be seen as conceptual models/order parameters (very similar to Alexander's patterns), which play an important role in encoding– decoding processes discussed in the literature of cognitive maps. On the third level, that of concepts and categories, one can speak of conceptual cognitive maps such as a city, a village or a country, and also of specific category cognitive maps such as Israel, France, USA, or New York, London and so on.

Edelman

Alexander and Lakoff draw their basic data from the external environment and their theories are based on a careful study of architectural and artistic artifacts, concepts, categories and the like. Edelman's view on cognition is similar in many ways to both, yet he arrives at that similar view from exactly the opposite direction: from a careful study of the anatomy and physiology of the brain, and from the perspective of embryology. Edelman – a Nobel Prize winner in physiology in 1972, and a student of neurobiology – attempts to reformulate neuropsychology and the issue of cognition in terms of biological evolution and his notion of *Neural Darwinism* (Edelman 1987).

Neural Darwinism starts by criticizing classical cognitive science for adopting the computer metaphor and its associated *instructive* viewpoint. The latter suggests that the brain, like a computer, acts by following instructions, which must be given by the *homunculus* – "the little man that one must postulate ... acting as an interpreter of signals and symbols in any instructive theory of mind". "If such an interpreter actually existed", writes Edelman (1992, 82), then "another homunculus is required in his head and so on, in an infinite regress".

In place of the classical theory of cognition Edelman has proposed the *Theory of Neural Group Selection* (TNGS). This theory, which encompasses all aspects of cognition, starting with the anatomy and biological operation of the brain, and ending with concept formation, categorization and consciousness, is based on three tenets (Figure 8): *developmental selection, experiential selection* and *reentrant mapping.*

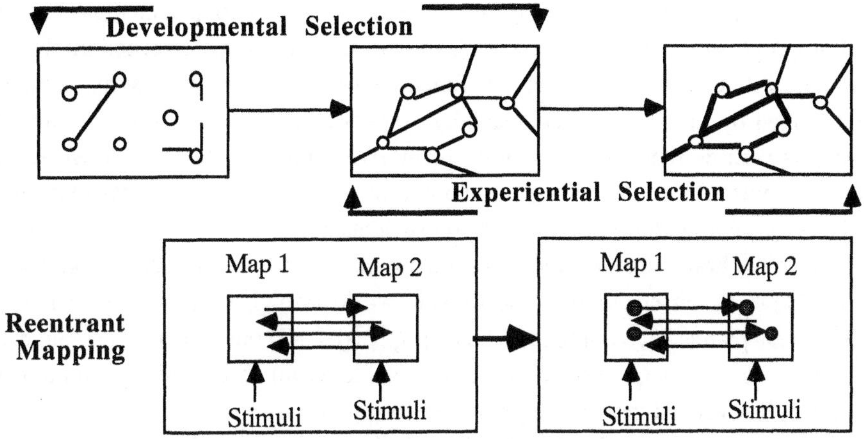

Figure 8: The three tenets of the TNGS (after Edelman 1992, 84).

The first tenet, developmental selection, is a process which leads to the formation of the neuroanatomical characteristic of a given species. Its essence is that groups or populations of neurons are engaged in a topobiological competition, yielding a certain anatomical network termed *primary repertoire* (Figure 8 top left).

The second tenet, experiential selection, concerns the selection of patterns of response. As illustrated in Figure 8 (top right), the process does not involve anatomical alterations, but a "selective strengthening or weakening of populations of synapses as a result of behavior" (ibid, 84). This process yields the *secondary repertoire.* Both the primary and secondary repertoires are unique to every individual.

Developmental and experiential selections form maps, and the *map*, according to Sacks (1993, 44), is one of the two most radical of Edelman's concepts (the second, as we shall see below, is *reentrant signalling*):

> Neurons can be anatomically arranged in many ways and are sometimes disposed into maps... Maps relate points on the two dimensional receptor sheets of the body (...the skin or the retina ...) to corresponding points on the sheets making up the brain... Furthermore, maps of the brain connect with each other via fibres... [The anatomy of the brain] is staggering in its intricacy and diversity. But it also has general organizing principles: It is made up of sheets that have topographic maps [connected to] sensory sheets and out to the

> muscles of the body. And maps map to each other... these maps are not fixed... there are major fluctuations in the border of maps over time. Moreover, maps in each different individual appear to be unique... Most strikingly, the availability of maps in adult animals depends on the available signal input... What is striking is that the ability to partition 'objects' and their arrangements depends on the functioning of the maps (Edelman, 1992, 19; 21-22; 27-8).

The third tenet, reentrant mapping (Figure 8, bottom), "is perhaps the most important of all proposals of the theory", showing how "the brain areas that emerged in evolution coordinate with each other to yield new functions"(ibid, 85). The process starts with the maps formed by the primary and secondary repertoires. These maps are connected in parallel as well as reciprocally, among themselves and with the outside world, thus enabling reentrant signaling and the mapping of one map onto the other (e.g., in the monkey's visual system there are over thirty such maps).

Maps in the brain receive signals from other maps in the brain and, through sensory maps (vision, touch etc.), from the outside world (which itself might be maps or environments planned according to some maps/plans). And consequently the brain makes maps from its own maps, as well as from outside-world maps and in a similar way categorizes its own and the outside world categorizations.

As noted, the essence of Edelman's TNGS is that the brain and its activity evolve and develop by means of Darwinian selection rather than by instructions. This implies, first, that every evolutionary and developmental stage comes into existence not by following some pre-existing, innate instructions, but is *selected* by means of reentrant signaling. Second, that the relevant units for selection are not individual neurons, but populations of neurons: neural groups, maps and concepts. The process of selection is implemented by a complex network of interacting neural groups and maps which, through competition and reentrant signaling, self-organize to form a certain neural network, response or behavior. Edelman notes at one point that "the brain is an example of a self-organizing system" (ibid, 25), and to this one might add that environmental selection is a typical process in such systems: input from the environment triggers an internal self-organization process of competition between different configurations of neurons, networks of neural cells, maps, reentrant maps (Figure 8) ... and so on. Cognitive capacities, even the most complex ones such as conceptualization, categorization, thinking, consciousness, etc., are thus selected exactly as the very complex creature known as the human being was naturally selected in the process of evolution. The various innate structures and capacities thus do not instruct the body and the mind what to do next, but set the potentialities and constraints. The final selector is the relevant environment. This is so with both the genetic and epigenetic cognitive capacities, the implication being that no cognitive capacity can be understood independent of its environment.

The whole body and its cognitive systems are thus seen as open, complex and therefore self-organizing systems, each interacting with its environment and creating an environment for its lower order systems. This view stands in opposition to the more prevalent classical view of cognition, for it implies, for example, that "to build syntax on the bases of grammar, the brain must have reentrant structures that allow semantics to emerge first (prior to syntax) by relating phonological symbols to concepts" (Edelman 1992, 130). In line with Macnamara, Edelman's theory proposes "that children are able to learn language *because* they first make sense of situations involving human interactions. Children make sense of things *first* and, above all, they make sense of what people do. ... It also seems that a child first makes sense of situations and of human intentions and *then* of what is said. This means that language is *not* independent of the rest of cognition" (ibid, 244-5). With this view we come full circle to our port of departure – to Vygotsky and his notion of internalization.

Edelman (1992) has recently summarized his theory in his book *Bright Air Brilliant Fire*, and closes the book with a "critical postscript" which criticizes classical cognitive theory at its hard core: its use of the computer metaphor, its attempt to isolate cognition from the outside world of environment, society and culture, and its tendency to consider the various cognitive faculties as independent entities. In particular he challenges Chomsky's generative grammar, and in so doing seems to fulfill Alexander's prediction (see above) that "Chomsky's work on generative grammar will soon be considered very limited". Edelman suggests to go beyond classical cognitive theory with its formal Chomskyian grammar and formal semantics, which "cannot account for the richness of reality", and to follow ideas such as Lakoff's (1987) as they appear in *Women, Fire and Dangerous Things.*

The picture which emerges out of Edelman's theory is that of a brain "full of maps", interactively related to each other, to other parts of the brain (i.e. hippocampus), as well as to objects in the external world via the senses. The information which flows through this interactive system by means of reentry entails what Edelman calls global mapping, which is the process responsible for cognitive capacities such as concept formation and categorization. Edelman's theory thus implies a new and surprising view of the general notion of mapping and cognitive mapping: mapping is a physiological and at the same time a psychological cognitive process. Ordinary mapping is therefore an external representation of a physiological-cognitive embodiment. This view is in line with Lakoff's notion of embodied experience in relation to concept formation and categorization. The notion of "map" in cognitive mapping is thus not only a metaphor as it is commonly perceived, but also a real material entity constructed in the brain in the processes of evolution, development and behavior; and global mapping is a process constructed out of the interaction between maps and between them and the external environment. A cognitive map is therefore the product of global mapping.

Concluding Notes

Edelman's TNGS takes us full circle back to our port of departure – to the notion of IRN as a framework to the study of cognitive maps, to the holomovie metaphor, to Bohm's orders and Haken's synergetics. Only now we come back to this port of departure with a richer conceptual and contextual background accumulated and elaborated in the previous section. That is to say, with Vygotsky's social formation of mind, his notion of internalization and the view that tools and signs are legitimate and integrative elements in the overall cognitive process; with Gibson's affordance which implies that as the structure and architecture of the brain/mind afford certain cognitive potentialities, so does the specific structure of the environment; with Rumelhart's et al. paper, which shifts the discussion of IRN to the domain of PDP and connectionist networks, introduces the notion of external representation, and describes thinking as a sequential interplay between internal and external representations, and Bartlett's scenarios of serial reproduction, which illustrate this interplay with an emphasis on remembering. Donald's evolutionary perspective adds the view that the current cognitive system is a network composed of a public, non-biological external memory, to which individuals with their biological memories are connected, whereas with Alexander's theory we further suggest that this external memory includes also the external artificial environment of buildings, cities, and other human artifacts, and that it is constructed by means of a pattern language. Lakoff, with his radial structure of categories and the spatialization of form hypothesis, puts forward the suggestion that there is a pre-linguistic experiential level from which concepts, categories and cognitive maps are derived, whereas Edelman's TNGS not only connects Alexander's and Lakoff's theories, but also relates them to the anatomy and biology of the brain. This is done by adopting "population thinking" and selectionism with the implication that the elements of cognition are not individuals, but groups – ranging from anatomical neural groups and maps, all the way to Alexander's patterns, Lakoff's concepts and "our" cognitive maps. These elements are constructed, that is to say, selected, by the complex interaction process of reentrant signaling.

The similarity between Edelman's theory and the perception of cognitive maps in terms of IRN and its tenets – the holomovie metaphor, the implicate order and synergetics – is striking. His model (Figure 8) can be seen as a multi-level interplay between Bohm's implicate, explicate and generative orders: the interactive IRN of neural groups can be interpreted as an implicate order enfolding multiple potentialities which, when realized, form the explicate order of the primary repertoire. The latter becomes an implicate-generative order, enfolding its own potentialities which, when realized, form the second repertoire and so on.

At the heart of the above process is Edelman's reentrant signaling, which is very similar to Haken's synergetics approach to self-organization and cognition. In both certain configurations of sub-systems (neurons, maps...) are formed and enter into a complex interaction and competition, at the end of which a unique configuration is eventually selected, that is to say, wins the competition. To this Haken's theory adds, that the wining or selected state is an order parameter which enslaves the system to its mode. Haken's order parameter and the slaving principle thus show how, despite the fact that every reentrant ends with a unique individual solution, the variable self-organized solutions are all enslaved by some global and slower order parameters. Thus, while individuals "create" their own unique set of words, concepts, categories, architectural-urban patterns and cognitive maps, they are all enslaved by collective languages which act as global order parameters.

From the above perspective cognitive maps can be seen as IRN, the elements of which are groups of neurons, brain-maps, maps of maps in different modalities, concepts, categories, patterns, "ordinary maps" as well as cultural and social human groups. They all exist in people's minds as well as in the outside world, and as such participate in forming the public collective memory; they all come into existence by this reentrant, synergistic, self-organization interaction between the internal and external elements of the IRN, and by the complex interplay between the implicate, explicate and generative.

Edelman's main data is drawn, as noted, from his studies in embriology and brain-anatomy. Most other studies surveyed and discussed above draw their data mainly from the interface between the individual, society, and the natural and artificial environments. As such they shed light on the various global scale facets of cognitive maps as IRN, as well as on their dynamics, and thus on the construction of cognitive maps. In particular, Bartlett serial reproductions suggest a simple and very effective method of experimenting with the issue. From the serial reproductions discussed above we have already seen that the process is sequential, involving an interplay between internal and external representations, and that its evolution takes a form typical to self-organizing systems as envisaged by synergetics. Indeed, while the aim of the present paper was to introduce the notion of IRN and elaborate its properties in the context of the cognitive theories we have surveyed, the next stage is to cast it into the formalism of synergetics. A first step towards this aim is suggested in the paper by Haken and Portugali (this volume), in which we start to operationalize the above conceptual framework and apply it to the various aspects of cognitive maps and their construction.

References

Alexander, C. et al. (1977). *A Pattern Language*, New York: Oxford University Press.

Alexander, C. (1979). *The Timeless Way of Building*, New York: Oxford University Press.

Bartlett, (1932/1961). *Remembering: A Study in Experimental and Social Psychology*, Cambridge: Cambridge University Press.

Bohm, D. (1980) *Wholeness and the Implicate Order*, London: Routledge & Kegan Poul.
Bohm, D. and Hiley, B.J. (1993). *The Undivided Universe*, London: Routledge.
Bohm, D. and Peat, F.D. (1987). *Science, Order and Creativity*, New-York: Bantam.
Couclelis, H. (1988). The truth seekers: geographers in search of the human world. In *A Ground for Common Search* (R.G. Golledge, H. Couclelis and P. Gould eds), Santa-Barbara: Santa-Barbara Geographical Press, pp. 148-155.
Donald, M. (1991). *Origins of the Modern Mind: Three Stages in the Evolution of Culture and Cognition*, Cambridge Mass: Harward University Press.
Donald, M. (1993). Precis of origins of the modern mind: three stages in the evolution of culture and cognition. *Behavioral and Brain Sciences* 16, 737-791.
Edelman, G.M. (1987). *Neural Darwinism: The theory of Neural Group Selection*, New York: Basic Books.
Edelman, G.M. (1992). *Bright Air Briliant Fire: On the Matter of the Mind*, London: Penguin Books.
Gardner, H. (1984). *The Mind's New Science*, New York: Basic Books.
Garling, T. and Golledge, R.G. eds. (1993). *Behavior and Environment*, Amsterdam: North-Holland.
Gibson, J.J. (1979). *The Ecological Approach to Visual Perception*, Boston: Houghton-Mifflin.
Giddens, A. (1984). *The Constitution of Society*, Cambridge: Polity Press.
Golledge, R. G. (1993). Geographical perspectives on spatial cognition. In *Behavior and Environment* (T. Garling and R.G. Golledge eds.), pp. 16-46. Amsterdam: North-Holland.
Gould, S.J. and Eldredge, N. (1977). Punctuated equilibria: the tempo and mode of evolution reconsidered, *Paleobiology* 3, 115-151.
Grabow, S. (1983). *Christopher Alexander: The Search for a New Paradigm in Architecture*, Stockfield: Oriel Press.
Haken, H. (1979). Pattern formation and pattern recognition – an attempt at a synthesis. In *Pattern Formation by Dynamical Systems and Pattern Recognition* (H. Haken, ed.), pp Berlin: Springer.
Haken, H. (1983). *Synergetics, An Introduction*, 3rd. ed., Berlin: Springer.
Haken, H. (1987). *Advanced Synergetics. An Introduction*. 2nd. print., Springer.
Haken, H. (1988). Synergetics in pattern recognition and associative action. In *Neural and Synergetic Computers* (Haken, H. ed), Berlin: Springer.
Haken, H. ed., (1990). *Synergetics of Cognition*, Berlin: Springer.
Haken, H. (1991). *Synergetic Computers and Cognition*. Berlin: Springer.
Heft (this volume)
Hiley, B.J. and Peat, F.D. (1987). *Quantum Implications*, London: Routledge.
Johnson, M. (1987). *The Body in the Mind: The bodily basis of meaning, imagination and reason*, Chicago: The University of Chicago Press.
Kruse, P. and Stadler, M. (1993). The significance of nonlinear phenomena for the investigation of cognitive systems. In *Interdisciplinary Approaches to Nonlinear Complex Systems*, (H. Haken and A. Mikhailov, eds.), Berlin: Springer.
Lakoff, G. (1987). *Women Fire and Dangerous Things: What Categories Reveal About the Mind*, Chicago: The University of Chicago Press.
McNamara, T.P. (1992). Spatial representation In *Geography Environment and Cognition* (J. Portugali ed.), a special theme issue *Geoforum* 23 (2) 139-150.
Mark, D.M. (1993). Toward a theoretical framework for geographic entity types. In *Spatial Information Theory* (A.U. Frank and I. Campari eds.), pp. 270-283, Berlin: Springer.

Mark, D.M. and Frank, A.U. (1989). Concepts of space and spatial language. In *Proceedings, 9th internationa Symposium on Computer Designed Cartography*, pp. 538-556, Baltimore, Maryland.

McClelland, J.L., Rumelhart, D.E. and the PDP Research Group (1986). *Parallel Distributed Processing, Explorations in the Microstructure of Cognition. Volume 2: Psychological and Biological Models*, Cambridge Mass: MIT Press.

Neisser, U. (1987). From direct perception to conceptual structure. In *Concepts and Conceptual Development: Ecological and intellectual factors in categorization* (U. Neisser, ed.), pp. 11-23. Cambridge: Cambridge University Press.

Portugali, J. (1990). Social synergetics, cognitive maps and enviromental recognition. In *Synergetics of Cognition* (H. Haken and M. Stadler eds.), pp. 379-392. Berlin: Springer Verlag.

Portugali, J. ed. (1992). *Geography Environment and Cognition* , a special theme issue *Geoforum* 23.

Portugali, J. (1993). *Implicate Relations: Society and Space in the Israeli-Palestinian Conflict*, Dordrecht: Kluwer Academic Publishers.

Portugali, J. (1993a). Frontiers, borders, centers and peripheries as cognitive maps. In *International Conference on Regional Development: The Challange of theFrontier*. THe Negev Center for Regional Development, Ben-Gurion University.

Portugali, J. (forthcoming). On the nature of world urbanization, *Progress in Planning*.

Portugali, J. and Haken, H. (1992). Synergetics and cognitive maps. In *Geography Environment and Cognition* (J. Portugali, ed.), a special theme issue *Geoforum* 23, 2, 111-130.

Pribram, K. H. (1990). Prologomenon for a holonomic brain theory. In *Synergetics of Cognition* (H. Haken and M. Stadler eds.) pp. 379-392. Berlin: Springer.

Rosch, E., Mervis, C., Gray, W., Johnson, D. and Boyes-Braem, P. (1976). Basic objects in natural categories. *Cognitive Psychology* 8, 382-439.

Rumelhart, D.E., Smolensky, P., McClelland, J.L. and Hinton, G.E. (1986). Schemata and sequential thought processes in PDP models. In *Parallel Distributed Processing, Explorations in the Microstructure of Cognition. Volume 2: Psychological and Biological Models* (J.L. McClelland, D.E. Rumelhart, and the PDP Research Group), Cambridge Mass: MIT Press.

Sack, O. (1993). Making up the mind. *The New York Review*, April 1993.

Stadler, M. and Kruse, P. (1990). The self-organization perspective in cognition research: historical remarks and new experimental approaches. In *Synergetics of Cognition* (H. Haken and M. Stadler eds), pp. 32-52. Berlin: Springer.

Tversky, B. (1992). Distortions in cognitive maps. In *Geography Environment and Cognition* (J. Portugali, ed.), a special theme issue *Geoforum* 23, 2, 131-138.

Vygotsky, L.S. (1978). *Mind in Society*, Cambridge Mass.: Cambridge University Press.

Wittgenstein, L. (1953). *Philosophical Investigations* (Translated by G.E.M. Anscombe). Oxford: Blackwell.

Juval Portugali
Department of Geography
Tel Aviv University
Tel Aviv 69978, Israel.

SYNERGETICS, INTER-REPRESENTATION NETWORKS AND COGNITIVE MAPS

Hermann Haken and Juval Portugali

Abstract: Synergetics and its concept of order parameters can provide a general theoretical framework for the study of cognitive maps. Close links between processes in the internal world of the individual and processes in the external environment have led to the notion of inter-representational networks (IRN). In this paper it will be shown how the concept of order parameters allows us to cast the notion of IRN into a mathematical form that in the present paper is based on graphical representations. After a short reminder of synergetics, cognitive maps and IRN we present a general model of IRN in terms of synergetics, where the mathematical basis is demonstrated. We then show how the above model can be applied to the construction of cognitive maps, in particular to environmental learning and transformations in cognitive mapping. Finally,we show, how collective cognitive processes may take place.

Introduction

In two previous papers it has been demonstrated how Haken's theory of synergetics can provide a general theoretical framework for the study of cognitive maps (Portugali, 1990, Portugali and Hakenw 1992). One of the central features of these discussions is that order parameters of the external environment (social, cultural, political...) play an important role in processes associated with cognitive maps of individuals and collectivities. These close links between processes in the internal world of the individual and processes in the external social and cultural environment have led to the notion of IRN (Inter-Representational Network). That is to say, the relevant network for the study of cognition and cognitive maps is not an internally represented network, but a network composed of some internally represented and some externally represented elements (Portugali, this volume). The present paper follows the above suggestions. In particular, it demonstrates how the above ideas can be cast into the mathematical formalism of synergetics. This new formalism enables us to investigate the various theoretical properties of cognitive maps as IRN and to relate them to some empirical findings.

The discussion below begins with an introductory section aimed to remind the reader of the basics of synergetics and their relations to cognitive maps, and of the notion of IRN. Then, in the next section, we turn to the formalism of synergetics and introduce a

J. Portugali (ed.), The Construction of Cognitive Maps, 45–67.

general synergetic model of IRN. The model is based on a new look at the relations between pattern formation and pattern recognition as suggested in the past by Haken (1979), and on Haken's (1991) demonstration, how the algorithm of pattern recognition can be realized on parallel networks. Having built the general model we use it to examine three issues which are central to the study of cognitive maps: Learning an environment and constructing a cognitive map of it; transformations in cognitive maps - mainly between verbal and visual representations, as they interact in way-finding and navigation, and collective cognitive processes related to cognitive maps and to the construction of cities.

A short Reminder of Synergetics, Cognitive Maps and IRN

Synergetics

Synergetics is an interdisciplinary field of research which deals with systems composed of many subsystems. By means of their interaction, the subsystems may spontaneously produce spatial, temporal or functional structures. Synergetics focusses its attention on those situations where new structures evolve. We briefly introduce some of its basic notations, which we will use in the following. First of all we have to describe the state of a system. We do so by means of a state vector

$$\mathbf{q} = (q_1, q_2, \ldots, q_j \ldots, q_n) \quad (1)$$

which possesses components q_1, q_2, ... To illustrate the meaning of the state vector, consider a simple example from population dynamics, where q_j denotes the number of people living at a location denoted by the index j. Thus (1) describes the distribution of a population over a certain area. This distribution will change in the course of time due to processes of birth, death and migration. We shall assume that such a change of the state vector over time is described by the so-called evolution equations

$$\frac{d\mathbf{q}}{dt} = \mathbf{N}(\mathbf{q}, \alpha) + \mathbf{F}(t), \quad (2)$$

where $\mathbf{N}$ is a nonlinear function of the state vector $\mathbf{q}$ and depends on control parameters α which describe the conditions of the environment. For instance, in the case of population dynamics, they may describe a general utility of an area. $\mathbf{F}(t)$ describes the influence of chance processes that are not accessible to a deterministic analysis. One of the prominent strategies of synergetics is to study the behavior of the state vector (1) when one or several control parameters are changed. In general, when a control parameter is changed, the state vector will adapt smoothly. There are, however, important situations in which the state vector changes dramatically. We then speak of the instability of a state vector. At those instability points, the newly developing state vector can be written as a superposition of new elementary vectors $\mathbf{v}_u$ with time-dependent coefficients $\xi_u(t)$. Thus the newly developing state vector can be written in the form

$$\mathbf{q} = \sum \xi_u \mathrm{v}_u + \text{small corrections}, \tag{3}$$

where we shall neglect the small corrections. As a general experience found in the field of synergetics, it turns out that close to instability points the dynamics of the growth of the new state vector $\mathbf{q}$ is determined by very few quantities ξ_u that are called *order parameters*. Since the vectors $\mathbf{v}_u$ describe whole configurations, we may state that the order parameters determine which new configurations evolve. To use a technical term of synergetics, the order parameters *enslave* the subsystems. Order parameters may compete so that only one order parameter survives in the course of time, or they may cooperate. In the former case, the final state vector is determined by one of the elementary state vectors $\mathbf{v}_u$. In general, the order parameters ξ_u obey a rather simple dynamics that is described again by equations of the type (2), namely now in the form

$$\frac{d\xi_u}{dt} = \boldsymbol{M}_u(\xi) + \boldsymbol{F}_u(t). \tag{4}$$

As has been shown by numerous examples both in the natural sciences but also in the context of cognitive maps, the concept of order parameters and enslavement is a powerful tool. We hasten to add that the mathematics that lies at the basis of synergetics is quite sizeable, but for the purpose of our present paper it will suffice to know the basic notions as order parameters and enslavement. In fact, the derivation of $\mathbf{v}_u$ and of (4) requires a considerable amount of mathematics which can, however, be circumvented in a number of applications.

A well-known example of the approach of synergetics to spontaneous pattern formation is provided by the *fluid dynamics paradigm*, related to a liquid in a vessel heated from below (Haken, 1983, 1987). At the initial stage of the process no macroscopic motion can be observed in the movement of the liquid. However, when the temperature difference between the lower and upper surfaces of the liquid exceeds a critical value, it can be observed that the motion of the liquid exhibits an ordered macroscopic pattern in form of rolls. According to synergetics, at the initial stage several roll configurations are formed and as the process continues, their order parameters enter a competition, which is eventually won by one order parameter which enslaves the subsystems. An important property of that process is that at the beginning the system is *multi-stable* in the sense that it enfolds many possible patterns; which pattern will eventually be realized depends on initial conditions.

A similar process takes place in pattern recognition. A typical example is face recognition: a person sees a few features of a known face and by means of associative memory he or she can recognize the entire face. According to synergetics, the process is analogous to pattern formation as described above: the cognitive system of the person is multi-stable as it enfolds, i.e. stores, many known faces. When the person sees a few features, or part of a face, several configurations of features and their order parameters are formed by means of associative memory. The order parameters enter into a competition which is, eventually, won by a certain order parameter.

The analogy between the process of pattern recognition by associative memory and the process of pattern formation as conceptualized by synergetics is illustrated graphically in Figure 1. It was first demonstrated by Haken (1979) in a seminal paper which has, in fact, launched the research domain today known as *synergetics of cognition*. Synergetics of cognition has since then become a most active field of research and much of the progress made in recent years in the theory of synergetics was connected to this issue. For a recent statement and updated bibliography see Haken's (1991) *Synergetic Computers and Cognition*.

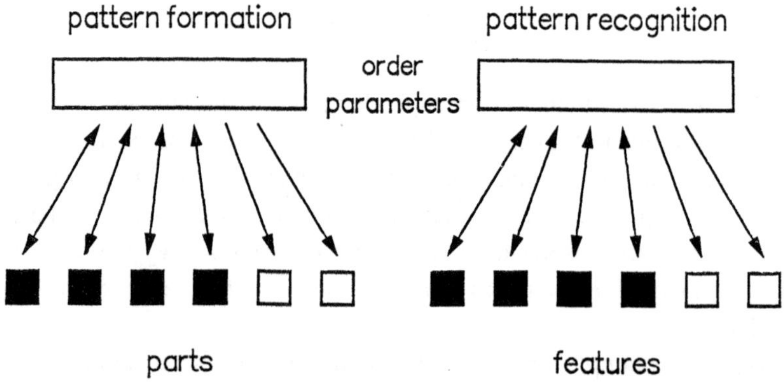

Figure 1: Analogy between pattern formation and pattern recognition. On the left: when some parts of a system are in an ordered state, they may generate the order parameter which, in turn, enslaves the rest of the system so that the total system is brought into an ordered state. On the right: when some features of a pattern are given, they generate their order parameter which, in turn, enslaves the total system (human brain or computer) and forces it to complement the rest of the features.

Synergetics and Cognitive Maps

The process of pattern formation and pattern recognition as theorized by synergetics offers an appropriate theoretical and conceptual framework for the study of cognitive maps. This possibility was first suggested by Portugali (1990) and was further elaborated by Portugali and Haken (1992). The suggestion is that similar to the synergetics' conceptualization of pattern recognition of faces, for example, the cognitive system associated with cognitive maps constructs or forms a whole pattern/map on the basis of only a partial set of features of it. In the language of synergetics one can say that a partial set of features of an environment shown to a cognitive system triggers a competition between several configurations of features and their order parameters, until one wins and enslaves the system so that a cognitive map is thus constructed.

There are, however, important differences between the processes of pattern recognition and cognitive map construction. In face recognition, for example, the aim is to use the partial set of features given to the system in order to recognize a face, out of a repertoire of known and stored faces. In cognitive maps the aim is to construct an initially unknown pattern/map out of a partial set of features of a certain environment. In pattern recognition we are usually dealing with one mode of cognition, in face recognition, for instance, it is vision. In cognitive maps we are usually dealing with several modalities. The reason is that because of size, the whole environment cannot usually be seen in its entirety and the mind has to construct the cognitive map not only by means of direct visual information (obtained by navigation, for example), but also by indirect visual (such as maps or photographs) and nonvisual (i.e. verbal descriptions) information. Now, much of this indirect information refers to the way other people see or imagine the environment – it is their cognitive map of it. Cartographic maps, for example, can be regarded as external representations of environments as shaped by a synergistic process of collective cultural, social or political processes. In fact, as noted in our previous study (Portugali and Haken, 1992), a person is born into an environment which is already self-organized and enslaved by some order parameters. Consequently some of the features out of which one constructs a cognitive map come up already enslaved by order parameters and it is quite likely that the individual will construct a cognitive map not by enslaving the given partial set of features, but by being enslaved by one or some of them.

Inter-Representation Networks

The above formulation of cognitive maps within the framework of synergetics leads to a conceptualization which gives much more weight to the external environment, and to external cognitive storage and representation in it, than is conventionally given in cognitive sciences. In fact, in concluding his *Synergetic Computers and Cognition*, Haken (1991) too hints at this direction This in turn has led to an exploration of the possibility that the cognitive system in general, and the one associated with cognitive maps in particular, extends beyond the individual's mind/brain into the external environment (Portugali, this volume). As it turns out, a survey of the relevant literature shows that although this idea was not very popular in cognitive science in the past, it was nevertheless always present in the work of scholars such as Vygotsky (1978) or Gibson (1979) and their followers. In recent years, however, the idea has become more acceptable in various domains, such as in PDP (Rumelhart et al., 1986), in relation to concept formation and categorization (Lakoff, 1987) and in various attempts to reconsider cognition from the theoretical perspective of evolution (Donald, 1991, Edelman, 1992). Thus Rumelhart et al. (1986) suggest that external cognitive representations play an important role in sequential thought processes and as such must be regarded as an external extension of the mind and its network. Or, from another perspective, Donald (1991) goes much further by suggesting that the emergence of external memory storages in the form of libraries, maps, computer data-bases and the like, marks the third and most recent stage in the evolution of human cognition.

Portugali (this volume) has surveyed the above ideas in some detail and used them as building blocks in developing the notion of IRN. That is to say that the cognitive system must be seen not as an internal network representing the external environment, but as an internal-external network, where some of the elements are internally represented or stored in the mind/brain, and some exist, are stored or externally represented in the outer environment. Portugali further suggested that some of the nicest experimental examples of the operation of IRNs are the so-called Bartlett scenarios conducted by Bartlett (1932/1961) as part of his studies on *Remembering*. The general structure of the Bartlett scenarios will help to convey the notion of IRN.

A typical Bartlett scenario evolves like this: a test person is given a text or shown a figure and is asked to memorize it. He or she is then asked to externally reproduce it out of memory (i.e. to rewrite the text or re-draw the figure, etc.). This external representation is given to another test person and so on. The usual result of such scenarios is that after several strong fluctuations in the reproduction, the text or the figure are stabilized and do not change much from iteration to iteration. The interpretation offered (ibid) is that what we have here is (i) a cognitive network composed of internal and external elements and representations, (ii) a sequential interplay between the internal and external elements of the system, and (iii) a typical synergetic process: this sequential interplay exhibits, first, strong fluctuations between competing configurations of texts or figures, which then lead to the emergence of a certain order parameter which enslaves both the external and the internal elements and representations of the system. Thus, instead of the usual process of pattern formation by which the order parameter(s) enslave(s) some external subsystems, and the usual process of pattern recognition by which the order parameter(s) enslave(s) some internal features, we have here an integrated process - the order parameter(s) enslave(s) both the externally represented subsystems and the internally represented features.

A General Model of Synergetics and IRN

In order to cast this integrative view into a graphic, and subsequently a mathematical form, we must remind the reader of the synergetic network model of pattern recognition as presented by Haken (1991) in *Synergetic Computers and Cognition.* According to these concepts, the synergetic computer can be realized by a three layer network as shown in Figure 2: The input layer with (model) neurons labeled by k, where $q_k(0)$ represents the initially given input activity of neuron k; the middle layer representing the order parameters ξ_j; the output layer with neurons labeled ℓ where $q_\ell(\infty)$ represents the final acticity of neuron ℓ.

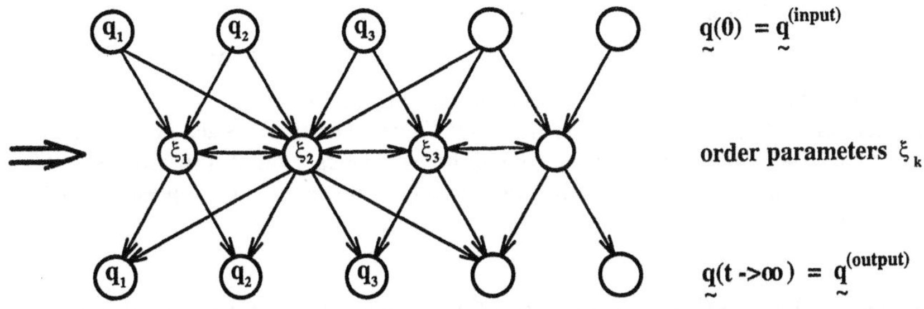

Figure 2: A three-level network of the synergetic computer. The first (upper) layer consists of model neurons that receive the input. This first layer projects on the second layer that represents the order parameters. The third layer represents the output from the order parameter layer. Though formaly similar to a neural computer arrangement, the algorithm of the synergetic computer is quite different, e.g. the model neurons are interacting by means of soft nonlinearities. Note that learned patterns are encoded in the connections between the first and second layers and those between the second and third layer. In the case of static patterns, the connections between the order parameters are of the same universal form, whereas in the case of dynamical patterns the order parameter connections may depend on the movement patterns to be generated. These remarks hold for all the following figures.

Note that in contrast to conventional neurocomputers, the numbers of neurons and order parameter cells are uniquely given. For what follows it will be convenient to look at the network of Figure 2 from the side, as indicated by the arrow. We then arrive at Figure 3. Now we are in a position to present our integrative view in a graphic representation (Figure 4). Here we have two kinds of inputs, $\mathbf{q}^{(\text{internal})}$ and $\mathbf{q}^{(\text{external})}$ and two kinds of outputs, again internal and external. The middle node symbolizes the brain, in which one or several order parameters of size ξ_j have been established. The index j differentiates between different order parameters. In the context of our paper it is important to note that the same order parameters ξ_j may govern quite different external outputs. For instance, the order parameter ξ_j may be connected with a specific output pattern $\mathbf{v}_j^{(e)}$. Such an output pattern may be a text or a drawing as in the Bartlett scenarios, or some other action, such as movements or writings, etc. that may lead to an external

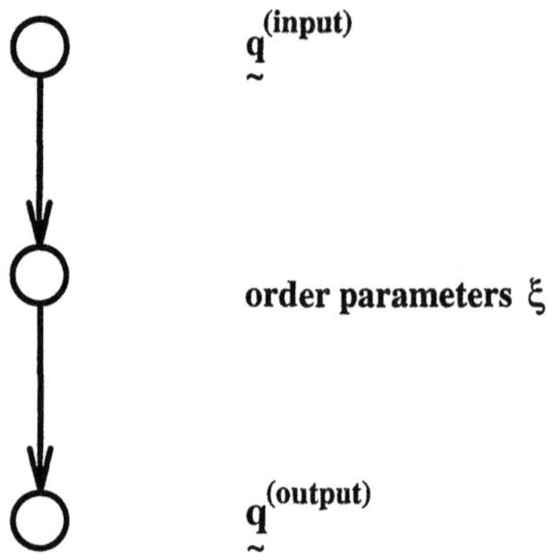

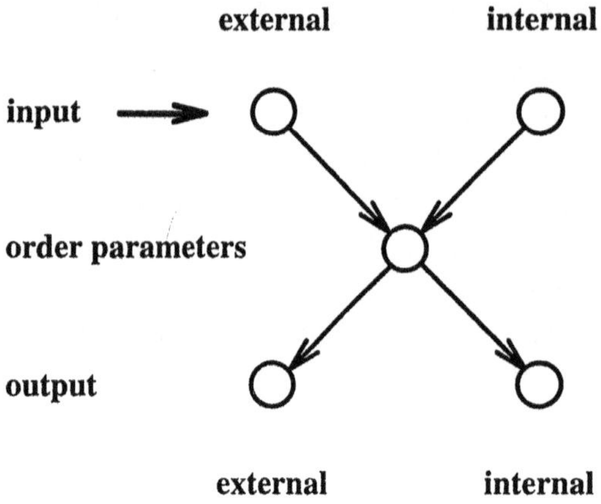

Figure 3 (Left): Figure 2 seen from the side as indicated by the arrow
Figure 4 (Right): The simplest cast of an IRN-model with its external in- and outputs and its internal in- and outputs. The middle area represents the order parameters. Note that in analogy to Figure 3, a network corresponding to Figure 2 is seen from the side so that each circle represents a whole set of model neurons

storage, say, in hand writings or in computers. The total set of possible drawings is then represented by the output vector $\mathbf{q}^{(e)}$

$$\mathbf{q}^{(e)}(t) = \sum_j \xi_j(t)\mathbf{v}_j^{(e)} \tag{5}$$

This output vector develops in the course of time. In general we will consider it for a large time so that the temporal change of the order parameters has finished. In an analogous fashion the order parameters ξ_j may govern the formation of *internal* patterns, such as internally stored images or learned patterns. These patterns are denoted by $\mathbf{v}_j^{(i)}$. The total set of possibilities is represented by

$$\mathbf{q}^{(i)}(t) = \sum_j \xi_j(t)\mathbf{v}_j^{(i)} \tag{6}$$

The reader should not be deceived by this notation in which patterns, both external and internal, are denoted by the same letter, $\mathbf{v}$. Note that the upper index may indicate quite different patterns, for instance, $\mathbf{v}_j^{(e)}$ may refer to spoken words, while $\mathbf{v}_j^{(i)}$ may refer to an internally stored image corresponding to that word.

In the next step of our analysis we have to consider the causes that generate the order parameters. To this end, according to Figure 4, we consider two different inputs to the order parameter level, namely an *external* and an *internal* input. The external input is denoted by $\mathbf{q}^{(e)}(0)$, the internal by $\mathbf{q}^{(i)}(0)$. The zero in brackets indicates that these inputs are taken at time t equal to zero. Again the inputs may have quite different modalities. The vectors $\mathbf{q}$ represent sets of data in different modalities and may have different dimensions. $\mathbf{q}^{(e)}$ is externally given via the senses, visual, auditory or tactile, for example. $\mathbf{q}^{(i)}$ is internally given, for instance, by vague ideas, phantasies, dreams, thoughts, etc.. An important step from the upper level, namely the input level, to the middle level dealing with order parameters is made by means of preprocessing. For instance, the given data may not be complete (compare Figure 5) or they are distorted or displaced in space, rotated or differently scaled (Figure 6). Thus the given patterns are checked against stored so-called prototype patterns denoted by $\mathbf{u}_j$. The preprocessing may play an important role, but we must be careful to distinguish between technicalities and essentials. For instance, it is a rather trivial task to shift a pattern in space or to slightly rotate it. A nontrivial task is, for instance, the rotation of a pattern by 180 degrees, where a computer has quite a

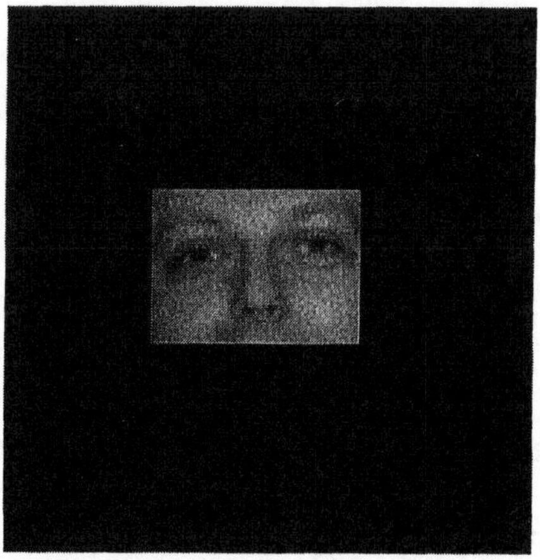

Figure 5 (Left): Example of an incomplete face.
Figure 6 (Right): A face that is rotated, displaced and scaled differently than the original.

different capability of recognizing a pattern from that of a human being, as is exemplified by Figure 7. An important subject of preprocessing is the removal of deformations. An example is shown in Figure 8. Quite evidently, the concept of Gestalt enters at this stage. In the computer one may either allow for rather general deformations or for limited deformations involving a cost function which limits the degree of deformations (Daffertshofer and Haken, 1994). These preprocessings may be performed either on $\mathbf{q}^{(e)}$ or on $\mathbf{u}_j$ or on both. By means of the prototype patterns $\mathbf{u}_j$, we may decompose the externally given data vector $\mathbf{q}^{(e)}$ according to

$$\mathbf{q}^{(e)} = \sum_j \xi_j^{(e)} \mathbf{u}_j + \mathbf{w}^{(e)}, \tag{7}$$

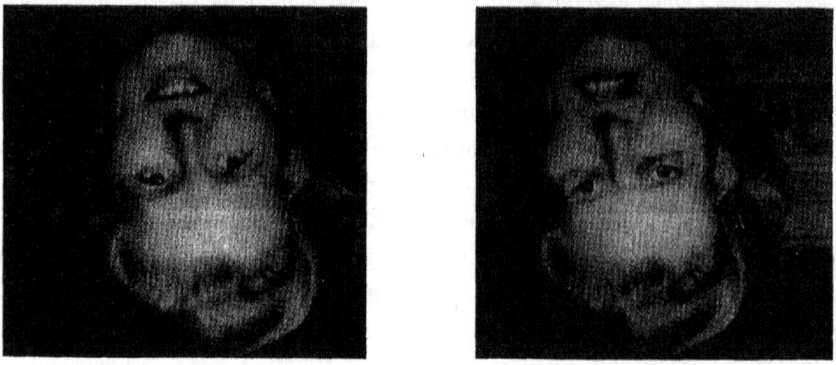

Figure 7: Two faces. Do you note any difference?

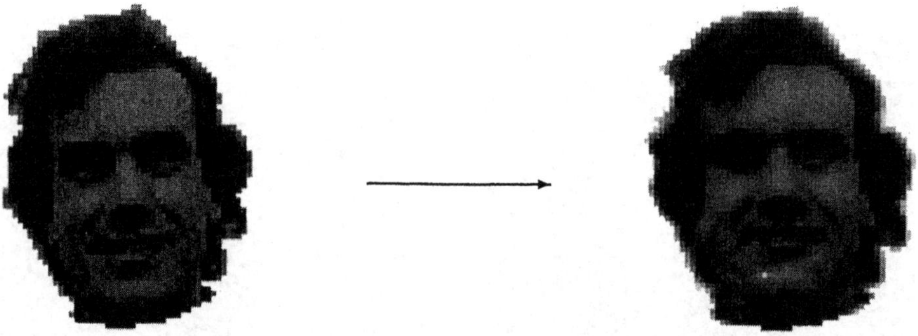

Figure 8: A deformed and undeformed face. The recognition procedure allows one to identify deformed faces.

where $\mathbf{w}^{(e)}$ is a rest term that need not be considered. An important concept is that of adjoint vectors $\mathbf{u}_j^+$ which obey the orthogonality relation

$$\left(\mathbf{u}_j^+ \mathbf{u}_k\right) = \delta_{jk} \tag{8}$$

where δ_{jk} is a Kronecker symbol = 1 for $j = k$ and = 0 for $j \neq k$. When we multiply eq. (7) by the adjoint vector, we immediately obtain

$$\xi^{(e)}(0) = \left(\mathbf{u}_j^{(e)+}\mathbf{q}^{(e)}(0)\right), \tag{9}$$

where the zero on both sides indicates that we take these values at time $t = 0$, i.e. at the beginning. In complete analogy to the externally given signal, the internally given signal may also be processed according to

$$\mathbf{q}^{(i)} = \sum_j \xi_j^{(i)}\mathbf{u}_j^{(i)} + \mathbf{w}^{(i)} \tag{10}$$

where, in general, we may assume that different criteria for the prototype patterns apply, or that the prototype patterns may have different modalities. That is why we distinguish the prototype patterns by upper indices e or i, respectively. In analogy to (9), we form

$$\xi^{(i)}(0) = \left(\mathbf{u}_j^{(i)+}\mathbf{q}^{(i)}(0)\right). \tag{11}$$

Now the question arises: how are the internal order parameters determined? To this end we define new order parameters for the total system, external and internal, i.e. the weighted superpositions of the order parameters (9) and (11) according to

$$\xi_j(0) = \alpha_j \left|\xi_j^{(e)}(0)\right| + \beta_j \left|\xi_j^{(i)}(0)\right| . \tag{12}$$

We then subject the order parameters ξ_j to a competition process that is well-known from pattern recognition by the synergetic computer. It means that in our approach we are dealing with a recognition of an internally or externally given pattern, in which different order parameters ξ_j with indices j compete and one order parameter eventually wins the competition, namely the one that obtained the highest value (12) at the beginning. This competition is described by the eqs.

$$\xi_j = \left(\lambda_j - (\boldsymbol{B} + \boldsymbol{C})\boldsymbol{D} + \boldsymbol{B}\xi_j^2\right)\xi_j , \tag{13}$$

where $\boldsymbol{D}$ is given by

$$\boldsymbol{D} = \sum_{j'} \xi_{j'}^2 , \tag{14}$$

and $\boldsymbol{B}$ and $\boldsymbol{C}$ are positive constants. Eq. (13) has the property that only one order parameter wins, or in other words, that the "winner takes all" strategy holds. One may think also of other mechanisms in which order parameters cooperate, but we shall not be concerned with this possibility in this paper.

Note that all the steps indicated above, including preprocessing, can be performed - and have been performed – by a computer so that our approach is entirely operational. This remark holds also for the rest of the present paper.

Some Applications to the Construction of Cognitive Maps

In the following we examine some central topics concerning the construction of cognitive maps by means of the above model. In particular, environmental learning, transformations between different modalities and collective cognitive processes will be considered.

Environmental learning

The question of environmental learning, that is to say, the way people learn an environment and in the process construct a cognitive map of it, was discussed in several research domains, including the development of spatial abilities, way finding, navigation and the like (see McDonald and Pellegrino, 1993 for a discussion and bibliography). A basic assumption in many of these studies is that the environment is *out-there*, and the task of the mind/brain is to learn the environment by encoding and internally representing what is already *out-there*. The notion of IRN, and its formulation in terms of synergetics in the above model, imply a different view. Environmental learning is seen as a process by which the individual is constructing patterns in the external environment and corresponding patterns in the mind. The result is an IRN cognitive map, part of the elements of which are constructed in the external environment, as external representations, and part in the mind/brain, as internal representations. The process as a whole evolves in line with the model we have presented above, as a synergistic interplay between external and internal inputs and outputs, ordered by the order parameters, which evolve in the process and enslave the interacting representational subsystems.

Consider, for example, a case of an individual who comes to a new environment, say a city, and makes several excursions in order to learn the area. Figure 9 illustrates graphically how the process develops in terms of our synergetic-IRN model. Starting from the left side of Figure 9 it can be seen that in the first excursion the individual's cognitive system is subject to two flows of incoming information: a flow of external input which comes in as the individual advances in space, and a parallel flow of internal input stemming from some initial previous environmental knowledge stored in the individual's memory. The interaction between the external and internal flows of input, ordered and enslaved by the order parameters which emerge in the process, entails also an interactive interplay between internal and external flows of output. In this sequential interplay between external and internal representations, objects and patterns in the external environment are being determined, marked, and internally represented as the IRN of the emerging cognitive map. The output of the first excursion is the first cognitive map of the area, and it provides part of the input for the second excursion, and so on in iterations. As can be seen in Figure 9, this map is composed of an external output $\mathbf{q}^{(e)}$ in the form of external landmarks, and an internal output $\mathbf{q}^{(i)}$ in the form of internally remembered and represented landmarks. The important feature of this process is that the cognitive system is not just photographing what is out-there in the environment, but it is actively constructing the external and internal network which makes the cognitive map.

It is important to note that at the start of the process the initial internal input need not be directly related to the new environment which is being learned. It is enough if the individual stores in memory some general environmental concepts and categories, such as

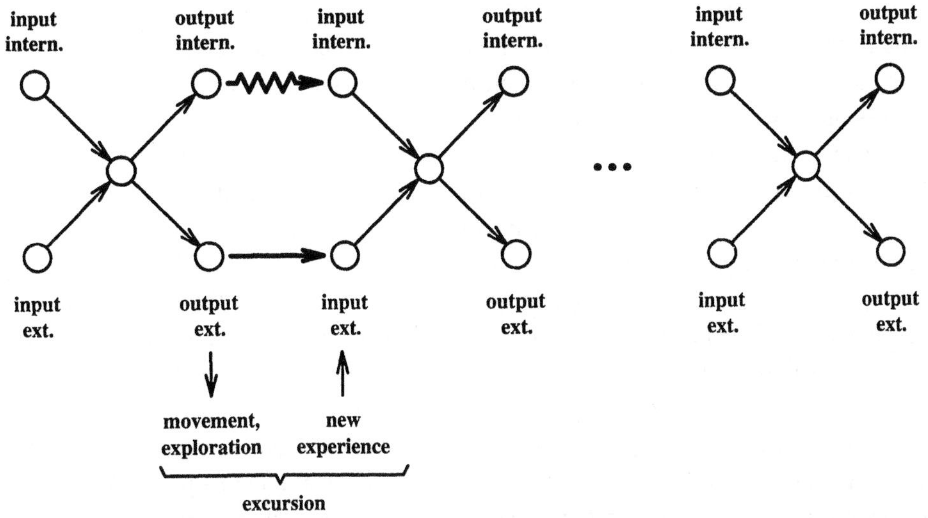

Figure 9: Graphic representation of a person exploring a neighbourhood. Each individual diagram represents the network within the person, and illustrates the different stages according to which the states of the network change.

"houses", "streets", "pavements", "traffic lights", "forest" and the like. Furthermore, non-environmental internal information can also play an important role in the process. For example, an Israeli making his first learning excursion in a European city is likely to be attracted by a Hebrew sign which will then become an externally and internally represented landmark of that individual's cognitive map.

Each excursion consumes the time and energy needed to walk, take notice, and (land)mark objects in the environment, and each excursion is also constrained by the time and energy available to the individual. The outcome of the first excursion is thus a partial and incomplete cognitive map: Only some objects and patterns in the environment are marked and become the elements of that individual's cognitive map. This externally and internally represented output then become the starting point, i.e. the input for the next excursion, and so on in a sequential process. Each new excursion adds more details to the previously constructed cognitive map and/or increases its perimeter. As conceptualized by our synergetics IRN model, in the first excursions there are likely to be marked differences between the cognitive maps reproduced from iteration to iteration. Such fluctuations are interpreted as a competition between the order parameters of the emerging cognitive map. From a certain point onwards, an order parameter of the area is established, enslaves the various externally and internally represented patterns, and brings the cognitive map to a steady state. Once this state is reached, new excursions do not change much the structure of the cognitive map of that area (see Figure 2 in Portugali, this volume).

The same mechanism will apply if the person is taking many excursions in various parts of the city: in the neighborhood, in the center, in recreation areas and the like. In each of these parts of the city and in their overall configuration, we will have a similar process: in all we have a situation where a certain order parameter of the neighborhood, the center .. the whole city, enslaves the internal and external incoming and interacting data. As just noted, according to the synergetic approach to cognitive mapping, at this stage the system, i.e. the cognitive map, enters a steady state, during which it does not change dramatically. This stable state is reached not necessarily when the cognitive map corresponds to the accurate cartographic map, but when it enables the person to survive and function in that specific environment. Thus, as was found in various studies (Downs and Stea, 1976), different people (children, taxi-drivers, pilots, and so on) tend to construct their own personal and specific cognitive maps.

The above description might lead to the impression that cognitive maps constructed as above are very personal and subjective. This is indeed so, but up to a point. As noted in previous studies and above (Portugali, 1990; Portugali and Haken, 1992; Portugali, this volume), many of the concepts, categories, schemata and patterns that we use in constructing our personal cognitive maps come already self-organized and enslaved by a complexity of inter-subjective collective order parameters. Ordinary spoken and written language has already been suggested as an example of such a collective order parameter (Haken, 1994). Another example from the domain of architecture and urbanism is Alexander's et al. (1977) Pattern Language (Portugali, this volume), whereas Lakoff's (1987) models, as derived from his "the spatialization of form hypothesis", might be interpreted as order parameters (Portugali, ibid) and thus as a third example related to cognition in general. These and other collective variables and order parameters play an important role in the synergetics-IRN process of cognitive map construction, with the consequence that the processes of environmental learning and cognitive mapping are associated with many collective variables and properties.

It is not difficult to cast the feedback mechanism described in this section into a mathematical form. Consider Figure 9 to this end. By means of the whole process an internal output $\mathbf{q}^{(i)}$ is eventually produced. This output may coincide with an already stored pattern, but it may also lead to the formation of a new pattern or a distorted new pattern. This new pattern may then serve as a new prototype pattern that is entering the preprocessing, i.e. it enters as a new $\mathbf{u}_j^{(e)}$

$$\mathbf{q}^{(i)}(\infty) = \mathbf{v}_j^{(i)} = \mathbf{u}_j^{(\text{new})} \tag{15}$$

This process can be modeled by the learning algorithm of the synergetic computer (cf. Haken, 1991). Note that the external output may involve movements and explorations, while the external input consists of new experience. In complete analogy to this process, the internal output may also determine a new prototype pattern that is used internally to check suggested ideas etc.. This new $\mathbf{u}_j$ again is defined by

$$\mathbf{q}^{(i)}(\infty) = \mathbf{v}_j^{(i)} = \mathbf{u}_j^{(\mathrm{new})} \quad . \tag{16}$$

Our approach may be generalized so that a cross-over modalities can also occur, so that one $\mathbf{u}_j$ in the output is replaced by a new $\mathbf{u}_j$ as a prototype pattern vector in the input.

Transformations in Cognitive Mapping

As afore said, because of size, large scale environments such as neighborhoods, cities and the like cannot be seen in their entirety, and consequently the mind/brain must construct their internal spatial representation by means of various forms of visual (navigation, map learning etc.) and non-visual (haptic, verbal etc.) means of information. Cognitive maps are thus by their nature multi-modal entities. Consequently, transformations of information from one modality to another play an important role in the construction of these maps. In fact, several of the studies in this volume elaborate in some detail such transformations (see Daniel et al. and Franklin on transformations from verbal to visual information, Golledge et al. and Ungar et al. on visually impaired people, and Lloyd and Cammack on transformations between various forms of visions). The suggestion here is that our model lends itself in a natural way to be a framework for the study of such transformations.

The synergetic IRN model for transformations is graphically shown in Figure 10. As can be seen, external input $\mathbf{q}^{(e)}$ is coming in one modality, say verbal, and internal input $\mathbf{q}^{(i)}$ in another modality, say, in the form of spatial images. An example is a situation in which one person is describing to another the way from location A to location B. The describer is producing an external verbal representation of his internally represented image of the route. This externally represented verbal flow becomes the external input for the listener. Parallel to this inflow of external verbal information, the listener is also subject to internal imagined visual flow of information. The latter is derived from the listener's memory, as described in the previous section on environmental learning. As in the latter case, here too, this information might be previous knowledge of the area under consideration, but need not be so. Some general environmental concepts, categories and schemata might suffice. The interaction between the incoming external verbal information and the incoming internal imagined-visual information gives rise to an order parameter which enslaves in the usual way the external and internal subsystems and features and governs the interaction between the subsystems and their output.

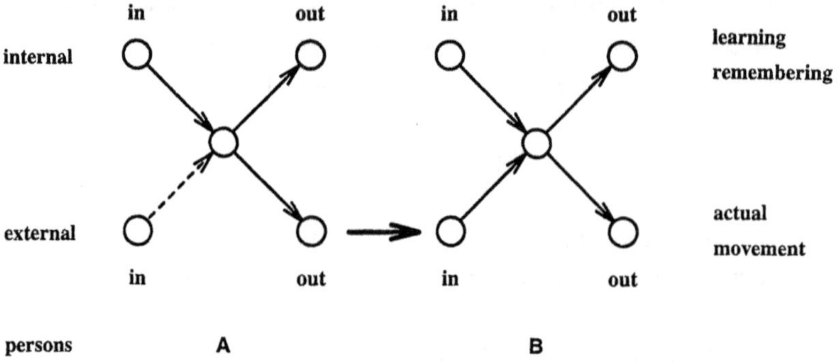

Figure 10: Communication between two persons A and B.

Collective cognitive maps

Portugali (this volume) has discussed at some length the implications to cognitive mapping of Donald's (1992) suggestion regarding the three major cognitive transitions in the evolution of the human mind: *The mimetic transition,* which for the first time enabled hominids to use their whole body as an external representation device; this stage entailed the emergence of mimetic cognitive maps which could represent the environment by means of mimesis. *The lexical transition,* which enabled internal and external speech and thus various forms of external verbal memory, including various forms of verbal cognitive maps. *The externalization of memory* is the third transition and it came into being with the appearance of writing. The essence of this stage is that some parts of human memory and thought could be put in the external environment, in archives, libraries and the like, and thus become independent of biological human memory. This externalized memory included also lexical cognitive maps, that is to say, written descriptions of buildings, cities, territories and other large-scale spatial entities. It is interesting to note, that some of the earliest known texts - the cuneiform found in the ancient city of Uruk – were detailed territorial descriptions, and that according to available data, these cognitive textual maps preceded the appearance of the drawn pictorial maps. As conceptualized by Donald (ibid), each sequential cognitive transition did not replace its previous one, but added to it new cognitive capacities, with the implication that each sequential transition entailed a more complex IRN, and a more complex external memory (Portugali, ibid).

One of the most important properties of the notion of IRN concerns the collective potential of externally represented elements of the network. Once a person constructs an external representation, it becomes a public domain. Other people might use it for various purposes, including as the external input to some of their IRNs. The externally represented elements of the individual thus enter various collective processes, cultural, social, and the like. This applies also to Donald's mimetic, lexical and textual forms of external representation; they enable the emergence of external collective representations, which are both biological (mimetic and lexical) and non-biological. The same with cognitive maps. On the one hand, they are subjective and personal, but on the other, because of their externally represented elements, they are engaged in collective cultural, social and environmental processes, which are themselves self-organizing systems (See Portugali, 1993; Haken, 1994). We thus have here an interaction between cognitive processes on two scales. One which is personal, where the individual interacts with the environment by means of internal and external representations as described above, and one which is collective, and its elements are the external representations of the individuals.

One of the main principles of the synergetic approach to self-organization is that the relatively slow order parameters tend to enslave the relatively faster ones, with the implication that when interacting with the environment the individual's cognitive system is likely to be enslaved by externally represented elements, shaped by collective cognitive processes. The individual thus interacts with an environment which is already self-organized and in the process internalizes elements which have been shaped by cognitive collective processes.

To see how the process works in terms of our synergetic-IRN model, we return to the Bartlett scenarios and reformulate them in line with our model. In order to complete the description we will start with a subjective *intra personal* scenario and proceed with an *interpersonal* collective scenario and then with a "more collective" *interpersonal with a common reservoir* scenario. As already noted, the first two scenarios were originally designed by Bartlett to illustrate how memory works internally, and by Stadler and co-workers to suggest that the cognitive system is a self-organizing system, which evolves in line with the synergetics of cognition. We use the various scenarios to illustrate the synergistic IRN nature of various cognitive processes and the way they are linked with externalized collective cognitive entities.

Intrapersonal subjective process

The externally produced output of a person may be first laid down, say by writing, and then reading this again can serve as an input to the same person. This leads to a description of Bartlett's experiments with respect to a single person. This iterative process is graphically described in Figure 11 and formally by

$$\mathbf{q}_{n+1}^{(e)}(t=0)=\mathbf{q}_{n}^{(e)}(t\rightarrow\infty),\qquad \mathrm{n}=1,2,\ \ldots \tag{17}$$

The index *n* refers to the number of iterations. Note that this process as indicated by $\mathbf{q}_n^{(e)}(t \to \infty)$ may be rather complicated because of cost functions. The reproduction of the pattern, for instance by drawing, may be imperfect. Rather, one has to consider a convergence, as shown experimentally by Bartlett, and as to be shown in a paper to be published by the present authors.

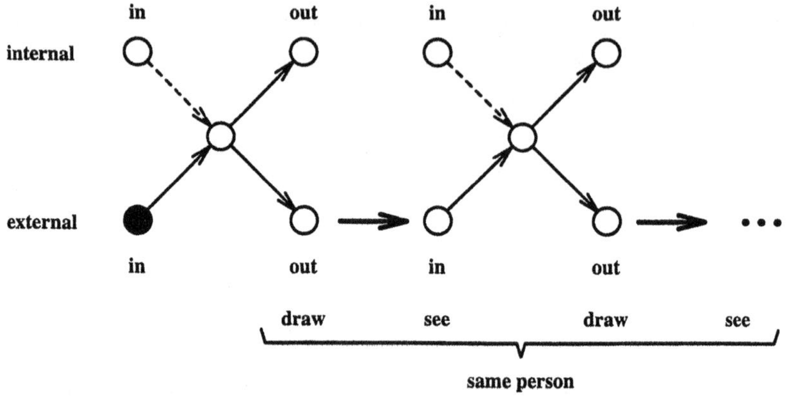

Figure 11: The same as Figure 10, but with respect to the same person

Interpersonal collective process

This is the classical Bartlett scenario as originally formulated by Bartlett and recently reproduced by Stadler and Kruse (1990; Kruse and Stadler, 1993). A typical experiment of this kind starts, as noted, with a given external input, such as a story, a drawing and so on, and proceeds with a sequence by which each person's externalized reproduction of the remembered input becomes an input to the next person to remember and externalize, and so on. Figure 12 illustrates such a sequence. As noted, the interesting result of such a scenario is that as in the intrapersonal case, here too, after several initial steps which exhibit major changes from one reproduction to the other, the story or the drawn figure stabilizes and does not change much from iteration to iteration. In the language of synergetics we would say that a certain order parameter has enslaved the system and brought it to a steady state. From the perspective of the synergetics of IRN, the interesting part of this experiment is that the system here is interpersonal. That is, several people are involved in generating these reproductions and their individual-subjective cognitive systems participate in producing a collective and externalized cognitive product. The results thus show how, as a consequence of the interplay between internal and external representations of individuals, an external collective cognitive product is constructed, without the individuals being aware of having participated in such a collective enterprise. As this sequential process evolves, and its collective product constructed, each individual's externally represented reproduction gradually becomes "more" collective and so does each individual's internally represented remembering. The individuals engaged in the process are thus being enslaved by the collective order parameter which emerges in the process. This interpersonal scenario can be described by means of the formula

$$\mathbf{q}_{n+1}^{(e)}(t=0) = \mathbf{q}_{n}^{(e)}(t \rightarrow \infty), \qquad n = 1, 2, \ldots \tag{18}$$

The index n refers to the person involved in the corresponding step before the transfer takes place. From a formal point of view the process in Figure 11 is indistinguishable from the process in Figure 12, provided that the preprocessing is governed by the same explicit laws for each person.

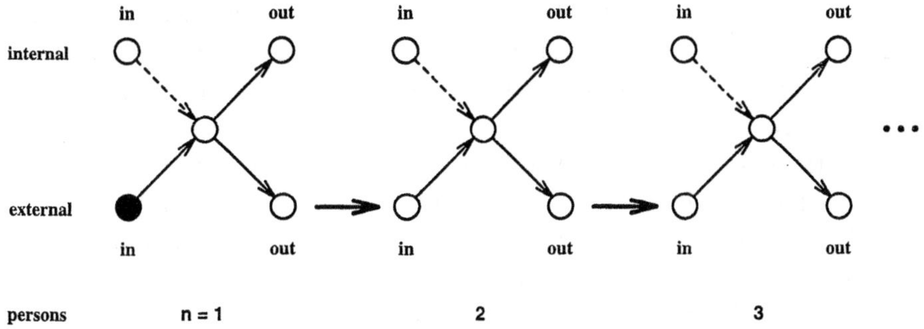

Figure 12: A sequence of communications between persons (compare text).

Interpersonal with a common reservoir

Note that the usual Bartlett scenarios and the above formulation are fully dependent on the biological memories of individuals. In order for a genuinely externalized memory and cognitive map, as defined by Donald (ibid), to emerge, we have to allow for some parts of the emerging external collective memory to become detached and independent from the biological memories of the individuals engaged in the process. According to Donald (ibid), this stage in human cognitive evolution became possible with the appearance, in the 3rd millennium B.C, of writing, and since then with the development of other means to externally store information. With these new means the individual does not need to remember the whole story or myth or landscape or territory, but just the code (i.e. language) to this common reservoir. Donald's externalized memory is thus restricted to what might be termed "pure" forms of information (texts, computerized data bases, maps, GISs and the like).

According to Portugali (ibid), such a restriction is neither needed nor necessary. First, many of the artifacts produced by humans have been created in a similar process of serial reproduction, that is, by an interplay between external and internal representations as formulated by our synergetic-IRN model. Second, many artifacts are not just functional tools, but they are also information carriers: Artifacts such as altars, sculptures, paintings, temples, office buildings, private houses, urban squares and whole city landscapes, transmit information about holiness, power, money, modesty, and most importantly, information about various forms of social, cultural, environmental or urban orders. The suggestion is that in the same the way people learn specific languages and thus gain access to externalized memories in the form of texts or computerized databases, they also learn environmental and urban languages and thus gain access to the externalized information enfolded in the environment and in the urban landscape. This suggestion is in line with Alexander et al. (1977) and his notion of *A Pattern Language* of architecture and urbanism, as well as with Lakoff (1987) and Edelman (1992), who seem to speak in line with Langacker's (1987) "cognitive grammar", which is beyond the grammar of ordinary lexical language. The more profound process of environmental learning is thus the process by which one learns the language of the environment - its semantics, its grammar and the order enfolded in them. As with ordinary language, once one learns the environmental or urban language, one can use it for the more specific tasks of navigation, way-finding and the like. This environmental language, with its environmental concepts, categories and schemata, was in fact one of the sources of the flow of internal input in the environmental learning model we have devised above.

The suggestion in this section is that this common reservoir of external, artificial and non-biological memory, in the form of texts, databases as well as of buildings, cities and environments, is created by a process the principles of which we have elaborated in our generalized synergetic-IRN model. As an illustration of the way the process develops, Portugali (this volume) has devised an experiment which is a new type of a Bartlett scenario. Its essence is a process of a sequential reproduction which is interpersonal, collective, and in addition also *public* – the participants observe the common reservoir as it develops. The experiment can be regarded as a game in which each player in turns is locating a building in the city (using mockups on a scale of 1:50). In a typical game (ibid, Figure 6) the players observe the city as it develops, and in the process also learn the order which is spontaneously emerging in the environment. After several initial iterations a certain urban order emerges, the participants internalize this emerging order and tend to locate their buildings in line with it. Such an experiment thus includes all the ingredients of the synergetic-IRN model: sequential interplay between internal and external representations, strong fluctuations at the start of the process, emerging order parameter(s), the principle of enslavement, and a relatively stable steady state.

Graphically, and in terms of our model, it can be described as in Figure 13, where each individual introduces his or her externalized output into a common reservoir and receives external input from that reservoir. From a formal point of view we have the relations

$$\mathbf{q}_k^{(e)}(0) = \boldsymbol{W}_k \mathbf{q}\text{common}, \quad k = 1,\ldots h \tag{19}$$

and

$$\mathbf{q}\text{common} = \sum_k c_k \mathbf{q}_k^{(e)}(t \to \infty)\,. \tag{20}$$

The index n enumerates the individuals. $\boldsymbol{W}_k$ is a "personal window" operator which selects part of the information stored in **q**common.

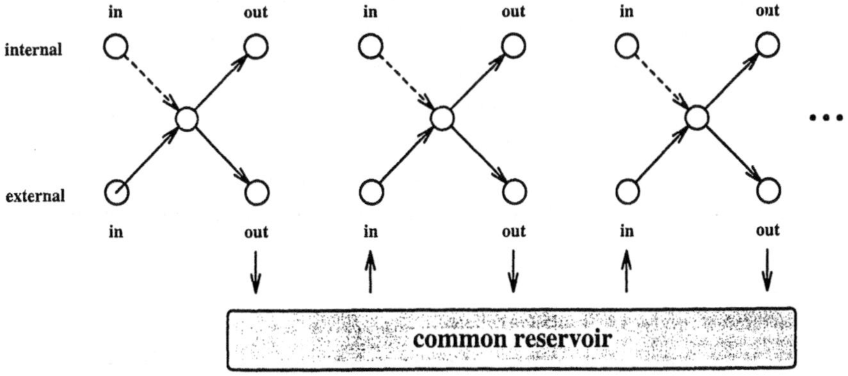

Figure 13: Communication between persons and a common reservoir, such as a library or even a city.

Concluding Remarks

In our paper we have shown how the notion of IRN may be cast into the conceptual framework and formalism of synergetics. This was done by means of a graphical representation which allowed for a simple visualization and provided us with a prescription for the mathematical model that followed. According to the latter, input data encoded in vectors **q** are preprocessed and transformed into order parameters and prototype patterns. The order parameters then govern new external and internal output patterns that may occur themselves in different modalities.

Casting the notion of IRN into the formalism of synergetics made the somewhat abstract concept of IRN operational, and as illustrated in the second part of the paper, ready to be applied as a general framework for the various aspects of cognitive mapping. The next stage would be to further elaborate and develop the model in conjunction with experiments and empirical data from specific case studies.

Reformulating IRN in terms of synergetics has added further insight to cognitive mapping and its relation to IRN. In particular, the nature of cognitive maps as self-organizing systems is now much clearer and so are the relations between the various modalities engaged in their construction and the complex dialectical tension between individuals' cognitive maps, collective cognitive maps and the physical environment. The construction of cognitive maps appears here as a process by which individuals actively construct their own external and internal environments, and in the process are engaged in the construction of collective cognitive maps and the artificial environment itself.

While our main focus was on cognitive maps, the notion of IRN, the model we have developed, and some of the issues we have elaborated, extend beyond the specific domain of cognitive maps. In particular, our synergetics-IRN model has proved to be an efficient tool for modelling sequential cognitive processes associated with general issues such as remembering, inter-modal transformations, and the self-organization nature of cognitive processes in general. Our model contributes also to the general theory of synergetics, as it genuinely integrates the processes of pattern formation and pattern recognition – an idea already suggested in the past in the claim that "pattern recognition is pattern formation" in the mind (Haken, 1991). In our synergetic-IRN model we show that the relations between pattern formation and pattern recognition are not only analogical, but homological – the relations between two interactive facets of a single cognitive system.

References

Alexander, C. et al. (1977). *A Pattern Language*, New York: Oxford University Press.

Alexander, C. (1979). *The Timeless Way of Building*, New York: Oxford University Press.

Bartlett, F.C. (1932/1961). *Remembering: A Study in Experimental and Social Psychology,* Cambridge: Cambridge University Press.

Daffertshofer, A. and Haken, H. (1994). A new approach to recognition of deformed patterns. *Pattern Recognition* 27 (12) 1697-1705.

Donald, M. (1991). *Origins of the Modern Mind: Three Stages in the Evolution of Culture and Cognition*, Cambridge Mass. and London: Harward University Press.

Downs, R.M. and Stea, D. (1976). *Maps in Minds*, New York: Harper & Row.

Edelman, G.M. (1992). *Bright Air Briliant Fire: On the Matter of the Mind*, London: Penguin Books.

Gibson, J.J. (1979). *The Ecological Approach to Visual Perception*, Boston: Houghton-Mifflin.

Haken, H. (1979). Pattern formation and pattern recognition – an attempt at a synthesis. In *Pattern Formation by Dynamical Systems and Pattern Recognition* (H. Haken, ed.), pp Berlin: Springer Verlag.

Haken, H. (1983). *Synergetics, An Introduction*, 3rd. ed., Berlin: Springer.

Haken, H. (1987). *Advanced Synergetics, An Introduction*, 2nd. print., Berlin: Springer.

Haken, H. (1991). *Synergetic Computers and Cognition*, Berlin: Springer.

Kruse, P. and Stadler, M. (1993). The significance of nonlinear phenomena for the investigation of cognitive systems. In *Interdisciplinary Approaches to Nonlinear Complex Systems* (H. Haken and A. Mikhailov, eds.), pp. 138-160. Berlin: Springer.

Lakoff, G. (1987). *Women Fire and Dangerous Things: What Categories Reveal About the Mind*, Chicago: The University of Chicago Press.

Langacker, R.W. (1987). *Foundations of Cognitive Grammar, Volume 1: Theoretical Prerequisites*, Stanford: Stanford University Press.

McClelland, J.L., Rumelhart, D.E. and the PDP Research Group (1986). *Parallel Distributed Processing, Explorations in the Microstructure of Cognition. Volume 2: Psychological and Biological Models*, Cambridge Mass: MIT Press.

McDonald, T.P. and Pellegrino, J.W. (1993). Psychological persperctives on spatial cognition. In *Behavior and Environment* (T. Gärling and R.G. Golledge eds.), pp. 47-82. Amsterdam: North-Holland.

Portugali, J. (1990). Social synergetics, cognitive maps and enviromental recognition. In *Synergetics of Cognition* (H. Haken and M. Stadler eds.), pp. 379-392. Berlin: Springer.

Portugali, J. (1993). *Implicate Relations: Society and space in the Israeli-Palestinian conflict*, Dordrecht: Kluwer Academic Publishers.

Portugali, J. and Haken, H. (1992). Synergetics and cognitive maps. In *Geography, Environment and Cognition* (J. Portugali ed.), a special theme issue *Geoforum* 23 (2), 111-130.

Rumelhart, D.E., Smolensky, P., McClelland, J.L. and Hinton, G.E. (1986). Schemata and sequential thought processes in PDP models. In *Parallel Distributed Processing, Explorations in the Microstructure of Cognition. Volume 2: Psychological and Biological Models* (J.L. McClelland, D.E. Rumelhart, and the PDP Research Group), pp. 7-57. Cambridge Mass: MIT Press.

Stadler, M. and Kruse, P. (1990). The self-organization perspective in cognition research: historical remarks and new experimental approaches. In *Synergetics of Cognition* (H. Haken and M. Stadler eds), pp. 32-52. Berlin: Springer.

Vygotsky, L.S. (1978). *Mind in Society*, Cambridge Mass.: Cambridge University Press.

Hermann Haken
Institute for Theoretical Physics and Synergetics
Pfaffenwaldring 57/4, D-70550 Stuttgart.

Juval Portugali
Tel Aviv University, Department of Geography
Ramat Aviv, Tel Aviv 69978, Israel.

NEURAL NETWORK MODELS OF COGNITIVE MAPS

Sucharita Gopal

Abstract: Neural networks - parallel systems consisting of highly interconnected simple neuron-like processing elements - have become a subject of intense interest to scientists spanning a broad range of disciplines including psychology, physics, mathematics, computer science, biology and neurobiology. A variety of connectionist or neural network models have been proposed, ranging from simplified models to more realistic models of learning, associative memory, and sensori-motor development. Neural networks hold a great promise in the field of spatial cognition since they are capable (in principle) of approximating any real valued function mapping, and have been used to solve complex problems in allied fields such as visual pattern analysis and robotic control. In addition, neural networks have biological relevance, a problem that has plagued the field of symbol processing Artificial Intelligence (AI) systems. Neural networks simulated based on known physiological and anatomical properties of the brain may reveal the process by which groups of neurons interacting according to some local rules undergo self-organization. This paper will examine specific problems in spatial cognition where neural networks have been used and have produced plausible models of spatial behavior. Neural networks have helped in understanding the types of computations that might be performed by the place cells in the hippocampus when the animal moves about and constructs an internal spatial representation. Other problems in spatial cognition such as recognizing places and locating goals further demonstrate the success of neural networks in this domain. A neural network model of route learning is proposed that can learn different routes in an environment and locate a goal given the route information.

Introduction: What are Neural Network or Connectionist Models

Neural Network or connectionist models are large networks of extremely simple computation units, massively interconnected and running in parallel. Formally, a neural network may be viewed as a dynamic system with the topology of a directed graph which can carry out information processing by means of its state response to continuous or episodic input (Hecht-Nielson, 1990). The nodes of the graph are called processing elements (PEs), and the directed links (unidirectional signal channels) are termed as connections. The PEs communicate with one another by signals that are numerical rather than symbolic. Neural network models are designed to perform a task by specifying the

J. Portugali (ed.), The Construction of Cognitive Maps, 69–85.

architecture: the number of processing elements, the network topology (i.e. the interconnections of the PEs), and the weight or strength of each connection via learning rules. These models can be applied on a large scale to model the whole brain system or, on a smaller scale, to model specific circuits in the brain. The principles underlying these models were first proposed by Hebb (1949) and were influenced by several researchers including Steinbuch (1961), Marr (1982), van Malsburg and Willshaw (1981), Kohonen (1984) and Grossberg (1982).

Are neural networks or connectionist models a viable alternative to traditional symbolic models of cognitive function? To answer this question one must examine the fundamental issues of psychological adequacy and neurobiological plausibility of connectionist models. In the first place, representation of knowledge in connectionist networks involve patterns of activity across many PEs. The behavior of connectionist networks emerges from the structural changes at the level of individual PEs, in terms of activation thresholds and connection weights. The distributed information storage in connectionist networks means that any given PE is involved at different times in the representation of many different events. In this context, associative memory is the ability to reinstate a previously active pattern given a related one or, more generally, to retrieve a previously active pattern given an incomplete piece or fragment of the original pattern (pattern completion or feature detection). The principles governing modification of connection strength and activation in fact represent the learning and memory procedures (Smolensky, 1988). This is in contrast to symbolic AI systems where, in general, semantically interpretable symbolic elements are transformed by a set of rules. These rules (or computer programs) are themselves represented symbolically. The distinction between program and computer (software and hardware) that is intrinsic in symbolic processing is absent in connectionist models.

Neural network models tend to have "soft" constraints and exhibit a graceful degradation in performance. Removal of some processing units leads to a gradual decline in performance but never a total break-down in the system. Thus connectionist models have some psychological relevance. Symbolic systems, on the other hand, tend to have "hard" constraints and are brittle. They are not fault-tolerant and may fail outside the exact environment for which they were developed. Removal of particular units imply a loss of specific functions resulting in abrupt degradation in performance.

The connectionist approach can be considered a theoretical approach to brain modeling in contrast to an empirical approach. The empirical approach attempts to delineate the micro-level details of biochemical and biophysical properties of neurons, the rules that determine their connectivity and the mechanisms through which their properties and connections are modified during the learning process. This approach leads to realistic brain models (Sejnowski, Koch and Churchland, 1988). For example, Hodgkin-Huxely model (1952) of the action potential in the squid giant axon is a realistic model at the level of a single neuron. The action potential is the transient, electrical event that propagates along an axon and is used for relaying information over long distances. The action potential is the direct result of voltage- and time-dependent properties of several types of membrane channels. Hence any attempt to model action potential must model the dynamics of membrane channels. Hodgkin and Huxley modeled the dynamics by a set of coupled, nonlinear differential equations that were solved numerically. Their model predictions of action potential was within 10% of the measured value. In this context, it is important to note that realistic models of this type are based on an extensive number of empirical experiments.

In many instances, it may be impossible to obtain information about the cellular structure or perform experiments in living tissue such as selective lesion of particular channels. The solution is to adopt a theoretical approach. The theoretical approach uses mathematical models in an attempt to simulate and synthesize known and hypothesized principles of brain functions. This approach may provide insight into the brain by showing how a particular function is in principle computable by neurons. Neural network models come under this approach. These models abstract from the complexity of individual neurons and the patterns of connectivity in exchange for analytical tractability (Hopfield and Tank, 1986). The most widely used architecture is the class of layered feedforward networks. This architecture is characterized by a hierarchical design consisting of an input layer, successive layers called hidden layers, and an output layer. Each layer consists of neurons. Information is coded as a pattern of activity that is successively transformed by the successive layers. The intermediate or hidden layers are capable of representing complex non-linear functions (Lippmann, 1987; Lapedes and Farber, 1988). The performance of the network depends on a number of critical parameters including the initial set of weights, input pre-processing, etc. These models have been useful in modeling problems in certain domains (Hinton and Anderson, 1981; Rumelhart and McClelland, 1986).

The theoretical approach is often labeled as a "dry" approach and the empirical approach as a "wet" approach to modeling. The work of the theorists and modelers result in new hypothesis to test experimentally, while the work of the empiricists provides new data for the improvement of the models. It should be noted that the integration of the two approaches can progressively lead to a greater understanding of the brain. In the following, we will first describe the nature of the cognitive map and then examine specific problems in spatial cognition where neural networks have been used and have produced plausible models of brain functioning and behavior. We will limit our discussion to a few models. This choice reflects the research interest of the author and in no way implies that other models developed in this area are not relevant.

Cognitive Map

In the last two decades, a variety of disciplines have investigated people's ability to learn about and navigate in large-scale spatial environment. A large portion of wayfinding research has been carried out explicitly or implicitly within the theoretical frameworks of environmental cognition put forward by psychologists. Siegal and White (1975) described the sequence of spatial representation developed by individual learning about a large-scale environment. The first stage is marked by an ability to identify objects or landmarks, followed by an integration of knowledge about the routes linking individual landmarks. The individual finally obtains an overall understanding of the spatial relationships between all the places within that environment. Such an understanding has been referred to by a variety of terms: *configurational knowledge* or *survey knowledge* or `vector knowledge' or more generally as a `cognitive map'. Thus the contents of the cognitive map evolves over time as its structure becomes increasingly complex. Cognitive maps include both spatial as well as nonspatial knowledge (Lynch, 1960). For example, certain areas of the city (spatial information) may evoke fear (nonspatial information) and may be avoided. Spatial knowledge consists of three levels -- sensori-motor, topological, and metric knowledge (Kuipers, 1978). Sensorimotor knowledge is the most primitive level of spatial knowledge and consists of a sequence of view-action pairs. The navigating individual is aware that a certain motor action (either a turn or move) will result in a certain view of the environment. A sequence of such view-action pairs can be stored in the cognitive map that can enable an individual to navigate a route or compute what can be seen if a certain action is executed at a particular location. The second level of knowledge is topological; proximity, closeness, dispersion or clustering relationships between objects (landmarks) in the environment are stored in the cognitive map. There is some evidence to show that individuals in new and novel environments show topological accuracy but not metric accuracy. The third level of knowledge is metric knowledge; both (absolute and relative) distance and direction information may be gained by the navigator. A sense of direction is more important than a sense of distance (Levine et al., 1982).

The cognitive map is used for a variety of wayfinding problems including-identification of current location, finding a desired destination in the environment and planning a path between current to desired destination, making inferences about spatial relations (e.g. distance and direction), and other kinds of problems. Performance outcomes in each of the above problems depends on how information is encoded into and retrieved from the cognitive map. (Golledge et al., 1985; Hirtle and Jonides, 1985; Levine et al. 1982).

Neurobiological Studies Relating to the Structure of the Cognitive Map

Before turning to a discussion of neural network models of the cognitive map, it is important to point out prior experimental and neurophysiological data that suggests spatial cognition may be supported by a number of distinct cognitive systems. For a review of neuroanatomical and functional divisions in the visual system of primates see Mishkin et al., (1983) and Parkinson et al., (1988). The role of the hippocampus in representing spatial information is implicated in a number of studies (O'Keefe and Nadel, 1978; McNaughton, 1989; Eichenbaum and Buckingham, 1990; Squire et al., 1989). Much of what is known about the neural basis for the cognitive maps has come from behavioral and neurophysiological studies on rat. Hence most of the neural network models of cognitive maps have also been built to model the navigational abilities of these remarkable creatures. However, the observations in rats may have correlates in the hippocampal functions of primates and humans.

O'Keefe and Nadel, in their seminal work on cognitive maps, make a distinction between two systems in rat brain that represent route and survey knowledge. The taxon system supports route knowledge and locale system contains the hippocampus which processes the spatial relations among multiple sensory cues and constructs a spatial or cognitive map of the environment. Mishkin and Appenzeller (1987) identified a possible neuroanatomical locus for route knowledge in the habit system that includes the striatum (a part of basal ganglia) as the link between the sensory and motor systems. The habit system is responsible for stimulus-response learning in monkeys. This stems from the fact that route knowledge is viewed as a sequence of stimulus-response-stimulus associations for how to get from an origin to a destination or goal.

O'Keefe and Nadel (1978) proposed that the hippocampus acts as a cognitive mapping system. This view is supported by activity of the hippocampal neurons during spatial memory tasks. O'Keefe and Nadel's studies of hippocampal neuron in freely moving rats indicate its role in the representation of spatial knowledge, i.e. the hippocampus acts as the cognitive map. Experiments reveal that certain locations in the external environment of the rat correlate to place field cells in the animal's hippocampus. Place fields fire at a maximum rate only when the animal is located at a particular location relative to a set of distal landmarks. Removing landmarks or altering their spatial configuration removes the place field responses of these units in the hippocampus (O'Keefe and Nadel, 1978). Zipser (1986) reports that many place field cells code places on environment-centered coordinates. It is also noted that the response of the place field cells is independent of the type of motor activity in which the rat may be temporarily engaged. Thus a location is recognized by the spatial configuration of landmarks around it. Place fields are also sensitive to viewer orientation as the rat has an ability to distinguish right from left. Place fields that fired at a particular location when the animal was headed in one direction do not fire when the animal is at the same location headed in the reverse or opposite direction. Zipser (1986) refers to these cells as view-field cells because they code a place in viewer-centered coordinates.

Other studies relating to the hippocampus indicate there may be many layers of the cognitive map (O'Keefe, 1989). Representations of these maps are probably distributed across the same set of synapses and neurons in the hippocampus. This means that maps that share a same set of landmarks share the same synapses and neurons. When an animal moves in a familiar or novel environment, a whole set of maps may be examined and compared to the current representation. Operations involving rotation, translation or, dilation may be performed to retrieve the map representing the best match. All the above findings suggests that the cognitive map has a complex structure and a variety of spatial information is encoded in it.

More recent studies by Squire et al., (1988, 1989) suggest that the role of the hippocampus is more general as it may be involved in the acquisition of new information, both spatial and non-spatial. Research on long term potentiation (LTP) suggests that the hippocampus plays a special role in associating inputs that occur simultaneously or in near succession. This finding has some relevance in the place recognition experiments described above. Whatever its role in spatial cognition, the hippocampus is in a unique position to receive highly integrative information from the neocortex. It compares and combines this information and then influences the neocortex. Squire et al. (1989) propose that the hippocampus is the seat of declarative knowledge. Although the issue of specificity of the hippocampal memory function is still open to debate, it has been widely studied by both empiricists and theorists.

Neural Network Models Relating to the Cognitive Map

Some excellent neurally plausible models have been built in the area of spatial cognition to model selected aspects of the cognitive mapping process and to provide a detailed understanding of its contribution to spatial knowledge representation. Though these models have immediate relevance in psychology and neurobiology, they are constrained since much of what is known about cognitive maps is obtained through observations of spatial behavior rather than an understanding of how cognitive maps are organized in the brain. The development and use of a cognitive map is a general problem that cannot be addressed by a complete model. Rather the existing neural network models are limited to a small set of spatial navigation problems including the following questions: How does the navigating observer recognize a place? How does the navigating observer locate a goal?

How does the observer infer path knowledge? These are just a subset of the general set of problems in this domain. Other problems are addressed in the area of robotics and control when the constraint of biological plausibility is not involved and unrestricted symbol manipulation is used to model the problems (for e.g., Seibert and Waxman, 1989; Bachrach, 1991).

Zipser (1986) proposed three computational models set in parallel distributed processing (PDP) framework to model place recognition, and path planning to a fixed goal. The first model is that of place fields which are based on extensive neurophysiological evidence of the activity of the place field cells in the hippocampus described above (O'Keefe and Nadel, 1978; Muller, Kubie and Ranck, 1987). The second and third models incorporate goal localizations and are somewhat more speculative since less is known about the underlying neurophysiology.

Place Field Model

The first model predicts the activity of place fields in the hippocampus of an observer. The neural network implemented by a set of processors consists of two layers and a sensory system. It is assumed that the sensory system detects the landmarks and generates descriptions of landmarks and location information (e.g. distance between the landmark and the observer). This information is input to the units in Layer I of the network (see Figure 1). Layer I consists of PEs representing one single landmark and its associated place field. The response of the PEs in this layer reach a maximum value when the current and stored location parameters are equal. Layer II acts as a place field neuron and computes the overall similarity between the current scene and scene as recorded at the center of the place field. The model assumes that the observer remains still until the scene is scanned. Hence PEs in Layer I, once computed, remain constant until the observer moves again. The pattern matching performed by the two layers are implemented by a set of mathematical functions (including a gaussian function) that use two parameters – matching criteria parameter and threshold parameter. The two layers of the neural network model represent the hippocampus. This simplistic model does not produce a specific output since the main objective of the model is to predict the activity of the place field cells in the hippocampus based on the observer's location in the environment in relation to a set of landmarks.

The neural network model simulation requires a scene consisting of a set of landmarks (whose size, position and number can be varied), the initial viewpoint representing the center of the place field, and the two parameter values. The program estimates the areas of the landmarks from the initial viewpoint. Then the observer is moved systematically through a number of locations in the simulated environment. The final product of the simulation is the location and shape of the place fields in the hippocampus. The simplified model predicts the effects of size, position and number of landmarks on the shape of the place fields of the hippocampus. In addition, the model successfully predicts the effects of dilation experiment. It can be concluded that a neural network model is able to successfully replicate some experimental findings relating to the activity of place field cells in the hippocampus of the rat.

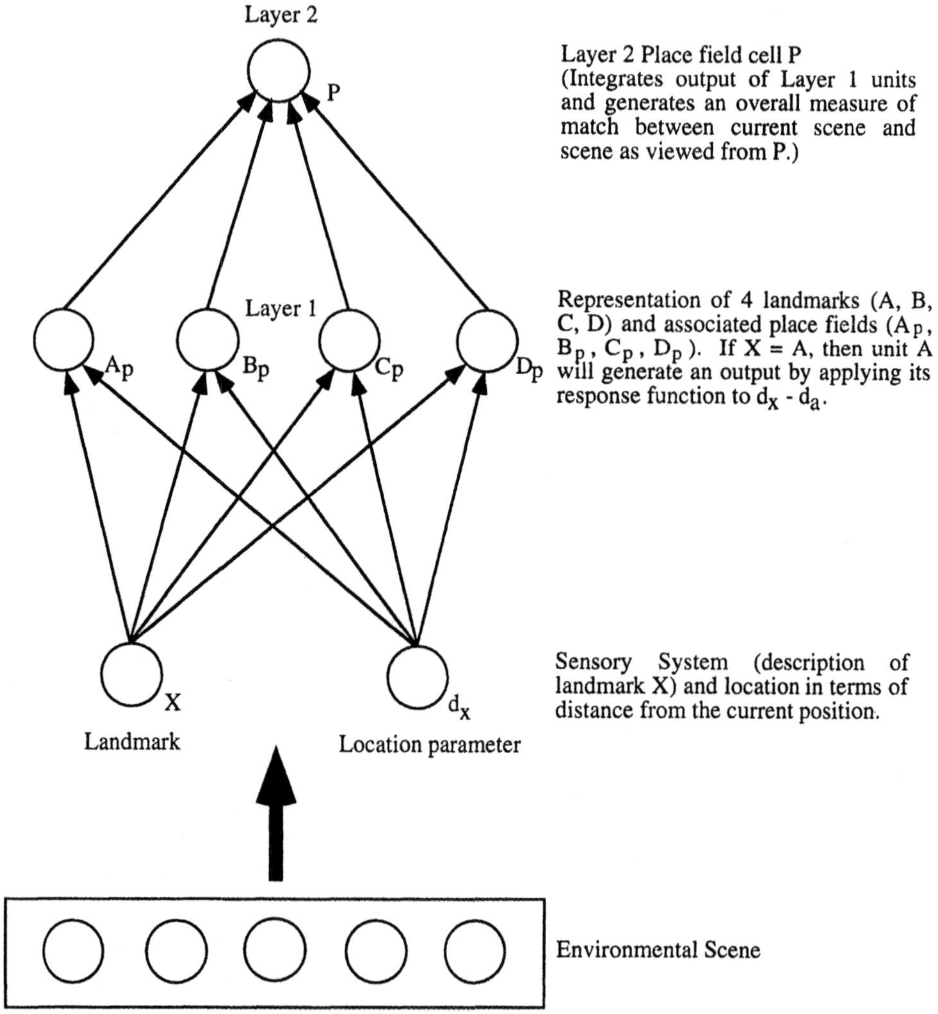

Figure 1: Place Field Model (adapted from Zipser 1986)

Rats in the O'Keefe and Nadel experiments could distinguish right and left. It was hypothesized that some place field cells in the hippocampus are orientation specific. But the place field model is unable to account for such findings since it ignores the problem of observer orientation and left-right asymmetry. Hence any modification of the place field model has to account for such effects.

Distributed View Field Model

Zipser's (1986) second model called distributed view field model incorporates orientation. This neural network model is a model of the cognitive map whose main purpose is to locate and direct the observer towards a goal (i.e. source of food or home). The model consists of three layers of PEs. The first layer consists of `object units' that encode three types of information relating to a landmark, its type, distance and orientation from the observer. The output of the visual system in this model consists of left, center and right sets of objects (i.e. what is to the right, center and left of the viewer oriented in a particular direction). The object units are also classified into three corresponding groups – right, center and left. The second layer consists of view field units (place field units along with orientation). The goal units form the third layer in the model. The location of the goal relative to the current position and orientation is encoded as goal units whose output is interpreted by the motor system as the direction of the goal.

The simulated observer can have two motivation states corresponding to exploration (curiosity) and homing (fleeing). The first state allows the observer to explore the environment at random and encode spatial information into view fields. The object units record information necessary to recognize and locate currently visible set of landmarks while goal units record the location of the goal to the current position and orientation. Subsequent goal directed travel during the second motivation state is dependent on the existing set of view fields unless some novel object item is seen in the environment. Thus the observer codes multiple viewer-centered representations of a place. The navigational strategy implemented is computationally simple but the drawback is that a very large number of view fields may eventually get stored as the observer moves around. In addition, the observer may sometimes end up thrashing in the vicinity of goal because of conflicting view fields.

Beta-Weight Model

The third model called Beta-Weight model overcomes the limitations of the distributed view model. Once again the problem is that the observer is located at an initial position recording the location of a set of landmarks and the location of the goal. At some later time, the observer moves to a new location from which the goal is invisible. The first step involves determining the location of landmarks from the new location. The beta-weight model implements a complex computational algorithm (inverse matrix of locations) to estimate the relationship between various landmarks from a single perspective. The algorithm enables the recovery of the structural configuration of landmarks from any location which the observer subsequently occupies and from which only a limited view is visible. Thus the model is able to build an environmentally centered representation of landmark interrelationships from a single view of the configuration of landmarks in the forward field. The beta-weight model attempts to explain the observed empirical data by proposing a novel representation and computational scheme. Whether it is biologically true could only be proved by experimentation.

The Hippocampus as a Cognitive Mapping System

O'Keefe's (1989) approach combines both the theoretical and empirical approaches discussed in Section 1 of the paper. He uses the neurophysiological evidence about the hippocampus to guide in the construction of a computational model of the hippocampus system. The structure of the hippocampus is compatible with a matrix-like system and this principle underlies the computational model. The computations performed by the hippocampus during navigation may be implemented as a set of simple functions including addition, multiplication, and trigonometric calculations as well as a set of more complex calculations involving vector rotation and matrix operations. Such complex computation can be implemented on parallel computers using systolic array processors and wavefront array processors (O'Keefe, 1989). Is such a computational model with matrix inversions and other mathematical operations biologically plausible? The answer to this question will emerge in future research. The model is described below.

O'Keefe's model assumes that the brain uses several systems for navigation. One system may be a representation of the metric coordinate space in which stimuli are located. Two other systems, an orientation system that enables the learning of a sequence of sensori-motor actions (stimulus-action pairs), and a guidance system that enables the animal to avoid or approach specific stimuli are included. The computational model has to account for the following navigational tasks -- representation of information about the stimuli, comparing the present sensory array with the information already stored in the cognitive map, and navigation from a current location to a desired location.

The first major function that has to be considered is the representation of environment within the cognitive map. As the animal moves around and explores its environment, it constantly has to recalculate the location of the stimuli in its environment i.e. update its position. The model assumes a cartesian coordinate system that is centered on the animal's head (see Figure 2). In terms of computation, the updating and representation of stimulus is easily accomplished by a set of matrix operations. Recall that the structure of the hippocampus is assumed to be matrix-like. Multiplication of the matrix representing the location of the stimuli by a transform matrix represents the rotation and translation produced as a result of the animal's navigation in space.

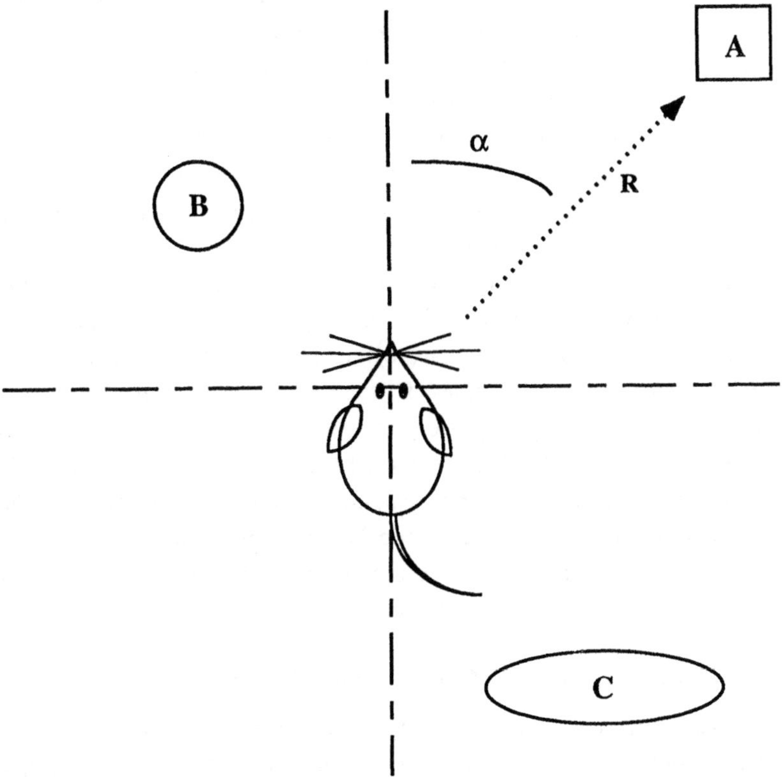

Figure 2: Cartesian axis framework centered on rat's head and location of objects A,B, and C within it. R represents the distance to object A in polar coordinates; α represents the angle (adapted from O'Keefe 1989).

During the exploratory phase, distance and angle of the cue impinging on the animal's senses represented in an egocentric head-centered coordinates has to be transformed into a non-egocentric cartesian coordinate system. The hippocampus then has to combine the coordinate information with the transform representing the animal's displacement to produce a prediction of the resulting coordinates in a new set of axes. Over a series of cycles, the process of updating successively leads to closer approximation of the estimated distance relative to the true distance. In this computational model, changes in angle between cues in the environment have large effects in the place fields and changes in size of the cue does not have a similar effect. This is true in Muller et al., (1987) experiment where cutting a cue card in half had little impact on the place fields of rat.

The second major function of the cognitive maps is to guide movement from current location to a goal. Much of navigation involves movement from point A to point B where point A may represent current location and point B the goal. The goal may be dictated by the motivational system and may be driven by instincts such as hunger or fear. The model assumes that goal locations are stored outside the hippocampus in other limbic areas. Goal information is retrieved and sent to the hippocampus; for example, information about a food source may be retrieved when the animal is hungry. The computational operation involves multiplication of the desired location (goal) matrix by the inverse of current location. There might be neural areas in the brain that may perform such computations and generate the required motor programs.

The next major function is to retrieve information stored in the cognitive map. How does an animal know that it has entered a familiar or novel environment? It is assumed that the activation of a map or set of maps will result in response to the input of stimulus information and the egocentric angles. Distances between stimuli are calculated and compared to the distances stored in the cognitive map. The map that has the closest match to the current sensory output is chosen and used to compute the navigational actions. The animal, upon entering a familiar environment, may activate a default version of the map of the environment which is oriented in a particular way in relation to the background cues. If this default map does not match the current map, the two maps are manipulated (rotated) until the two are in correspondence.

Spatial Transition Matrix to Model a Cognitive Map

McNaughton (1989) views the hippocampus as an associative memory network and proposes a computational model called the spatial transition matrix. His computational model deals with spatial representation in the hippocampus. This model is based on some aspects of the hippocampal activity but simplifies the structure of the PEs, weight modulation rule and connectivity between the network modules. The model has two classes of inputs corresponding to the actual inputs received by the hippocampus from inferotemporal cortex and parietal cortex. The former signals the nature of objects in the visual field of the animal and the latter provides information about the movement or motion in space. The animal forms a link of conditional association between local views of the environment and specific movements. Given a starting location, orientation and information about a movement, the animal may recall a representation of a resulting view. A sequence of such local views and sensori-motor actions would enable the animal to move forward. Such a memory would take the form of a spatial transition matrix supporting associations between movements and local view representations. The model should be able to generate a representation of the local view that would result from a specific movement in that location. At the very least, the computational model should discriminate between left turn, right turn and forward motion since these are essential movements in navigation.

There is an obvious constraint in the model. How does the animal plan novel trajectories or paths to a particular goal if it has not previously experienced such paths? McNaughton describes three possible solutions to this problem of path planning. The first solution is the animal can use spatially invariant landmarks and learn association between goal objects and such landmarks. This solution obviously fails if the route that is to be followed lacks such visible landmarks. The second solution is more sophisticated in that if two previously learned routes overlap, a new trajectory or path can be computed at the point of overlap. For example, if we assume two routes form two sides of a triangle, the third side of the triangle representing a new trajectory can be easily computed. A third solution to the problem of path planning is that the system can be provided the ability to compute movement equivalences (for example, a 270° left turn is equivalent to a 90° right turn or a right turn in the forward direction is equivalent to a left turn in the opposite direction). Thus if there is any path in the transition matrix linking two locations, spatially equivalent movement sequences can be used to compute other routes.

From the review of the models, it can be seen that the cognitive map can be modeled as an active global representation involving encoding of landmark information, distance and orientation between different landmarks in some sort of viewer-independent framework (O'Keefe, 1989). The alternative approach is to completely dispense with the notion of a global representation and assume that the cognitive map consists of local association. That is form a link of conditional association between local views of the environment and specific movements (McNaughton, 1989). While Zipser's models are all implemented as PDP models and address specific issues in navigation, O'Keefe (1989) takes a broader perspective and suggests a plausible model to support his earlier theory of the hippocampus as a cognitive mapping system. O'Keefe also points out how such a neural network model could be implemented on a parallel processing machine. More recently, Burgess, Recce, and O'Keefe (1994), Bachelder and Waxman (1994), Hetherington and Shapiro (1993) and Wan (1993) have proposed models that provide useful insights into the possible biological mechanisms underlying hippocampal navigation.

A Simple Model of Route Learning and Assimilation in the Cognitive Map

For the purposes of our discussion on building a neural network model of cognitive map, a simplified model of route learning and assimilation is proposed based on previous research (Gollege et al., 1985; Gopal et al., 1989). The steps in building the model of *learning and assimilation* is as follows:

1. Start at any location in the environment. Assume this location as the origin O with a fixed orientation i.e. North or 0°.
2. Perceive the landmarks visible from O. Note the distance and direction to each visible landmark if they have never been perceived before. Create a link between O and each landmark L. The link or edge ε (consists of distance (d_i) and direction (α_i) information. The representation of routes from O may be: ($O - L_1 : d_1, \alpha_2, O - L_2 : d_2, \alpha_2, O - L_3 : d_3, \alpha_3, \ldots O - L_n : d_n, \alpha_n$).
3. Move towards a new location based on a random exploration strategy or the commands issued by an external module. Let this location be A. The coordinates of A are fixed with respect to O. Establish the edge ε between them and store the distance and direction of ε ($O - A : d_{OA}, \beta_{OA}$).
4. At A, check to see if any landmark is visible. If yes, check to see if it a landmark that has already been stored. Store new link between A and L in the cognitive map.
5. Repeat step 2-4 at each new location.
6. At the end of navigation trial, recompute the position of all landmarks from O. The set of landmarks that is stored can be limited to those which are above a certain level of perceptual threshold.
7. The output of the model is the route that is learned and the relative positions between the selected set of landmarks.

A neural network model is being built that will use the procedure outlined above. Each input set to the neural network model represents a route that is learned in a large-scale environment. Each input unit, in turn, consists of the location, direction and distance of each landmark along the route. All landmark information is computed from an origin. The output unit consists of the location, direction and distance of the goal unit from the origin. The model can learn to associate different inputs with different goals and the representation of the relevant route knowledge. Initially, a supervised learning strategy can be used to teach the model different routes in the same environment. The weights connecting the input units to the predicted goal should represent the route information and the model should be able to compute the location of the goal given the route. The actual details of the architecture are currently being worked out.

Conclusion

This chapter presented neural network models of the cognitive map and the kind of computations that may be performed in order to solve some way-finding problems. Part of the appeal of the neural network approach is that it is strongly linked to the neural basis of cognition than symbolic or other approaches. The models examined in this chapter attempt to build cognitive maps at a level of abstraction somewhat higher than the level of individual neurons and connections. In such models, the connection between part of the brain structure and particular elements of the models cannot be fully specified, since the mapping is only approximate. Neural network models of cognitive maps are somewhat limited due to the fact that neural mechanisms underlying the cognitive mapping is not fully known. In addition, most of the neurophysiological experiments in this field relate to the cognitive abilities of rats and some primates. Therefore caution must be exercised in generalizing the results. In spite of all these limitations, the class of models presented in this chapter explore the computational properties of the cognitive map. Such models are important for understanding complex biological systems and, at the same time, can provide real insight in designing artificial systems (robots) with brain-like capabilities. They also raise interesting questions on certain aspects of cognitive map functioning that may lead to specific testable predictions.

Acknowledgements: This research is supported by a grant from National Science Foundation (SBR-9300633).

References

Bachelder, I. A. and Waxman, A. (1994). Mobile robot visual mapping and localization: a view-based neurocomputational architecture that emulates hippocampal place learning. *Neural Networks* 7, 1083-1099.

Bachrach, J.(1991). A connectionist learning control strategy for navigation. In *Neural Information Processing Systems*. (R. P. Lippmann, J. E. Moody and D. S.Touretzky, eds.), pp. 457-464. San Mateo, CA:Morgan Kaufmann Publishers.

Burgess, N., Recce, M., and O'Keefe, J. (1994). A model of hippocampal function. *Neural Networks* 7, 1065-1081.

Eichenbaum, H. and Buckingham, J. (1990). Studies on hippocampal processing. experiment, theory and model. In *Learning and Computational Neuroscience: Foundations of Adaptive Networks* (M. Gabriel and J. Moore, ed.), pp. 171-233 Cambridge, MA: Bradford book, MIT Press.

Golledge, R.G. Smith, T.R. Pellegrino, J.W. Marshall, S.P. and Doherty, S. (1985). A conceptual model and empirical analysis of children's acquisition of spatial knowledge. *Journal of Environmental Psychology* 5, 125-152.

Gopal, S., Klatzky, R., and Smith, T. (1989). NAVIGATOR: A psychologically based model of environmental learning through navigation. *Journal of Environmental Psychology* 9, 309-331.

Grossberg, S. (1982). *Studies of Mind and Brain*, Boston: Reidel Press.

Hebb, D. O. (1949). *The Organization of Behavior*. New York: Wiley.

Hecht-Nielsen, R. (1990). *Neurocomputing*, Reading, MA: Addison-Wesley Pub.

Hetherington, P. A. and Shapiro, M. L. (1993). A simple network model simulates hippocampal place fields: II. Computing goal-directed trajectories and memory fields. *Behavioral Neuroscience* 107, 434-443.

Hirtle, S. C. and Jonides, J. (1985). Evidence of hierarchies in cognitive maps. *Memory and Cognition* 13, 208-217.

Hinton, G. and Anderson, J. A. (eds) (1981). *Parallel Models of Associative Memory*, Hillsdale, NJ: Lawrence Erlbaum Associates.

Hodgkin, A. K. and Huxley, A. F. (1952). Currents carried by sodium and potassium ions through the membrane of the giant axon of Loligo. *Journal of Physiology* 116, 449-472.

Hopfield, J.J. and Tank, D.W. (1986). Computing with neural circuits: a model. *Science 233*, 625-633.

Kohonen, T. (1984). *Self-organization and Associative Memory*, New York: Springer-Verlag.

Kosslyn, S. and Schwartz, S.P. (1977). A simulation of visual imagery. *Cognitive Science* 1, 265-295.

Kuipers, B. (1978). Modeling spatial knowledge. Cognitive Science, 2, 129-153.

Lapedes, A. and Farber, R. (1988). How neural nets work. In *Neural Information Processing Systems*. (D. Z. Anderson, ed), pp. 442-443. New York: American Institute of Physics.

Levine, B., Jankovic, I.N. and Palij, M. (1982). Principles of spatial problem solving. Journal of *Experimental Psychology: General* 2, 157-175.

Lippmann, R. P. (1987). An introduction to computing with neural nets. *IEEE Acoustics, Speech and Signal Processing* 4, 4-22.

Lynch, K. (1960). *Image Of The City*, Cambridge, MA: MIT Press.

Marr, D. (1982). *Vision: A Computational Investigation into the Human Representation and Processing of Visual Information*, San Francisco: W.H. Freeman.

McNaughton, B. L.(1989). Neuronal mechanisms for spatial computation and storage, In *Neural Connections, Mental Computation* (L. Nadel, L.A. Cooper, P. Culicover, and R.M. Harnish, eds.), pp. 285-351. Cambridge, MA: The MIT Press.

Mishkin, M. and Appenzeller, T. (1987). The anatomy of memory. *Scientific American* 256, 80-89.

Mishkin, M., Ungerleider, L. G. and Macko, K. A. (1983). Object vision and spatial vision: two cortical pathways. *Trends in Neuroscience* 6, 414-417.

Muller, R. U., Kubie, J. L., and Ranck, J. B. (1987). Spatial firing patterns of hippocampus complex-spike class in a fixed environment. *Journal of Neuroscience* 7, 1951-1968.

O'Keefe, J. and Nadel, L. (1978). *The Hippocampus as a Cognitive Map*, New York: Oxford University Press.

O'Keefe, J. (1989). Computations that the Hippocampus might perform. In *Neural Connections, Mental Computation* (L. Nadel, L.A. Cooper, P. Culicover, and R.M. Harnish eds.), pp. 225-285. Cambridge, MA: The MIT Press.

Parkinson, J. K., Murray, E. A., and Mishkin, M. (1988). A selective mnemonic role for the hippocampus in monkeys. memory for location of objects. *Journal of Neuroscience* 8, 4159-4167.

Rumelhart, D.E., McClelland, J.L. and the PDP Research Group (eds.) (1986). Parallel *Distributing Processing, Explorations in the Microstructure of Cognition. Volume 1.* Foundations, Cambridg, M.A: The MIT Press.

Sejnowski, T.J., Koch, C. and Churchland, P.S. (1988). Computational neuroscience. *Science* 241, 1299-1306.

Seibert, M. and Waxman, A. (1989). Spreading activation layers, visual saccades, and invariant representations for neural pattern recognition systems. *Neural Networks* 2, 9-27.

Siegel, A.W. and White, S.H. (1975). The development of spatial representations large-scale environments. In Advances In *Child Development And Behavior* (H.W. Reese, ed.), pp. 9-55. New York: Academic Press.

Smolensky, P. (1988). On the proper treatment of connectionism. *Behavioral and Brain Sciences*, 11, 1-74.

Squire, L R., and Zola-Morgan, S. (1988). Memory. brain systems and behavior. *TINS Special Issue* II, 170-175.

Squire, L R., Shimamura, A. P. and Amaral, D. G. (1989). Memory and the hippocampus. In *Neural Model of Plasticity* (J. H. Byrne and W. O. Berry, eds.), pp. 208-239. New York: Academic Press.

Steinbuch, K. (1961). Die Lernmatrix. *Kybernetik* 1, 36.

van der Malsburg, C. and Willshaw, D. (1981). Co-operativity and brain organization, Trends in *Neurosciences* 4, 80-83.

Wan, H. S., Touretzky, D. S., and Redish, A. D. (1993). Towards a computational theory of rat navigation. In *Proceedings of the 1993 Connectionist Models Summer School* (M. Mozer, P. Smolensky, D.S. Touretzky, J.L. Elman, and A. Weigend eds.), Hillsdale, NJ: Erlbaum Associates.

Zipser, D. (1986). Biologically plausible models of place recognition and goal location. In Parallel Distributed Processing. *Explorations in the Microstructure of Cognition, Volume 2. Psychological and Biological Models* (J.L. McClelland, D.E. Rumelhart, and the PDP Research Group, eds.), pp. 432-470. Cambridge, MA: The MIT Press.

Sucharita Gopal
Department of Geography
Boston University
Boston, MA 02215

CONNECTIONIST MODELS IN SPATIAL COGNITION

Thea Ghiselli-Crippa, Stephen C. Hirtle and Paul Munro

Abstract: This chapter reviews neural network approaches to the study of spatial cognition and spatial language with a focus on the representations and processes that are used by humans for encoding, storing, accessing, and referencing spatial knowledge. Such processes are used for recall of spatial information, for navigation through space, for spatial decision making, and for generating spatial descriptions. Connectionist models for representing the structure of cognitive maps and understanding the language of spatial relations are discussed in detail, using the representation adopted by the model to evaluate the usefulness of each approach. In addition, the functional differences within neural network models, such as distributed versus local representations, are discussed. Finally, an agenda for future development of connectionist models, including the exploration of alternative network structures and data types, is proposed.

Introduction

Research on human spatial cognition during the past two decades has followed several distinct research paradigms, including empirical data gathering, theoretical speculation and computational modeling. Computational models require the building of systems that can mimic human processing within a specific domain. Such systems require the explicit statement of assumptions, representational codes, and processing dynamics. Thus, a computational model can facilitate the development of theories by focusing on the representation and on the processing dynamics (Hirtle and Heidorn, 1993). The predominant computational approach for the study of spatial cognition has been based on advancing a symbolic, or artificial intelligence (AI), framework (e.g., Gopal, Klatzky and Smith, 1989; Kuipers and Levitt, 1988; Yeap, 1988). The aim of this chapter is to examine more recent approaches using a connectionist, or neural network, framework to the modeling of spatial cognition.

Despite the explosion of research using neural networks that has occurred during the past decade in the many areas of cognition, there have been relatively few advances in the area of spatial cognition. Instead, much of the research on spatial processing using neural networks has been directed either towards low-level aspects, such as place recognition, goal location, or landmark learning (e.g., Touretzky, Redish and Wan, 1993; Zipser,

J. Portugali (ed.), The Construction of Cognitive Maps, 87–104.

1986), or towards optimization algorithms for specific spatial problems (e. g., Angéniol, Vaubois and Texier, 1988). There has also been a small body of research directed towards spatial applications in marketing (e.g., Halmari and Lundberg, 1991) and for many problems inherent within geographic information systems, such as image analysis or choice analysis (e.g., Fischer, 1994).

In this chapter, we examine and review research limited to the higher level domain of spatial cognition. By this we mean the representations and processes that are used by humans for encoding, storing, and accessing spatial knowledge (Hirtle and Heidorn, 1993). Such processes are used for recall of spatial information, navigation through space, and for spatial decision making. As such, the representation adopted by the model will prove to be an important factor in evaluating the usefulness of a non-symbolic, connectionist approach.

The chapter begins with a brief motivation to connectionist models. This is followed by detailed discussions of connectionist models for representing cognitive maps and connectionist models for the language of spatial relations. Finally, proposals for future developments and conclusions are discussed in the final two sections.

Connectionist Models

In general, connectionist approaches to modeling high-level cognitive functions focus on accounting for emergent properties generated by the interaction of many simple neuron-like computational units. In these models, knowledge is stored in the strengths of the connections between units, which are modifiable and change with exposure to the environment.

Connectionism has evolved in many directions, so that it has come to indicate various kinds of models, which all share the basic ideas, but which can be quite different along some lines. One distinction that will prove useful in our review of connectionist models in spatial cognition is the one based on the kind of representations used. Given a network of computational units and some entities that need to be represented in it, the choice is usually between using one computational unit for each entity and using many computational units whose various patterns of activity represent the various entities. In the first case, we have local representations, whereas in the second case, we have distributed representations.

Networks that use local representations, where each of the entities is represented by one computational unit, are appealing because they are easy to understand, since "the structure of the physical network mirrors the structure of the knowledge it contains" (Hinton, McClelland and Rumelhart, 1986). Also, in this case, the knowledge about an entity is stored only in the connections of a unique computational unit dedicated to that entity. In networks that use distributed representations, each entity isre presented by a pattern of activation over many computational units and each computational unit is involved in the representation of many entities. The network structure does not reflect the structure of the knowledge it contains and the knowledge about each entity is stored in the connections among the many computational units involved in its representation. The use of distributed representations can lead to automatic generalization, content addressability, and learning new concepts without modifying the network structure. Both local and distributed representational models are described in more detail below in reviewing their application for the study of spatial cognition.

Connectionist Models of Cognitive Maps

Models based on Local Representations

One of the first implementations of connectionist modeling is a series of models based on a theory of cognitive maps outlined by Kaplan and Kaplan (1982). According to Kaplan and Kaplan (1982), a cognitive map is created when a collection of isolated representations of elements of the environment are connected together using an associative mechanism. An individual's experience of an environment determines which locations are experienced and in which sequence. The sequences can be combined into a network of associations, which form the cognitive map. The network provides an efficient way to store sequences of experiences and allows for the generation of new sequences that have not been experienced before.

Under this model, each element in the network is the representation of a location and so is itself a network of features, which allows for recognition of each particular location. In terms of the cognitive map, since each network of features only represents one location, it can be considered as a single unit dedicated to that location, thus generating a local representation.

The representations originally proposed by Kaplan and Kaplan (1982) are implemented by Kaplan, Weaver and French (1990) as active symbols, where the active symbol is a recurrent network in which activity can persist over time and which is used to represent concepts; it finds its origins in the Hebbian cell-assemblies. The characteristics of the active symbols make them such that their activation does not depend exclusively on environmental stimuli and they become "the building blocks of more complex internal models of the environment" (Kaplan et al., 1990). In this framework, the cognitive map network is then recast in terms of a connectionist, spreading activation network with two control mechanisms: inhibition, to dampen the activity in other units, and fatigue, to dampen activity over time.

The structure of the network, the map of the sequences experienced in the environment, corresponds to the structure of the environment in the sense that sequences in the environment correspond to internal sequences, intersections also correspond and what is near in the experienced environment is also near, in terms of association, in the cognitive map (Kaplan et al., 1990). The authors also suggest that the cognitive map, the model of the environment, can be seen in the context of a layered, hierarchical system, in which more abstract models of the environment, or, more exactly, models of the model of the environment, can be developed, to provide a higher power of abstraction to cognitive processes.

Testing of the effectiveness of Kaplan and Kaplan's (1982) original idea of cognitive maps as a form of knowledge representation was part of the dissertation work performed by Levenick (1985), who developed the Network Activity Processing Simulator (NAPS) to study the effect of the choice of knowledge representation on the performance in a wayfinding task. Although the author stresses that NAPS is not intended to be a model of cognition, but only a tool to investigate the importance of knowledge representations, it is reviewed here, as it can provide insights on how knowledge representation is related to the performance on wayfinding tasks. The knowledge representations simulated by NAPS are all activity-based associative network structures and they vary mainly along two dimensions: the nature of their nodes and of their links, either of which can be binary or have variable activity. The parameter space of the simulator ranges from marker-passing semantic networks, with binary links and binary nodes, to what he calls cognitive maps, with variable strength links and variable activity nodes. Other parameters to choose include fatigue and three kinds of inhibition.

The NAPS model has two operating modes, one for building the network and one for testing the network. In the building operating mode, a set of parameters and a set of routes, specified as sequences of locations, are input to NAPS; the system generates the knowledge representation network specified by the input parameters by creating a node for each location and links between adjacent locations, as defined by the input sequences; the link strengths are determined by how often two adjacent locations appear in the input sequences. In the testing operating mode, NAPS is given a start node and a goal node; the system then uses the information in the network, by letting the activity from the start and goal nodes spread until it intersects, to try to find a subgoal node. To overcome the problems encountered with arbitrarily large, variable activity networks, in which activity from two sources will not always converge and cause enough activation of a subgoal node, Levenick (1985) introduces the concept of hierarchy in his system. If a group of locations is such that the locations are always experienced in the same sequence, a new unit, one level up in the hierarchy and connected to all of the location units in the group, can come to represent the group of locations; when start and goal are too far on the regular level, the system can jump up one level in the hierarchy and make use of the upper units for a more efficient search, as seen in Figure 1.

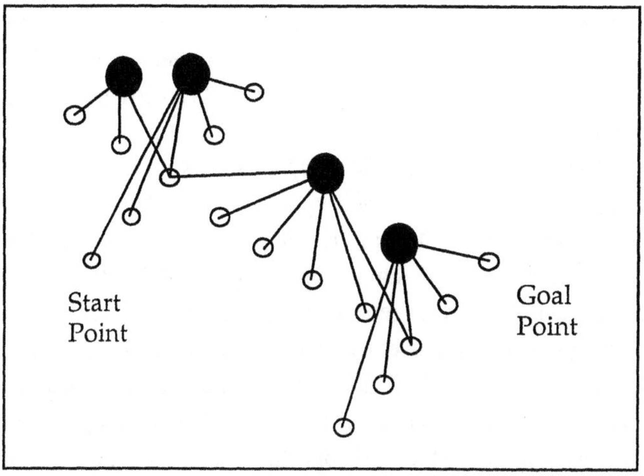

Figure 1: Schematic diagram of the hierarchical structure of units. Adapted from O'Neill (1991).

By examining NAPS's simulations, Levenick (1985) concludes that the marker-passing semantic network at one end of the parameter space provides a simple and efficient mechanism for knowledge representation, but that their binary nature, their inflexibility, the need to erase old marks, and the need for additional control structures in complex situations make it a less than appealing choice in any model of cognitive processes. At the other end of the parameter space, the cognitive maps, with the addition of the hierarchical structure necessary for their correct operation, can use the same kind of search mechanism independently of the complexity of the situation, thus providing flexibility and adaptability. So, in the context of wayfinding, Levenick's cognitive maps, as representations of the knowledge of an environment, appear to provide the best representation for the task.

The problem of modeling people's cognitive maps is directly examined by O'Neill (1991), who emphasizes how any theory of cognitive maps, and the representations and processes it defines, is influenced by the choice of the underlying metaphor of cognition. He argues that Kaplan and Kaplan's (1982) cognitive map is based on a biological metaphor, since it uses physiologically plausible mechanisms, it is functionally analog to the experienced environment and its structure can be altered by new experiences. The model is also in agreement with the psychological data which indicates hierarchical structuring of spatial information (Hirtle and Jonides, 1985). O'Neill further claims that the creation of a hierarchical structure in cognitive maps allows for the development of a survey representation, "since the links between high level nodes allow the representation of the spatial relationships between distant locations within the environment" (O'Neill, 1991). However, since the structure of the cognitive map reflects only the way in which the environment has been experienced, distant locations in the cognitive map might happen to correspond to close locations in the environment, if indeed these were never experienced close together; so, it appears that it would be difficult for a novel shortcut to emerge from the representation.

The implementation of the cognitive map model is an adaptation of the computer simulation system previously developed by Levenick (1985). The model is tested by comparing the performance of the model on simulated wayfinding tasks to human performance in the corresponding actual environments. Within the limited environments used in the experiments, the author reports no significant differences between the model performance and human performance in actual wayfinding and in path selection on sketch maps. The model is therefore able to generate decision making behavior in wayfinding tasks which is similar to the behavior exhibited by human subjects and it does so by constructing a representation which mirrors the structure of the environment, learning from an initial set of sequences through the environment, and generalizing to new searches between any two points. Learning from new searches and the introduction of salience in the network nodes, to simulate the effect of landmarks, appear as issues worth exploring with this kind of model.

The aspect that O'Neill (1991) stresses the most with regard to his system is that its wayfinding behavior is accomplished without programming any behavioral rules into the system. As O'Neill states, the biological metaphor approach "describes lower level physiologically based mechanisms that model some of the psychological processes of spatial cognition. Thus, any 'rules' governing psychological processes are emergent from the activity of the system, rather than specified as explicit rules as in the case for computational models" (O'Neill, 1991). The focus is on the structure of the system, on the physiological constraints, and, as opposed to symbolic models, the behavior of the system is emergent instead of being programmed into it as rules. Apart from wayfinding, though, it is not known how this model would perform with respect to other spatial tasks which make use of cognitive maps. In particular, it might be interesting to compare its performance with the model proposed by Munro and Hirtle (1989), described in the following section, in the context of priming studies concerning locations along routes.

One clarification might be necessary with respect to O'Neill's work and his description of a "physiologically plausible model of the cognitive map" (O'Neill, 1991). One has to be careful with the use of the term plausibility and its qualifiers: we take physiological plausibility to imply that the mechanisms used in a model are based on the mechanisms used by living organisms in their normal functioning; biological plausibility goes beyond just mechanisms, to include all aspects of a living organism. Only after all the biological aspects are known it is possible to develop a truly biologically plausible model: when a lot is known about, say, the neurobiology of place recognition, then a biologically plausible model of place recognition can be built; but, when only the mechanisms are known and they are employed in the model of some aspect of cognition for which the actual complete neurobiology is not known, it would be more appropriate to define such a model as only physiologically plausible. O'Neill is not trying to actually model the neurobiology of cognitive maps, since the current status of understanding of the neurobiology of spatial cognition is very limited: he is therefore correct in defining his account as biologically based but only physiologically plausible. In this respect, this model stands in contrast with other connectionist models which want to remain biologically plausible and which therefore only attempt to account for more elementary, better known aspects of spatial cognition, such as place recognition in rats (Zipser, 1986).

Models based on Distributed Representations

The two models described in this section rely on distributed representations: the network structure does not necessarily reflect the structure of the environment. Instead, each location in the environment is represented as an activity pattern over a set of units, in either the input layer or in the hidden layers. Such representations allow for the spatial properties to emerge from the interaction of units, rather than through the direct encoding of the spatial primitives.

A model for spatial priming

Munro and Hirtle (1989) propose a connectionist architecture for the structure of cognitive maps, which is used to model the spatial priming effects reported by McNamara (1986) and McNamara, Ratcliff and McKoon (1984). Priming studies have shown that spatial judgments are influenced not only by spatial location, but also by the cognitive structures imposed on the space. Such structures include routes, regions, and other clusters defined by subjective perceptions of neighborhoods or of semantic relatedness. To account for these two factors, the network, shown in Figure 2, includes two sets of units: a set of grid units, which represent the environment by means of a uniform rectangular grid, and a set of category units, which represent the clusters in the environment, four regions in one experiment and six routes in another experiment. Another set of units, the place units, correspond to locations in the environment; they are connected to the category units by means of bidirectional binary connections and to the grid units by means of bidirectional real connections. The grid units represent the location of a place using a form of coarse coding (Hinton et al., 1986): each place is represented by a Gaussian peak over the grid units centered at its location in the environment. There are no connections among units in the same set and between grid units and category units. In this network, the connection strengths are preset and not learned; given an input stimulus, which is the activation of one of the place units, the activity of all the units in the network is iteratively updated, following the interactive activation scheme proposed by McClelland and Rumelhart (1981).

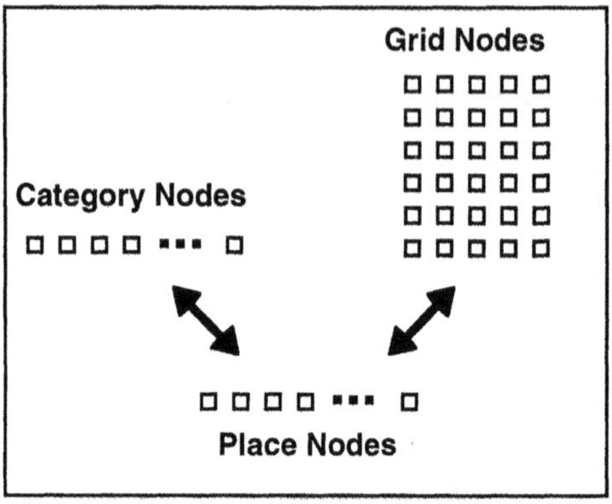

Figure 2: Network architecture used by Munro and Hirtle (1989) to model effects of spatial priming.

Munro and Hirtle (1989) simulate the spatial priming experiments by activating the priming place unit at a constant level for a fixed number of iterations, letting the network activity decay for a relaxation period, and then activating the target place unit, also at a constant level and for a fixed number of iterations. Reaction time is simulated by counting the number of iterations required for the target place unit to reach a criterion level of activity. In both experiments (with regions and routes), the results of the simulation show a good correspondence with the priming effects obtained by McNamara and his collaborators. The authors report some minor discrepancies, which involve the interaction of alignment with distance in the region experiment and the effect due to distance in the route experiment. Overall, the model accounts well for the experimental data and provides an alternative to the partially hierarchical, spreading activation, computational model that McNamara himself proposes for his data (McNamara, 1986).

The model proposed by Munro and Hirtle (1989) indicates that the systematic distortions found in the experimental data can be accounted for by considering several representations of the spatial knowledge relative to a given environment, which interact with each other when the process of retrieval acts upon them. This account is in agreement with the hierarchical theory of spatial knowledge, which proposes that spatial knowledge is represented in memory using hierarchical structures. In the cases described, the hierarchical structure has just one level (regions or routes); it would be interesting to study how the model could be extended to the cases where the hierarchical structure is more complex and multi-level. One of the drawbacks of the model is that, since it does not include a learning algorithm, it does not explain how the different representations (spatial location and cluster membership in the experiments described above) can be developed; it only explains how they interact to generate the spatial priming effects observed in the experimental data. The fact that the hierarchical structures can be considered either embedded in the environment, such as in the case of routes and regions, or superimposed on the environment by the person experiencing it, such as in the case of neighborhoods or when perceptual divisions are not present, would have to be considered carefully when trying to make this model adaptive.

A model of spatial learning

Considerations of learning from subjective experience with the environment guided the development of the model proposed by Ghiselli-Crippa and Munro (1994). This model focuses on how a global representation of the environment, a cognitive map, can develop from the integration of local spatial associations: the inspiration is provided by considering any exploration task, in which, at any step, only a minor portion of a new environment is available for perception. The goal of the authors is to study the internal representations developed by the network and determine if the structure of the internal representations captures the global properties of the environment.

The authors use a variant of the encoder architecture to associate each location in a two-dimensional grid environment with the set of its immediate neighboring locations, as shown in Figure 3. As in the model by Munro and Hirtle (1989), the input units only specify the location to be considered; there is no structure in the input domain that could be exploited in constructing the internal representations. In their main experiments, the authors use a feed-forward network with two layers of hidden units of which the first layer, identified as the topographic layer (T-layer), consists of just two hidden units. These two units provide distributed representations of the locations in the environment and it is these two-dimensional representations at the T-layer which determine if the global structure of the environment has been captured. Most of the results the authors report are for grid environments; although, after convergence, the networks tend to form representations in the T- layer which reflect the global structure of the environment, in some cases the correspondences between representations and environment are less obvious. The authors report a few techniques which were found valuable in promoting this correspondence between representations and environment. One technique involves the introduction of noise in either hidden layer; another technique involves the use of landmarks, that is, training the network on only a subset of locations at first and then introducing the remaining locations in a later training phase; yet another technique involves the combination of the previous two. The manipulation of these factors affects the degree to which the internal representations match the structure of the environment; they may be thought of as introducing large-scale constraints which are intrinsic to the environment and essential for the development of useful representations, but which are not necessarily reflected in the local neighborhood relations.

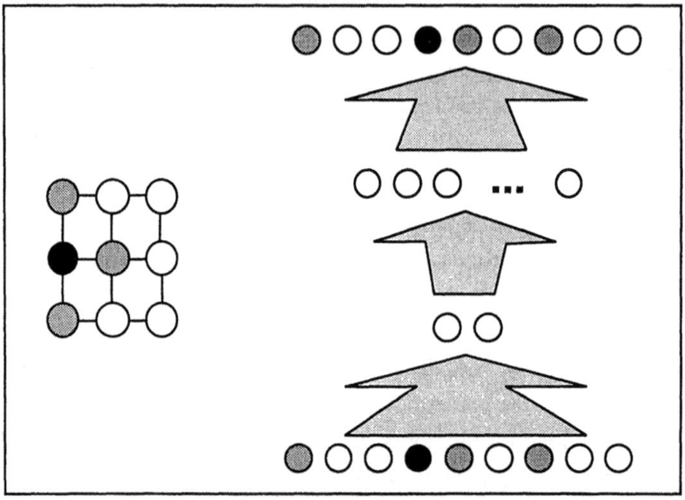

Figure 3: Network architecture for encoding spatial knowledge used by Ghiselli-Crippa and Munro (1994) .

The main aim of the work of Ghiselli-Crippa and Munro (1994) is to show how it can be possible to develop a representation which captures the structure of an environment using only local information and some large-scale constraints. One interesting extension of the work, as they suggest, would be to study if and how the representation of the environment can be affected by the exploration strategy, that is, to determine which of the two, systematic exploration or random exploration, provides a better representation or allows to reach a correct representation sooner. When learning a new environment, the information we take in is usually very redundant, but the redundancy is useful in speeding up the formation of our representation of the environment; in this respect, the model presented by the authors could also be used to investigate the importance of redundancy in learning spatial configurations. It would be interesting to study the quality and the development time of the representations which emerge when the redundancy is removed from all the local information, or when the local information at some locations is incomplete, like, for instance, when not all neighboring locations are specified.

Connectionist Models of the Language of Spatial Relations

When processing spatial information, the relations present in the environment, after having been perceived, need to be represented in some way, for immediate or later use. In many instances, the use involves language, like describing the spatial arrangement of some set of objects which we have perceived. The important research issues in this area involve the exploration of the connections between language, spatial relations, and the mental representation of these relations. For instance, what kind of mental representation is formed when a spatial relation is described using language only? How can language be used to describe a spatial relation? The connectionist models described in the following attempt to shed more light on these issues.

When language is used to describe spatial relations, people normally use spatial prepositions, also called locative prepositions, which are concise and also allow for additional flexibility on the part of the speaker, since they can be used in a deictic, intrinsic, or extrinsic way (Retz-Schmidt, 1988). The expressions which describe the spatial relation between two objects by means of a spatial preposition are called locative expressions (Herskovits, 1986). As many spatial relations can be described by these kinds of expressions, they have been chosen by many researchers as the starting point in their study of language and spatial relations.

From Spatial Relations to Locative Expressions

In his discussion of the connection between spatial relations and locative expressions, Wender (1989; see also Wender and Wagener, 1990) describes a series of levels which correspond to different representations of a spatial relation, from perceptual input to its expression in natural language. Proceeding through the levels, the representation becomes progressively more abstract; since the end points of this series of representational levels correspond to analog and propositional representations, Wender also suggests that there must be a level, which he calls spatial working memory, in which the analog representation is translated into a propositional representation, suitable to be handled by language.

The model proposed by Wender (1989) is inspired by a series of experiments he conducted (recognition tasks in spatial environments and comparative judgment tasks for spatial relations), which provide evidence for an analog representation of spatial relations. His purpose is to show how the model he proposes, appropriately trained to learn spatial relations, can account for the experimental results, which showed a spatial priming effect and an equivalent of the symbolic distance effect in linear orderings. The model described in Wender (1989) is limited to only one-dimensional representations and only two of the possible one-dimensional spatial relations (left-of and right-of). It consists of a cascade of three connectionist modules: a serial position store, a working memory, and a module for the detection of spatial relations. The spatial location store is usually preset for a particular configuration of objects, while the other two modules are trained on examples of spatial relations consistent with the given configuration.

Wender (1989) reports that the model successfully learns to verify spatial relations between two objects, that is, to verify if the locative preposition used to describe the spatial relation between two objects is consistent with the stored configuration, and it generalizes over positions of the objects in the configuration and over objects. Although verifying the spatial relations could be easily accomplished using simple rules, these rules are not coded into the network, which only sees examples of correct performance and still ends up behaving as if it knew these rules. The author also reports that the model is able to qualitatively exhibit the spatial priming effect.

Of interest is whether or not the model can be extended to two-dimensional and three-dimensional situations, with the full array of possible locative prepositions, including whether complicated relations, such as the one expressed by "around-the-corner-from" could be captured using the same approach (Wender, 1989). Another aspect which needs to be considered is the effect of context on the locative preposition used in describing a spatial relation between two objects; for example, the same locative preposition could describe different spatial relations depending on the objects which it relates (e.g., *the house on the lake* versus *the cup on the table*). This issue is explored further in the model described in the following section.

Locative Prepositions

When language is used to describe spatial relations, locative prepositions are very important, because they express the essence of the spatial relations. However, since it is usually necessary to distinguish among many different spatial relations and there are only a few locative prepositions, the context in which a preposition appears usually modifies its default meaning, to more accurately convey the intended spatial relation (Heidorn and Hirtle, 1993; Landau and Jackendoff, 1993). Although the overall context of a sentence could influence the meaning of a preposition, it is usually the two nouns connected by the preposition, which constitute its immediate context, which are mostly involved in determining its final meaning. To study the process of context-dependent mapping between a locative preposition and its meaning, Munro, Cosic and Tabasko (1991) have proposed a connectionist model, which is inspired by a theory on the use of language to express spatial relations put forth by Herskovits (1986). The theory is based on the idea that each locative preposition has an ideal meaning, a meaning based on the prototypical use of the preposition, and that this ideal meaning is modified by the nouns in the immediate context, with the modification usually influenced by the physical and geometrical properties of the noun referents. Given a set of semantic primitives characteristic of spatial relations, the semantic representation of a locative preposition would then change on the basis of which nouns constitute its immediate context; if the semantic representation of the preposition differs from the ideal meaning, then it is called a use type for the particular preposition. Herskovits defines the task of mapping a preposition to its semantic representation as decoding and the task of mapping the semantic representation of a spatial relation to its appropriate locative preposition as encoding.

To show the potential of the connectionist approach to learning the encoding and decoding of locative prepositions, the authors (Cosic and Munro, 1988; Munro et al., 1991) use a restricted but meaningful set of locative expressions: they use concrete expressions of the form noun-preposition-noun, in which the preposition is not used in an abstract or metaphorical sense; they only consider the five prepositions *at, in, on, under*, and *over*; and they limit the semantic representations to combinations of a set of ten semantic primitives representing topological and physical relationships between objects. The purpose of the authors was threefold (Munro et al., 1991): (1) to determine if the network could learn to encode and decode locative expressions; (2) to determine the context-free meaning of each preposition and compare it with the ideal meaning suggested by Herskovits (1986); and (3) to determine the similarity structure of the internal representations of the nouns used in the locative expressions and compare it with the grouping of nouns suggested by the preposition use types listed by Herskovits (1986), which indicates the importance of the physical/geometrical properties of the noun referents. The network, shown in Figure 4, is a feed-forward network with three banks of hidden units and is trained with backpropagation to learn the associations between locative prepositions and their semantic representations (decoding) and between semantic representations of spatial relations and the appropriate locative prepositions (encoding), using different immediate contexts.

The networks perform well on the encoding task, while the performance is significantly worse on the decoding task: this indicates that decoding from locative prepositions to semantic representations is considerably more difficult, due to the higher ambiguity in going from prepositions to the richer space of semantic primitives which describe spatial relations. The networks find context-free semantic representations for the locative prepositions which are fairly stable across the simulations and the representations of the five prepositions correspond nicely with the ideal meanings suggested by Herskovits (1986). Finally, the similarity structure of the internal representations of the nouns used in the locative expressions, examined using cluster analysis, tends to only weakly support Herskovits's suggestion: in fact, it reflects loose semantic categories, which aren't always organized along the physical/geometrical properties of the noun referents.

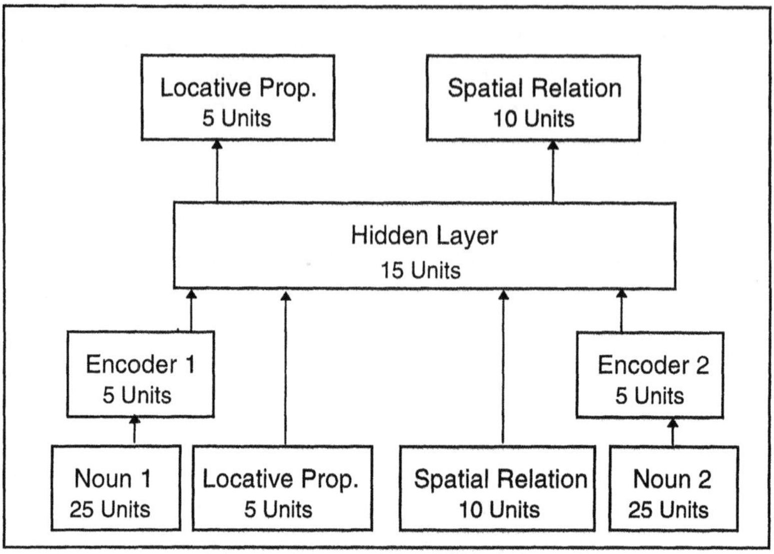

Figure 4: Network architecture for locative prepositions from Munro et al. (1991).

Translation of Locative Prepositions

The locative prepositions model has been extended further to handle the problem of translating locative prepositions between different languages (Munro et al., 1991; Munro and Tabasko, 1991). Two networks of the kind described above and trained using locative expressions from two different languages can be paired to provide translations between the languages. The encoding-decoding capabilities of the networks can be exploited to this avail: a locative expression in the first language is transformed into a semantic representation (decoding in network 1), which is then transformed into the corresponding locative expression in the second language (encoding in network 2).

The authors assume that the nouns in each expression are easy to translate and that, as opposed to the prepositions, their translation is not dependent on context. Also, as with the locative prepositions model, they only use concrete locative expressions, of the form noun-preposition-noun, in which the preposition is used in its concrete sense and not in an abstract sense. The major problem in translating locative prepositions is given by the fact that the correspondence between prepositions is often not one-to-one and the context information is therefore crucial. In their experiments, the authors use the English and

German languages, which provide a one-to-one correspondence for the English prepositions *in, under,* and *above*, but provide a one-to-many correspondence for the English prepositions *at* and *on*. Two networks are trained, an English network and a German network, using the same set of locative expressions, each in the corresponding language.

The results reported, in term of correct translations, are in general rather good, both for translations of expressions used to train the networks and of novel expressions. The performance of the networks in the translation tasks is limited by their performance during training and especially by their performance on the decoding task, which was the most difficult task to learn in both languages, and on the encoding task, for which learning was not perfect in the German network.

The approach to translation used in this system, which relies on the use of intermediate semantic representations, is known as the interlingua approach. The strong points of the system, the capability of learning bidirectional mappings between syntax and semantics from a set of example expressions and the capability of generalizing this knowledge to novel expressions, certainly warrant further exploration.

Future Directions

The inherent multi-dimensionality of spatial knowledge adds to the complexity of modeling spatial knowledge. In particular, it is apparent from past research that semantic and temporal knowledge (McNamara, Halpin and Hardy, 1992) are integrated with pure spatial knowledge, which itself is at least two dimensional. Several recent advances in neural networks might prove useful for this domain.

To build a spatial map through exploration requires an associative mechanism that operates in the temporal domain. A standard feed-forward network has some limited utility for forming temporal associations, even though it computes a static transformation. For example, the mapping can be trained to predict items in a sequence given the previous item; the network described by Ghiselli-Crippa and Munro (1994) can be trained in this way, without changing the results significantly. This approach does not represent the history of the sequence, except for the most recent item (this is analogous to a first order Markov process).

Temporal associations over a longer range can be learned by networks with recurrent connections; i.e., networks that are not feed forward, at least in the strict sense (Elman, 1990; Jordan, 1986; Pearlmutter, 1989; Williams and Zipser, 1989). The general idea is that certain units in the network receive activation from units in the previous learning trial as part of the new input; thus the influence of a stimulus can, in principle, have an effect on network processing that lasts arbitrarily long. Network models are expected to show improved performance acquiring spatial knowledge through exploration if they can represent the long term history of sequences. Such networks could most likely play an important role in modeling the development of spatial cognition and wayfinding performance.

Conclusions

In reviewing the literature on connectionist modeling in spatial cognition, a few issues appear prominent. In all the connectionist models of cognitive maps presented in this review, the emphasis is on modeling the final product, the representation of the environment that a person has developed. Even the issues of learning are mostly driven by the necessity to generate a plausible representation which can account for the experimental data. By focusing on this aspect, we might be overlooking another very useful aspect of connectionist models, which is the fact that, besides being used, after training, to model psychological processes based on stable representations, they can also be used during training, to model the evolution of those representations. As such, connectionist models can then be used not only to account for experimental evolutionary data, but also as tools to help develop theories of how representations evolve, which can then be proved or disproved by the experimental data. In the context of spatial cognition, more emphasis could then be put in their use to study spatial learning in children and adults (Siegel and White, 1975).

At the neuroscience end of the spectrum, much can be learned by studying low-level aspects of spatial cognition (such as place recognition, goal location, landmark learning) in lower animals and in developing appropriate models. Connectionist models have already been proven very useful in this field (e.g., Touretzky, et al., 1993; Zipser, 1986) and, being biologically inspired, appear to be the model of choice. Overall, the study of the issues related to animal navigation can help shed light on which kinds of representations are actually being used and this, as a consequence, can prove very useful in studying human cognitive maps. Thus, it is clear that the integration of both the biological aspects and the psychological aspects is essential for developing an account of human spatial cognition.

Acknowledgments: Preparation of this chapter was supported in part by a Department of Education Title IIb Fellowship to the first author. The paper was written while the second author was on sabbatical leave at the Department of Geoinformation at the Technical University of Vienna and their support and encouragement is gratefully acknowledged. In addition, we wish to thank Andrew Frank, Dan Montello, and Manfred Fischer for their comments and discussions concerning the issues presented in this paper.

References

Angéniol, B., Vaubois, G., and Texier, J.-Y. (1988). Self-organizing feature maps and the travelling salesman problem. *Neural Networks* 1, 289-293.

Cosic, C., and Munro, P. W. (1988). Learning to represent and understand locative prepositional phrases. In *Proceedings of the 10th Annual Conference of the Cognitive Science Society*, pp. 257-262. Hillsdale, NJ: Erlbaum.

Elman, J. L. (1990). Finding structure in time. *Cognitive Science* 14, 179-211.

Fischer, M. M. (1993). Expert systems and artificial neural networks for spatial analysis and modelling: Essential components for knowledge based geographical information systems. Geographical Systems 1, 221-235.

Ghiselli-Crippa, T. B., and Munro, P. W. (1994). Emergence of global structure from local associations. In *Advances in Neural Information Processing Systems 6* (J.D. Cowan, G. Tesauro, and J. Alspector, eds.), San Francisco, CA: Morgan Kaufmann.

Gopal, S., Klatzky, R. L., and Smith, T. R. (1989). Navigator - A psychologically based model of environmental learning through navigation. *Journal of Environmental Psychology* 9, 309-331.

Halmari, P. M., and Lundberg, C. G. (1991). Bridging inter- and intra-corporate information flows with neural networks. Paper presented at the Annual meeting of the Association of American Geographers, Miami, April 13-17.

Heidorn, P. B., and Hirtle, S. C. (1993). Is spatial information imprecise or just coarse? *Behavioral and Brain Sciences* 16, 246-247

Herskovits, A. (1986). *Language and Spatial Cognition.* Cambridge: Cambridge University Press.

Hinton, G. E., McClelland, J. L., and Rumelhart, D. E. (1986). Distributed representations. In *Parallel Distributed Processing: Explorations in the Microstructure of Cognition. Volume 1: Foundations* (D. E. Rumelhart, J. L. McClelland, and The PDP Research Group), pp. 77-109). Cambridge, MA: MIT Press.

Hirtle, S. C., and Heidorn, P. B. (1993). The structure of cognitive maps: Representations and processes. In *Behavior and Environment: Psychological and Geographical Approaches* (T. Gärling and R. G. Golledge, eds.), pp. 170-192. Amsterdam: North-Holland.

Hirtle, S. C., and Jonides, J. (1985). Evidence of hierarchies in cognitive maps. *Memory and Cognition* 13, 208-217.

Jordan, M. I. (1986). Attractor dynamics and parallelism in a connectionist sequential machine. In *Proceedings of the Eighth Annual Conference of the Cognitive Science Society*, 531-546.

Kaplan, S., and Kaplan, R. (1982). *Cognition and Environment: Functioning in an Uncertain World.* New York, NY: Praeger.

Kaplan, S., Weaver, M., and French, R. (1990). Active symbols and internal models: Towards a cognitive connectionism. *AI and Society* 4, 51-71.

Kuipers, B. J., and Levitt, T. S. (1988). Navigation and mapping in large-scale space. *AI Magazine* 9, 25-43.

Landau, B. and Jackendoff, R. (1993). "What" and "where" in spatial language and spatial cognition. *Behavioral and Brain Sciences* 16, 217-265.

Levenick, J. R. (1985). Knowledge Representation and Intelligent Systems: From Semantic Networks to Cognitive Maps. Unpublished Ph.D. Dissertation, Department of Computer and Communication Sciences, University of Michigan.

McClelland, J. L., and Rumelhart, D. E. (1981). An interactive activation model of context effects in letter perception: Part 1. An account of basic findings. *Psychological Review* 88, 375-407.

McNamara, T. P. (1986). Mental representations of spatial relations. *Cognitive Psychology* 18, 87-121.

McNamara, T. P., Halpin, J. A., and Hardy, J. K. (1992). Spatial and temporal contributions to the structure of spatial memory. *Journal of Experimental Psychology: Learning, Memory, and Cognition* 18, 555-564.

McNamara, T. P., Ratcliff, R., and McKoon, G. (1984). The mental representation of knowledge acquired from maps. *Journal of Experimental Psychology: Learning, Memory, and Cognition* 10, 723-732.

Munro, P. W., Cosic, C., and Tabasko, M. (1991). A network for encoding, decoding and translating locative prepositions. *Connection Science* 3, 225-240.

Munro, P. W., and Hirtle, S. C. (1989). An interactive activation model for priming of geographical information. In *Proceedings of the 11th Annual Conference of the Cognitive Science Society*, pp. 773-780. Hillsdale, NJ: Erlbaum.

Munro, P. W., and Tabasko, M. (1991). Translating locative prepositions. In *Advances in Neural Information Processing Systems 3* (R.P. Lippman, J. E. Moody, and D. S. Touretzky, eds.), pp. 598- 604. San Francisco, CA: Morgan Kaufmann.

O'Neill, M. (1991). A biologically based model of spatial cognition and wayfinding. *Journal of Environmental Psychology* 11, 299-320.

Pearlmutter, B. A. (1989). Learning state space trajectories in recurrent neural networks. *Neural Computation* 1, 263-269.

Retz-Schmidt, B. (1988). Various views on spatial prepositions. *AI Magazine* 9, 95-105.

Siegel, A. W., and White, S. H. (1975). The development of spatial representations of large-scale environments. In *Advances in Child Development and Behavior* (H.W. Reese, ed.), pp. 9-55. New York: Academic Press.

Touretzky, D. S., Redish, A. D., and Wan, H. S. (1993). Neural representation of space using sinusoidal arrays. *Neural Computation* 5, 869-884.

Wender, K. F. (1989). Connecting analog and verbal representations for spatial relations. Paper presented at the 30th Annual Meeting of the Psychonomic Society, Atlanta, GA.

Wender, K. F., and Wagener, M. (1990). Zur Verarbeitung räumlicher Informationen: Modelle und Experimente (The processing of spatial information: models and experiments). *Kognitionswissenschaft* 1, 4-14.

Williams, R. J., and Zipser, D. (1989). A learning algorithm for continually running fully recurrent neural networks. *Neural Computation* 1, 270-280.

Yeap, W. K. (1988). Towards a computational theory of cognitive maps. *AI* 34, 297-360.

Zipser D. (1986). Biologically plausible models of place recognition and goal location. In *Parallel Distributed Processing: Explorations in the Microstructure of Cognition. Volume 2: Psychological and Biological Models* (J.L. McClelland, D.E. Rumelhart, and The PDP Research Group), pp. 432-470. Cambridge, MA: MIT Press.

Thea Ghiselli-Crippa, Stephen C. Hirtle, and Paul Munro
Department of Information Science
University of Pittsburgh
Pittsburgh, PA 15260 USA

THE ECOLOGICAL APPROACH TO NAVIGATION: A GIBSONIAN PERSPECTIVE

Harry Heft

Abstract: From an ecological perspective, a basic form of navigating is wayfinding which involves the control of travel through perceiving temporally-structured visual information. This information consists of an optical flow of perspective structure generated by a perceiver moving along a path of travel. The generation of visual information through action, which in turn is controlled by that information, is indicative of the on-going, reciprocal interaction between the perceiver and the environment. It is suggested that the perspective structure consists of a sequence of transitions between successive vistas which uniquely specifies a route to a destination. Also, like the information specifying other types of events, this information can be described as a nested hierarchical structure that unfolds over time. A series of experiments are reviewed that employ dynamic displays of paths in order to examine this approach to way- finding. In addition, it is proposed that in the process of traveling paths through the environment, invariant information specifying the overall layout of the environment is revealed to a perceiver. In this way, the panorama of the environment is apprehended. The role of the affordances of places in navigational processes is also briefly considered. Overall, this ecological analysis suggests a need to reexamine our standard assumptions about the nature of perceiving and its role in navigation.

Introduction

> The Aboriginals, he went on, were a people who trod lightly over the earth. . . . each totemic ancestor, while traveling through the country, was thought to have scattered a trail of words and musical notes along the line of his footprints. . . . 'Providing you knew the song, you could always find your way across country'. Bruce Chatwin (1987), *The Songlines*

Few behavioral functions are more essential for mobile organisms than being able to navigate or travel systematically through the environment. Navigating to the location of functionally important resources and then being able to find a way back 'home' is a fundamental and vital achievement. How is this accomplished? What is the process by which we and other animals systematically find our way?

These questions are challenging for at least two fairly obvious reasons, and a third less apparent one. First, the information needed to address questions concerning animal navigation is not always easy to come by. On the one hand, this information can be obtained through a careful and sensitive analysis of behavior in natural settings – a

J. Portugali (ed.), The Construction of Cognitive Maps, 105–132.

daunting task under any circumstance, whose difficulties are exacerbated by the selection of a problem in which one does *not* want the animal to stay in one place. Alternatively, researchers can strive for carefully crafted investigations in controlled experimental settings that preserve essential correspondences with natural habitats – a standard that assumes one understands a great deal about the natural habitats in the first place.

A second reason that these are difficult questions to answer relates to the hallmark of animal-environment relations: adaptation. Because animals occupy numerous ecological niches and have evolved strategies suited to exigencies of their specific habitat, it is improbable that there is a single navigational process. Rather there are a multiplicity of processes that support navigational activities among mobile animals. The complexity of the problem increases further when we recognize the degree of flexibility some animals have in choosing alternative navigational strategies. Obviously, this issue is particularly acute in the study of human navigation.

In addition, there is a third and less obvious reason why the investigation of these questions can be difficult. The study of navigation is subtly structured by unstated assumptions that researchers bring to the examination of psychological processes. We need to remind ourselves continually that contemporary analyses of behavior and psychological processes, like all scientific analyses, do not begin *tabula rasa*. They reflect a long history of thought about the nature of psychological processes and about the relationship between individuals and the environment. In very concrete and discernable ways, the contemporary analysis of navigation has been shaped by ideas derived from long-standing philosophical traditions. Concepts stemming from these traditions can determine the kinds of questions that are raised and, as a result, can constrain the explanatory picture that emerges.

To take a concrete example of this influence, investigators of navigational processes often uncritically adopt the standard distinction employed by psychologists between perception and cognition. Accordingly, it seems natural to suppose that perception is based on a comparatively simple set of sensory processes, is limited to the momentary stimulation impinging on receptors, and as a result, often needs to be supplemented if not actually driven by more complex (i.e., "top-down") cognitive processes. The latter higher-order processes enable the perceiver to go beyond momentary sensory input. Armed with this tacit distinction between perception and cognition, investigators of navigational processes have often drawn a parallel distinction between route knowledge and configurational or survey knowledge. By comparison, route knowledge is considered to be simpler, often involving little more than 'correlating' movements with sensory stimulation along a path, although sometimes requiring additional contributions by higher-order processes. Configurational knowledge is based on these more complex, higher-order cognitive processes, the most critical aspect of which is a mental representation of the environment (Allen, 1987, Golledge, 1987).

This sort of distinction between perception and cognition, and by extension, between route knowledge and configurational knowledge, may only look intuitively obvious and unquestionably natural to many researchers because it reflects long-standing ways of thinking in our intellectual tradition about processes of knowing. In fact, this type of approach is derived from centuries old assumptions about the nature of perception, and importantly, assumptions which developed in a different intellectual milieu than that of today. For this reason, some of these views may be in need of reexamination.

Specifically, much work in perception and cognition conducted in this century bears the imprint of the Cartesian metatheoretical framework developed centuries *before* the formulation of *evolutionary theory*. Since the momentous appearance of an evolutionary perspective, perceptual and cognitive theorists have for most part attempted to incorporate Darwinian insights by "fine-tuning" the received views of perception. Few theorists have adopted the position that evolutionary theory *requires* that the received accounts of perception and cognition be more than amended, but instead that they be *reformulated* on ecological grounds.

Gradually over the course of his career, James J. Gibson (1904-1979) attempted to do just that. In effect, his approach to the analysis of perception, and ultimately to all cognitive processes, is an attempt to take a fresh approach to the study of psychological processes from an evolutionary perspective (Heft, 1988; Lombardo, 1985; Reed, 1988). The major aim of this chapter is to examine the topic of navigation from the point of view of Gibson's ecological theory. This discussion will primarily explore some of the theoretical implications of examining navigation from an ecological perspective. In the course of this presentation, we will see that many traditional claims about perception and cognition are called into question.

What is An Ecological Approach?

The defining idea in any ecological analysis is the *interrelatedness* among natural entities. From this starting point, an ecological approach to the study of psychology would be one which takes as its central concern the *adaptive fit* between an individual animal and the environment.

In addition, to be ecological a theory must not only take adaptation as its central theme, but importantly, it must maintain a focus on the *environmental conditions* to which a species has adapted. This claim must be sharpened further by emphasizing that the environmental focus assumed in an ecological approach is that of a *relational* and a *reciprocal* perspective.Such a perspective means that just as the animal's structural features and functional capabilities need to be viewed as reflecting adaptation to properties of the environment, and thus should be viewed relative to these properties; so too the pertinent properties of the environment in an ecological analysis need to be identified, from among the range of potential environmental properties, in relation to these characteristics of a particular animal (Heft, 1989; Michaels and Carello, 1981). Indeed, sometimes the environment is even modified by an animal's activities, both in subtle and substantial ways. Identifying those features of the environment that bear a functional relationship to an animal provides a description of that animal's *econiche.*

To specify the econiche for an animal is to consider the environment in relation to an animal considered at a *molar level* of analysis. Tolman (1932), who later would introduce the notion of cognitive map, championed this molar approach throughout the first half of this century. He claimed that the central concern of psychology should be the whole animal, rather than some molecular neurophysiological process or some isolated component of behavior, and examining the whole animal requires attention to its *purposive, goal-directed* actions. The econiche of humans includes such functionally meaningful environmental features as supportable surfaces that can be walked upon, supportable surfaces that can be sat upon, graspable objects that can be used as tools, enclosures that can be used as shelters, and so forth (see Gibson, 1979, pp. 33-42). According to Gibson's ecological framework, environmental *affordances* such as these are perceived through the detection of perceptual information specifying the functional properties of the environmental feature relative to an individual perceiver (E. Gibson, 1982; J. Gibson, 1979; Warren, 1978). These ideas will be developed further by comparing the ecological approach to more standard formulations of perceiving.

Two Approaches to Environmental Perception

Let us examine a classic problem in visual perception. Consider a cube resting on a surface. How is its three- dimensional shape detected? This question has traditionally been an especially difficult one for perception theorists to answer because of the way the problem has usually been formulated. Theorists have long assumed that the initial stage of visual perception is the projection of an image of an object, such as the cube, through the cornea and lens onto the surface of the retina. Whether one takes this 'image' simplistically in an iconic fashion as some sort of picture, or as an array of receptor firings, the fact remains that at this level of analysis we are considering a two-dimensional stimulus. Thus, according to the received metatheory, the initial stage of visual perception is a projective transformation of a three dimensional object in the environment into a two-dimensional stimulus array. Conceptualized in this way, it follows that as the perceiver adopts different vantage points relative to the cube, the projected form of the object on the receptor surface will change. In fact, there are a multitude of two-dimensional forms that are projective transformations of the cube and that correspond to various observation points the perceiver can adopt relative to it.

The significance of this formulation of the initial conditions of perceiving for theories of knowing becomes apparent once one realizes that a projected form of an object on the retina does not correspond uniquely to the one three-dimensional shape that is its source in the world. Rather this projected form can be derived from a large number of possible sources – a perceptual equivocality referred to as *the problem of equivalent configurations*. Therefore, if we take some projected form on the retina as our starting point, we can talk only about the probability that this image corresponds to a particular three-dimensional shape in the world. What perception must then involve, according to most versions of the received view, is *constructing* the most probable object that could be the source in the world for a particular retinal image.

This constructivist explanation, which comes in a variety of forms (see discussions by Heft, 1981; Lombardo, 1988; Turvey, 1977), has at least two significant consequences for the present discussion. First, it means that when one is perceiving a feature of the environment, what is *in fact* being perceived is a psychological construction based on equivocal sensory data. Thus, one does not perceive the world as such, but rather an idiosyncratic, privately constructed representation of that world. Second, this analysis means that all knowing is at best *probable*. As a result, animals must function adaptively in the environment in the face of unremitting uncertainty. From a functional or adaptive point of view this all sounds extremely precarious; but perhaps we must make do as best we can under the circumstances. Or perhaps the apparent precariousness of even our most basic visual accomplishments might prove to be less so if we approach this problem in a different manner.

Reexamining perceptual processes from the point of view of evolutionary theory, Gibson (1966, 1979) formulated an alternative to the preceding constructivist approach to vision – an alternative built, in the case of vision, on an *ecological optics*. Gibson saw as the first question to be addressed by a perceptual theory, what is the nature of the "environment to be perceived?" From the standpoint of a molar analysis, one is asking "what is the econiche for a particular animal?" In pursuing this question, he proposed that from a functional point of view an animal's econiche is set of affordances. These functionally significant environmental features are specified by information in an ambient array of light which is reflected from, and hence structured by, the surfaces of these features. This ambient array of light is the medium for visual perception.

From a single point in the ambient array, the information specifying the shape of an particular object may be equivocal, as we saw above. A stationary or fixed observation point, however, is in fact a limiting case. Gibson (1966) examined instead what happens when perceivers engage, as they normally do, in continuous, exploratory movements of the eye, head, neck, and entire body – that is, of their *visual perceptual systems*. These activities lead the perceiver to adopt a continuous series or a flow of observation points in the visual array; and importantly, generated by such actions from the point of view of the perceiver is a continuing transformation of the stimulus information that may be picked up.

To be more specific, the moving perceiver, by continually changing the point of observation, simultaneously generates two different types of information in the medium that is the ambient array: *perspective information* corresponding to the perceiver's own movements, and *invariant information* corresponding to persisting properties of environmental features. Gibson (1979) describes the situation this way:

> The optic array *changes*, of course, as the point of observation moves. But it also does not change, not completely. Some features of the array do not persist and some do. The changes come from the locomotion and the nonchanges come from the rigid layout of environmental surfaces. Hence, the nonchanges specify the layout and count as information about it; the changes specify locomotion and count as another kind of information, about locomotion itself. (p.73)

Returning to our example, as the perceiver moves with respect to the stationary cube, the continuously changing views of the object – the perspective information – that are generated serve as visual information to the perceiver about her own movements relative to the pbject. That which does not change in the ambient light, that which is constant across these transformations, specifies the invariant properties of the object. In other words, through exploratory movements of the perceptual system invariant information is separated from the flux of varying information. As a result, information specifying the object is revealed, and the individual perceives the object by detecting or picking up this invariant information. In the case of the cube, the invariant in question can be expressed mathematically in terms of geometric cross-ratios in the ambient array (Johansson et al., 1980).

In contrast to the consequences of the constructivist view previously discussed, note that from an ecological perspective, the perceiver is not mentally constructing a subjective world of probable objects. And unlike the probabilistic relationship between retinal stimulation and the environment, invariant information specifies layout features. Thus, in detecting invariant information in the ambient array, the individual is directly perceiving environmental features (Glotzbach, 1992). This is not to say that invariant information is always detected, even correctly. The perceiver certainly needs to be looking in the appropriate direction, often needs appropriate perceptual experience (E. Gibson, 1969), and even then, on occasion, errors can be made (J. Gibson, 1966). However, in contrast to the constructivist view, what we perceive is a structured environment, not a subjective realm constructed from probabilities. In other words, to invoke Wohlwill's (1973) memorable aphorism, "the environment is not in the head!" (Heft, in press a). May it soon become a mantra for psychology. Let us now apply this ecological framework to the problem of navigation.

Ecological Optics and Navigation: A Theoretical Framework

Environments are cluttered with features. This seemingly obvious observation is masked by the ubiquitous use of the terms "space" and "spatial" in the environmental cognition area – terms which connote environments as empty containers. But this observation has important implications for beginning an analysis of navigation. Because environments are cluttered with features, any view of the layout of the natural environment from a particular vantage point will be distinctive and unique, and similarly, any path through this layout will be unique.

When an individual moves along a particular path of travel, and thereby adopts a continuous series of observation points, she generates an optical flow of perspective structure. As discussed above, this flow of information specifies to the perceiver self-movement relative to environmental layout. There will be a unique flow of *perspective structure* specific to each particular path of travel, and different paths of travel will give rise to distinct flows of perspective structure specific to each route.

Can we be more precise about the structure of this flow of information? Two aspects of this perspective flow can be distinguished – *vistas* and *transitions*. A vista is an expanse, a layout of environmental features presently visible. It is "a semienclosure, a set of unhidden surfaces. [It] is what is seen from here" (Gibson, 1979, p. 198). As one travels a path within this expanse, some new information is revealed. Objects that may have been previously hidden from view by occluding surfaces become visible. Concurrently, previously visible objects become concealed behind occluding surfaces as one proceeds along the path. This uncovering and covering of environmental features are local changes occurring within the vista. One is still located at some position within the same expanse, among the same layout of environmental features.

At some point along the path of travel, however, a much more substantial change occurs in the field of view other than merely local, within vista changes. A different vista, a different expanse adjacent to the previous one, begins gradually to come into view replacing the latter. What separates a succeeding adjacent vista along a path of travel from the preceding one are visual barriers, such as a stand of trees, a building, or a crest of a hill. These visual barriers occlude adjacent vistas from sight. Accordingly, as the individual travels a sufficient distance along a path relative to an occluding barrier, or makes a turn around a barrier, a new vista comes into view. The new vista emerges gradually over time at the occluding edge of the visual barrier as the individual travels the path. This portion of the route where a previously occluded vista gradually comes into view constitutes a transition.

Neither vistas nor transitions are located at any one specific point or at any one instant of experience along the path of travel. They are both experienced over time along some portion of the path. However, because a vista consists of a single expanse, much of it can be viewed from a single observation point. For this reason, one can readily take a still photograph that is representative of a vista. However, it is more difficult to capture a transition in a photograph because transitions are most readily visible over time. Moreover, transitions are comparatively more salient portions of a route relative to vistas because it is those portions of a path of travel over which the individual can begin to survey the next adjacent vista. For this reason, transitions are functionally significant segments of the path of travel. They afford looking ahead.

This description of a path of travel as specified by a flow of perspective structure emphasizes *the temporal character of navigating*. For a variety of reasons, most notably the influence of static images such as cartographic maps on our thinking, it is this aspect of perceptual experience that has frequently been neglected in psychological considerations of navigation. The temporal structure of the environment as experienced when navigating has received slightly more attention in the environmental design literature, however (Appleyard et al., 1964; Cullen, 1961; Lynch, 1960, 1984; and Thiel, 1970, in press). For example, on the first page of his seminal work *The Image of the City*, Kevin Lynch describes city design as "a temporal art." And in a later essay he writes:

> City design can focus on *journeys* by which people actually experience cities. . . . It is routine to design streets, bridges, tunnels, and sometimes street facades, but only occasionally are they treated as sequential experiences: as comings out and comings in, as arrivals, glimpses, risings, fallings, a winging around, a sudden view – as approaches, progressions, or foretellings (Lynch, 1984, p. 503).

A consideration of the structure of perspective information that is generated by following a path through the environment suggests a form of navigation based on the perception of temporally structured visual information. This form of navigation will be called here *wayfinding*. In the following section, a number of experiments will be briefly reviewed which were designed to explore the approach to wayfinding just proposed.

Empirical Studies of Transitions and Vistas as Navigational Information

Over the past decade, my students and I have carried out two series of experiments to explore some ideas about wayfinding suggested by the ecological framework. It was hypothesized initially that the perceptual information utilized in wayfinding consists of a sequence of transitions perceived over time that connect successive vistas. In the first series of experiments, several claims concerning the saliency and the nature of transitions along a path of travel were explored. For these studies, a 16 mm color film was prepared, which consisted of a continuous dolly shot through a residential neighborhood. In the initial study (Heft, 1983, Study 1), two edited versions of this film were prepared: a transitions film, which consisted only of the transitions in the route separated by ten second intervals; and a vistas film which consisted of only the vistas, each separated by ten seconds. Participants in the study viewed three times in succession either one of these edited films or the complete unedited film. Subsequently, participants were transported nearby to the start of the route and were asked to walk it. The results indicated that performance accuracy of participants who viewed the transitions film was at a comparably high level to those who viewed the complete film, whereas the vistas film participants were significantly less accurate than the complete film group. These findings were taken as initial support for the view that the sequence of transitions is functionally more valuable information in learning a route than are the vistas.

A second experiment (Heft, 1983, Experiment 2), which was conducted in the laboratory, involved presenting to participants the complete film of this same path, followed by four edited films containing only the transitions. These edited films differed among themselves in the degree to which the order of the transitions corresponded to their order in the complete film. Participants' were asked to rate the similarity of each edited film to the complete film, and their resulting ratings reflected a sensitivity to the actual order of the transitions along the path. These findings indicate that when viewing a path of travel through the environment – in the present case, as displayed in the complete film – perceivers do detect the order or sequence of transitions in the route.

The final two studies in this initial series of investigations (Heft, 1985) explored the hypothesis that the information at the transitions is most readily detected over time. This claim was tested by examining the wayfinding performance of participants as they walked the route after having viewed: a) an edited film (or videotape) containing only the sequence of transitions in route; b) the complete route presented in a series of photographic slides (i.e., "stills") taken at regular, overlapping intervals; or c) a videotape version of the sequence of transitions, with "freeze frames" edited into the tape at four intervals during the course of each transition. The purpose of these latter two manipulations was to disrupt the temporal continuity of the transition information. These comparisons indicated that disrupting temporal continuity resulted in a degrading of subsequent wayfinding performance. Although the size of the effect was modest, perceivers who viewed the edited film of *continuous* transitions between vistas walked the route more accurately than those in the other two conditions.

Each of the studies in the initial series of experiments examined what perceivers *had* learned in prior exposure to a path of travel. In this sense, this work may have been more of an examination of what perceivers remembered about a path of travel than what they experienced as they were traveling the route. To explore navigation in a temporally more immediate way, a procedure was devised for a second series of studies (Heft and Kent, 1993) that permitted perceivers to respond as they were experiencing a path. Furthermore, these studies utilized unedited route presentations, thereby allowing perceivers greater intentional control over what information they might choose to attend. The research paradigm employed in this series was adapted from Newtson's "break point" method – a procedure originally designed for the study of the perceived structure of social action (Newtson and Enquist, 1976; Newtson et al., 1977; Newtson et al., 1989).

The methodological paradigm in this series of experiments was as follows: Participants viewed a videotape of a path of travel through a complex environment, and while doing so they were asked to press a computer key to mark places along the route according to specific task instructions. Thus, they responded to the route as they were perceiving it. A record of the time of each response from the start of a trial was automatically generated.

For the initial experiment in this series, a route along a rural road was videotaped from a moving vehicle. The videotape presented a route that included a number of curves, hillcrests, and two ninety degree turns. The landscape through which the path wound consisted of wooded areas as well as open fields, with a few houses and barns scattered at irregular intervals along the roadsides. Two judges identified places along the route that met the definition of a transition (as discussed above). Only those places selected independently by both judges were designated as transitions for the purpose of data analysis.

Participants were instructed prior to a second viewing of the videotape to indicate with a keypress those places in the route that were "most important for finding your way." The number of perceivers responding at each second from the start of the videotape is presented in Figure 1. The brackets along the abscissa of the figure mark places in the route where transitions occurred. It can be seen that many of the places along the route where the highest number of perceivers responded were at the transitions, although clearly there were many responses to other aspects of the route (such as road signs, unusual trees, even a passing car and a cyclist).

Frequency of responses during transition and non-transition intervals of the route were directly compared by calculating a response rate in each type of interval. The total number of responses during transition intervals and during non-transition intervals were divided, respectively, by the total number of transition interval seconds and the total number of non- transition interval seconds. A comparison of these response rates indicated that perceivers were responding at a significantly higher rate during the transitions in the route, even though in combination transition intervals comprised only 16% of the total of

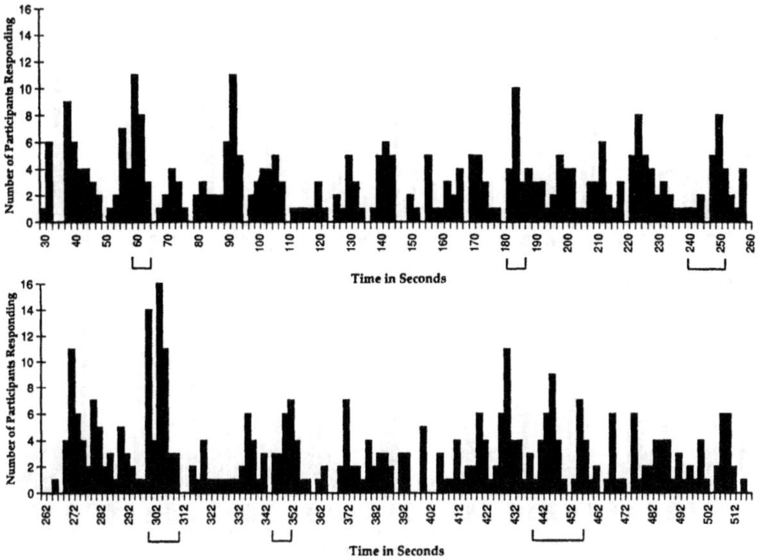

Figure 1: The number of perceivers' (n = 25) combined responses over time to the rural route. Brackets below abscissa indicate transitions.

the route. These findings indicate that transitions as experienced during a path of travel are designated as salient information by a perceiver, although other environmental features also prompted participants' responses. A more direct examination of these latter environmental features relative to transitions was the focus of a second study. Further, in subsequent experiments we modified the prior task instructions because it seemed likely that the somewhat dispersive pattern of responses was due to the nonspecific nature of the task instructions.

For the second investigation in this series, a videotape was prepared by walking with a shoulder-held video camera along a complex route on an architecturally diverse university campus. As in the previous experiment, only transitions independently identified by two judges were designated as transitions for the data analysis. The participants were assigned to one of two task conditions: a fine grain analysis or a large grain analysis condition. Participants in the fine grain condition were instructed during a second viewing of the videotape to segment the route, by pressing a key, into "its smallest meaningful and natural units." Alternatively, in the large-grain condition they were instructed to segment the route into "its largest meaningful and natural units."

Perceivers' responses in each condition are presented in Figures 3a and 3b. In the small grain condition (Figure 2a), responses tended to cluster at the transitions, but at other numerous places as well. The pattern of responses in the large grain condition

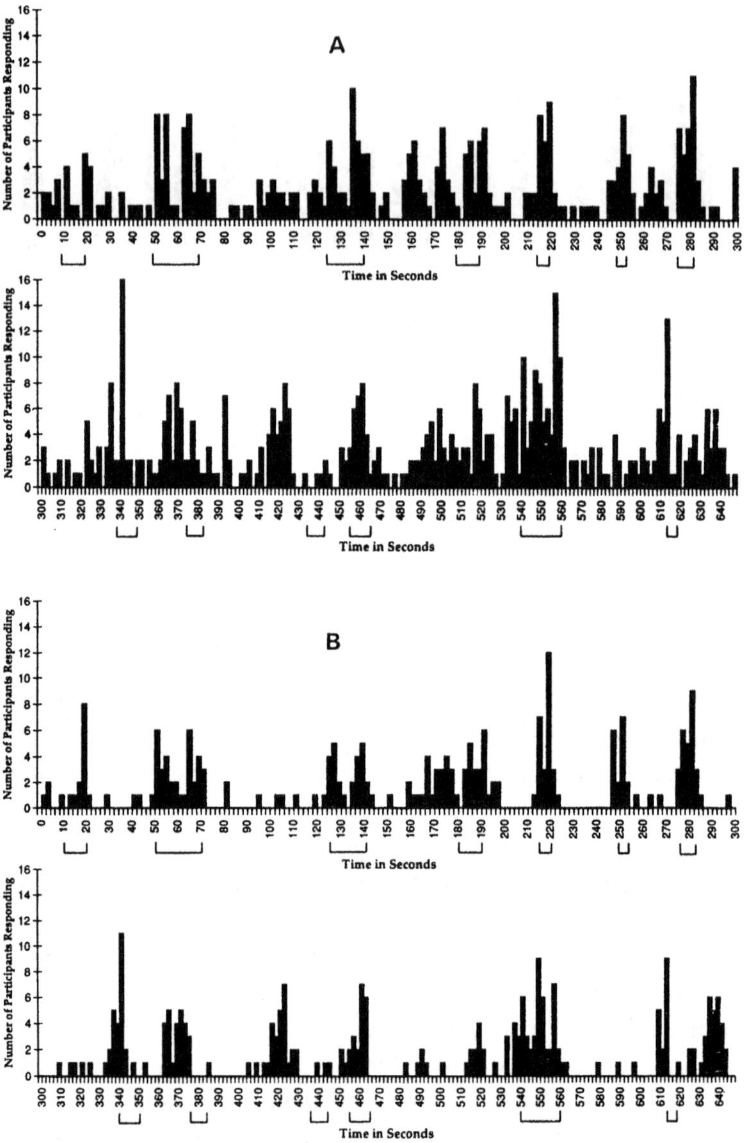

Figure 2: a. The number of perceivers' (n = 26) combined responses over time to the campus route in the "fine-grain" condition (second viewing). Brackets below abscissa indicate transitions.
b. The number of perceivers' (n = 26) combined responses over time to the campus route in the "large-grain" condition (second viewing).

(Figure 2b) indicated that perceivers were primarily responding at some point during each of the transitions in the route (marked by brackets). The performance between these two groups can also be compared by correlating for each condition the number of responses occurring in the transition and non- transition intervals (point biserial correlations). Among participants in the large grain condition, the correlation between number of responses and the type of interval in which responses occurred was substantial (r=.54). Thus, when participants were asked to mark large meaningful and natural units in the route, they reliably indicated places within the transitions. However, when perceivers were asked to divide the route into fine units this correlation was negligible (r=.08). Responses to features other than transitions that appear in the fine-grain condition may reflect a lower scale of analysis of route information. (We will return to this point in the next section.) In sum, these results add further support to the claim that when perceivers are viewing a path of travel, the transitions in the route are seen as its most salient features [also see, Allen and Kirasic (1985), which is based on a different theoretical perspective].

The findings of the two series of experiments reviewed above are consistent with the wayfinding framework developed previously. In general, the results indicate that perceivers can reliably identify transitions between vistas, that transitions may be most readily detected over time, and that the sequence of transitions may play a functionally important role in way- finding. The next step in developing this ecological approach to navigation is to consider how the sequence of vistas and transitions along a route is structured over time.

The Hierarchial Structure of Path Information

Ecological approaches in psychology have been proposed for the study of perception (Gibson, 1979), behavior in social settings (Barker, 1968; Schoggen, 1991), and social development (Bronfenbrenner, 1979). In spite of some important differences among this work, they all share a number of common features. One salient commonality is the structural claim that natural features occur at different levels of scale and that these levels are *hierarchically structured*, with subordinate units *nested* within superordinate units. For example, a leaf is nested within the superordinate feature - tree, and a tree in turn is nested within the more superordinate unit - hillside.

Each natural feature (such as the tree in the previous example) has a dual character, being simultaneously a superordinate or a higher-order unit and a subordinate or embedded unit or, to use Barker's (1963) terminology, both a circumjacent and an interjacent natural unit.

One important consequence of the fact that natural features are hierarchically nested is as follows:

> Hence, for the terrestrial environment, there is no special proper unit in terms of which it can be analyzed once and for all. There are no atomic units of the world considered as an environment. Instead, there are subordinate and superordinate units. The units you choose for describing the environment depends on the level of the environment you choose to describe (Gibson, 1979, p. 9).

Not only features but also *events* in the environment are hierarchically structured. The successive events of my day, such as 'this morning's breakfast', 'commuting to campus', and 'teaching my perception class', are all nested within a higher- order event unit, 'morning activities'. This superordinate event unit, along with the commensurate level unit 'afternoon activities' are both nested within the higher-order unit, 'events of the day September 19, 1994' (Figure 3a). Obviously, this event structure can be extended, in principle, indefinitely in the direction of more inclusive, superordinate units (e.g, third week of the semester/ the fall semester/ the academic year) or in the direction of more subordinate, nested units (e.g., discussion of the visual system in today's perception class/ the neuroanatomy of the visual system/ the retina). At any level an event is both interjacent relative to a superordinate event, and circumjacent relative to a subordinate event.

Traveling a path through the environment can also be described as a temporally-structured, hierarchically-nested event. If we view a path as temporally structured into a

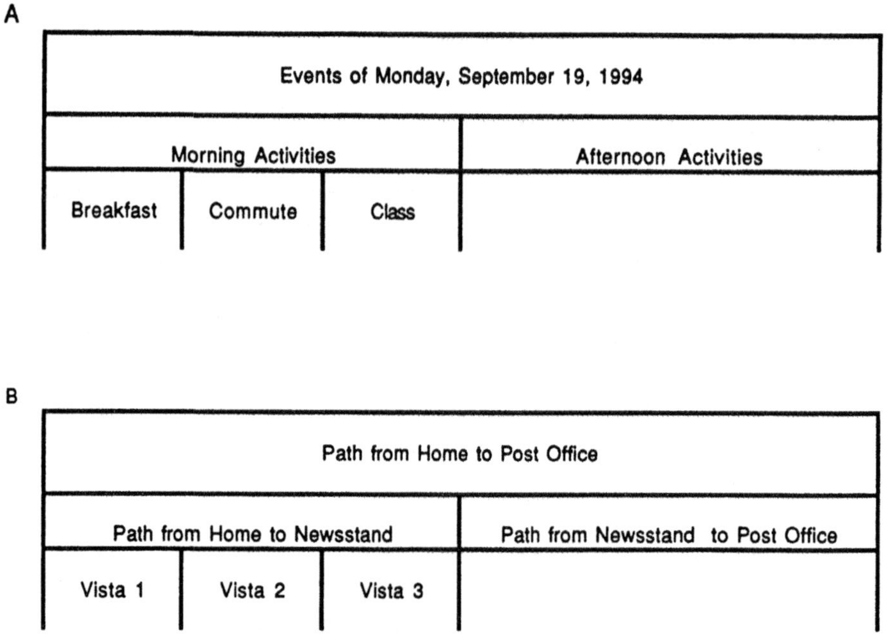

Figure 3: a. A nested hierarchy of events during part of one day. b. A nested hierarchy of path units.

succession of distinguishable vistas connected by transitions, movement through each vista – bounded by an opening and a closing of a transition – would be a basic event unit in the route. And a particular series of vistas would be nested within some higher-order unit. For example, as I follow a path of travel beginning at the front door of my house, I continue through a succession of adjacent vistas through my neighborhood, eventually arriving at my destination – the newsstand. The path through each of these successive vistas is embedded with the superordinate event unit 'path to the newsstand' (see Figure 3b). This event unit is followed by another unit at a commensurate level, 'path to the post office', within which is nested a succession of vistas, and so on. As noted above, there is no "special proper unit", but rather there are event units at varying degrees of circumjacency and interjacency. In fact, as will be seen below, the scale of the event unit that is the focus of travel may well shift with experience in traveling the route.

This type of structural argument has been made in several domains of event perception, the most formalized of these analyses concerning the perceived temporal structure of music (Jones and Boltz, 1988). An examination of the temporal structure of music will lead us to a richer understanding of the temporal structure of the visual information utilized in wayfinding. That an extension of music perception to other domains of event perception is warranted finds support in Jones and Boltz's assertion that their analysis is not intended to be limited to music.

> Events contain many nested time periods whose beginnings and ends are intrinsically marked by various structural changes... In auditory events, they involve onsets of unusual frequency or amplitude changes and are termed accents... In visual events, changes in direction and velocity serve similar functions. In any context, non-temporal information has some potential for carving out meaningful time intervals within and between events (p. 463).

Jones and Boltz (1988) describe a variety of ways in which musical rhythms can be hierarchically structured, and they report experiments indicating how event structure affects perceiving. With regard to the latter, they demonstrate that the temporal coherence or predictability of this structure affects how well perceivers can anticipate structural changes. Jones and Boltz refer to this anticipatory process as *dynamic attending*, which "reflects the attender's tacit use of an event's dynamic structure" (p. 473). Highly coherent time structures, that is, time structures with a recurrent periodicity, enable perceivers to anticipate the occurrence of rhythmic accents and changes in tempo, and thus to "look ahead" in anticipation of the change. However, when the perceiver has not yet discriminated the temporal structure of the music, or if this structure is somewhat incoherent, the perceiver cannot anticipate coming structural changes. Under these circumstances, the perceiver must maintain a more limited or local temporal focus.

Although Jones and Boltz employ the term "attending", they are not referring to an intra-organismic process. In their view, the structure of environmental information controls attending: event perception is the attunement of the perceiver to the temporal structure or the "environmental rhythmicities" of the event.

> In our view, the biological basis for responses to event time takes the form of attunement rhythms that selectively entrain, that shift over nested levels, and eventually are *shaped by the event itself.* (Jones and Boltz, 1988, p. 486; italics added)

This analysis of perceiving music is consistent with the view that a fundamental form of navigation, wayfinding, is a process of perceiving a temporally structured visual event, namely, a path of travel. The environmental information supporting wayfinding is a nested hierarchy of information specifying a path of travel. Further, learning a route is viewed here as a process whereby the perceiver becomes attuned to its hierarchically nested event structure.

With these ideas in mind, we can now define wayfinding more completely as follows: wayfinding refers to a process whereby the perceiver engages in goal-directed actions that generate information specifying a path of travel with a unique hierarchical event structure; and *reciprocally* and *concurrently*, the information that is revealed over time by these actions controls the path of locomotion.

Let us explore some of these issues in more detail by considering a few more of our research findings in navigation. One prediction that grows out of the preceding discussion is that as perceivers gain increasing exposure to a path of travel, they will detect higher-order event units. In other words, initially perceiving path information will be marked by a fine-grained focus and a relatively short time frame perspective in anticipating event structure, that is, in "looking ahead." With increased experience, the perceiver will detect higher-order units of route structure and can therefore look farther ahead in anticipation of changes or transitions in the event structure. Gibson (1979) makes the point this way: "Perceiving gets wider and finer and longer and richer and fuller as the observer explores the environment" (p. 255).

With regard to these issues, recall in a previously discussed experiment that when perceivers were asked to segment a videotape of a path of travel into either fine-grained units or large-grain units, they were able to make this differentiation reliably. Their ability to make this distinction independently reflects, at least, the presence of event structure at two levels of analysis. These findings lend support to the reality of temporally structured units at different levels of scale.

In addition to examining if perceivers can segment a route into either fine or large grain units, we can also consider how their segmentation of a route changes with increasing exposure to it, holding the scale of analysis constant. Accordingly, a experiment was

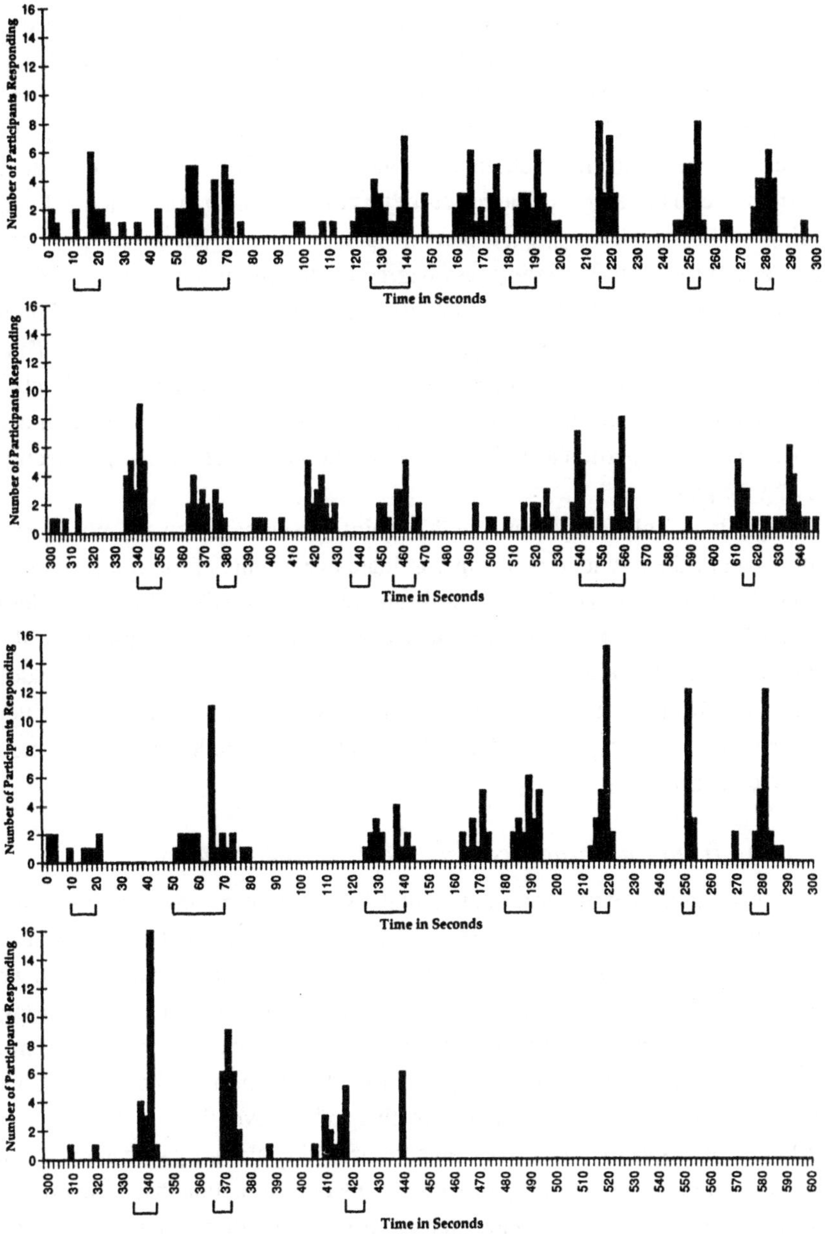

Figure 4: a. The number of perceivers' (n = 21) combined responses over time to the campus route in the "large-grain" condition (first viewing). Brackets below abscissa indicate transitions.
b. The number of perceivers' (n=21) combined responses over time to the campus route in the "large-grain" condition (third viewing). In this condition only the initial 440 seconds of the videotape were presented in order to shorten the time required to run the session.

conducted in which perceivers were asked to segment the videotape of the campus route "into its largest meaningful and natural units" either during a first viewing or a third viewing. It will be recalled that the perceivers in the large grain condition of the previous experiment segmented the route during the second viewing. Thus, the data from these three conditions will allow us to compare perceivers' segmentation of the route in response to "large-grain" instructions when the amount of prior exposure to the route is varied from one to three times. The results of this experiment are presented in Figure 4.

During an initial viewing, perceivers' responses tend to cluster at the transitions (Figure 4a), but only to a slightly less degree of definition than observed during the second exposure (Figure 2b). By comparison, during a third viewing (Figure 4b), responses are sharply limited to a smaller number of places along the route. Thus, with increased experience, perceivers are clearly segmenting the route more selectively. These places which are selected following more exposure are the most prominent transitions. They are marked by particularly distinctive structural changes in the perspective flow specifying the path.

Although it is only conjecture at this point, presumably these more prominent transitions mark the boundaries of superordinate event units of the route, within which the subordinate units, marked by the less distinctive transitions identified in earlier views, are nested. The fine-grain analysis, conducted previously, may be identifying still further subordinate, nested event features. Support for these claims will require additional research, including perhaps an examination of the extent to which perceivers can anticipate structural changes in the flow of path information following varied amounts of exposure to the route.

wayfinding from an Ecological Perspective: A Summary

From an ecological perspective, navigating by wayfinding involves the control of locomotion, or other forms of travel (e.g., driving), by perceiving temporally-structured visual information. This information consists of a flow of perspective structure which is generated by a perceiver moving along a path of travel. Each path through the environment is uniquely specified by a particular sequence of transitions connecting adjacent vistas. Moreover, this information, like that supporting the perception of any event, can be described as a nested hierarchy that unfolds over time.

wayfinding to a specific destination involves traveling along a particular route so as to generate or recreate the temporally structured flow of information that uniquely specifies that path to the destination. The generation of visual information through action, which in turn is controlled by that information, is indicative of the on-going, *reciprocal* interaction between the perceiver and environmental structure. It is not the case that the actions precede the information, *nor* that the information elicits the action in an S->R fashion. What is being described here is a continuous loop of perceiving and acting (Dewey, 1892; Gibson, 1966; and Neisser, 1976).

With additional experience, perceivers are able to follow the path with reference to increasingly superordinate or higher- order units of structure, and in so doing may be better able to "look ahead" and anticipate distinctive transitions. This latter claim needs to be qualified, however, because the ease in identifying higher-order units of information will be also determined by the structural complexity of any particular path and by the perceptual skills of the individual in wayfinding endeavors.

The notion that knowledge such as that based on path information can be temporally structured is inconsistent with the tendency of modern theorists to "spatialize" all knowledge (Bergson, 1910) – that is, the tendency to transpose temporal phenomena into spatial and thus atemporal terms. In the navigation literature, this tendency is reflected in the pervasive use of "cognitive map." Such hypothetical "spatial" representations are usually assumed to be essential for any complex navigational skills. However, as Pick (1993) has persuasively argued, many actions (e.g., making locational and directional inferences) which might seem to require a "spatial" representation can be accounted for in other ways. For example, Pick reviews evidence suggesting that knowledge of the relative location of environmental features may involve continuous updating based on the calibration of visual information (e.g., optical flow) and biomechanical (i.e., movement-produced) stimulation (Rieser et al., 1986; Rieser and Rider, 1991; see also findings of Hazen et al., 1978; Heft, 1979). Thus, the hegemony of representational explanations may be unwarranted on empirical grounds. Moreover, representational explanations may also be, in large measure, logically unnecessary after one has articulated a sufficiently rich description of the environmental information available to be perceived (Gibson, 1979; Heft, 1980). With respect to navigation, such a description would certainly include attention to the temporal structure of stimulus information. This temporal approach requires a departure from standard ways of thinking about navigation – a shift made easier if instead of drawing a parallel between navigational knowledge and perceiving a pictorial map, we recognize that a more appropriate parallel may be between perceiving route structure and perceiving musical structure.

Environmental Layout as Invariant Information: Being Here is Being Everywhere at Once

Our apprehension of paths of travel is only one type of knowledge about the environment. Another very evident type of environmental knowledge – indeed, the type that has received the most attention in the navigation literature – is awareness of the overall configuration or layout of the environment. What is the origin and nature of this form of knowing? A commonly held view is that knowledge of environmental configuration must be based on cognitive operations that construct a mental representation, or a "cognitive map", from discontinuous perceptual encounters. Such a constructivist account seems to be required because the overall layout cannot be perceived from any single location in the environment. (e.g., Kaplan and Kaplan, 1982; Thomson, 1987; Weisman, 1981).

In contrast to this position, the most radical aspect of Gibson's treatment of navigation is his claim that by following paths through the environment, eventually one does come to perceive the overall layout of the environment.

> When the vistas have been put in order by exploratory locomotion, the invariant structure of the house, the town, or the whole habitat will be apprehended. The hidden and the unhidden become one environment... One is oriented to the environment. It is not so much having a bird's-eye view of the terrain as it is being everywhere at once (Gibson, 1979, pp. 198-199).

Although this assertion might initially seem rather peculiar, it does follow directly from the ecological framework developed earlier. The claim that one can come to perceive the overall structure of the environment – and hence be "everywhere at once" – is consistent with a central thesis of the ecological approach that persisting features of the environment, including its layout, are perceived through the pick up of invariant information over time.

As discussed previously, invariant information is revealed in the context of a changing array of information in a flow of perspective structure. The perspective flow specifies to a perceiver self-movement relative to features of the object or the environmental layout in question. The invariant structure that is revealed specifies the object or the environmental layout independent of any vantage point a perceiver might momentarily adopt. The invariant structure specifying object shape is that information in the ambient light which is constant, which does not change (e.g, cross-ratios), regardless of where one begins examining the object or the order in which one examines its various surfaces. Likewise, the invariant structure specifying the overall environmental layout is that information which is constant about its layout, such as the relative position of layout features, regardless of where one begins exploring the layout or the order in which one travels various paths through it. To apprehend this invariant information is to "be everywhere at once."

Before considering these ideas further, it is interesting to note that Gibson was not the only student of perception to recognize that we perceive objects and environmental layout from all sides at once. The phenomenologist Merleau-Ponty (1962), whose work has much in common with Gibson's (Glotzbach and Heft, 1982), independently made essentially the same claim:

> I see the next-door house from a certain angle, but it would be seen differently from the right bank of the Seine, or again from an aeroplane: *the house itself is none of these appearances*; it is, as Leibnitz said, the flat projection of these perspectives *and* of all possible perspectives, that is, *the perspectiveless position from which all can be derived*, the house seen from nowhere (p. 67; italics added).

Of course, a house seen from everywhere at once is a house seen from nowhere in particular; thus, Gibson and Merleau-Ponty are in agreement. They also agree that this knowledge is attained by perceiving invariants in the context of change: "If the object is an invariable structure, it is not one *in spite of* the changes of perspective, but in that change or *through* it" (Merleau-Ponty, 1962, p. 90, italics added).

A similar idea has been expressed by Lynch (1960). Based on his ground-breaking work reported in *The Image of the City*, Lynch is deservedly well-known for promoting the importance of imagery in comprehending the layout of cities. However, in a few passages toward the end of this book he begins to sketch an alternative conceptualization of environmental knowledge somewhat akin to what we have been examining here:

> Considering our present way of experiencing a large urban area, however, one is drawn toward another kind of organization: that of sequence, or temporal pattern. . . . . Intuitively, one could imagine that there might be a way of creating a whole [urban] pattern that would only gradually be sensed and developed by sequential experiences, reversed and interrupted as they might be... The principal quality would be a sequential continuity in which each part flows from the next – *a sense of interconnectedness at any level or in any direction...* the region would be *continuous, mentally traversable in any order* (Lynch, 1960, pp. 113-115, italics added).

Presented with such ideas, one might object that it is simply implausible that the overall layout of the environment can be perceived by uncovering an invariant structure over time as one travels paths through the environment. The primary basis for this objection is that perceiving is obviously something that only occurs in "the present". Therefore, a *fortiori* one cannot possibly *perceive* invariant structure over extended periods of time. It is primarily on these grounds that theorists have distinguished between perception, which is assumed to be limited to the present, and cognition, which does not have this constraint.

But this characterization of perception, and the standard perception-cognition distinction that follows from it, is not so problem-free, and hence so intuitively obvious, as it might first seem. The previous argument rests on having achieved some conceptual clarity about the notion of the present. But to borrow one of Gibson's favorite expressions, this notion is "a muddle." As William James argued, the idea of the present as an instant in time is a fiction: "There is literally no such datum as that of the present moment... except as an unreal postulate of abstract thought" (James, 1895, p. 158). James (1890; Ch. 15) observed instead that the "sensible present" has *duration*, rather than a being "razor's edge," encompassing simultaneously what was present and is now past, and what is future and will soon be present. Its boundaries are not sharp and certainly not fixed.

If it is impossible to delimit the 'present' in any clear-cut and consistent way, it seems quite arbitrary where one might choose *a priori* to draw a time limit within which one can *perceive* invariant structure. The invariant information specifying the shape of a small object might be detected over a few seconds duration as we turn it in our hands. The invariant information specifying the shape of a larger object, such as a statue, or even a larger object, such as a building, might be detected over a minute or several minutes, respectively, as we walk around it. Over what duration should we say invariant structure can no longer be detected? Might there not be invariant structure that is typically detected only after several hours of exploration, (e.g., a complex building interior) or several days of exploration (e.g., a small town), or even several years (e.g., a large city)?

Where do these considerations leave us concerning the distinction between perception and cognition? Let us consider the following possibility. Instead of delimiting perceptual processes with respect to a temporal criterion, we might claim that perceiving refers to those cognitive processes involved in the *detection of information in the ambient array*. It is this on- going pick up of stimulus information that is hallmark of perceiving and that distinguishes it from other cognitive processes, such as remembering and imagining, which are not based on stimulus information pick up (see Gibson, 1979; Reed, 1987). From this perspective, it is justifiable to say that one can *perceive* the overall layout or configuration of the environment because this invariant structure can be detected or revealed as information in the context of changing perspective structure (i.e., paths of travel) over extended durations of time. As this claim suggests, perceiving may operate at high levels of complexity. It differs from other cognitive processes then, not with respect to its relative complexity, but rather because it is based on the detection of stimulus information specifying environmental features.

Finally, the claim that perceiving can extend over long durations is consistent with the functional perspective that lies at the heart of the ecological position. To know where one is in the environment, that is, to be *oriented* requires more than seeing an arrested moment or position in a path of travel. It requires a panoramic awareness of the overall layout of the environment (Gibson, 1979, pp. 112-114). Some recent experimental evidence (Beer, 1993) provides initial support for the claim that visual perception is panoramic and that visual perception extends beyond the information in the present field of view. [Although see Hochberg (1986) for an alternative interpretation of findings such as those of Beer (1993).]

Place Affordances and Navigation

Up to now, the discussion has concerned itself with "how" perceivers find their way around. But it has not addressed "why" they do so. Indeed, in contrast to investigations of migratory processes in birds, insects, and non-human mammals (Gallistel, 1990; Waterman, 1987), the "why" of navigation has received very little attention in the literature on human navigation.

Most of the time, navigation is purposive and goal-directed. Places in the environment typically have functional significance for us, and we travel to places to utilize and engage their affordance possibilities. There is considerable qualitative evidence indicating that functionally meaningful places are the most salient features in individuals' recollections of previous environmental experiences (Chawla, 1992; Cooper-Marcus, 1992; Hart, 1981, Lukashok and Lynch, 1956; Moore, 1986). Paths through the environmental layout leading to functionally significant places are reported to be salient aspects of the environment by children (Heft, in press b).

Only in a few instances, however, has navigation been studied in the context of an individual carrying out some functional goal (Cohen and Cohen, 1982; Cohen, Cohen, and Cohen, 1988; Gärling, Böök, and Lindberg, 1984). One reason for this circumstance is a characteristic of much psychological research historically: Psychological processes tend to be studied in a decontextualized manner. Thus, purpose of an action, its "why", if it is considered at all, is seen to have little bearing on "how" it is performed.

There is an additional reason, however, that is more specific to the study of navigational processes. Much of the initial research used knowledge of Euclidean geometric relations as a standard for assessing performance. In this work, individuals' proficiency in making judgments or performing actions that reflected an adequate appreciation for appropriate geometric principles was assumed to have a bearing on how well they could find their way around the environment. Thus, tasks utilizing table-top models, miniature rooms, and model towns were employed, especially with children, to assess geometric knowledge and, by extrapolation, their skills in environmental cognition. A concern expressed by some researchers reviewing this work was that children often appeared to be less competent in these experimental settings than they are in their everyday lives (Heft and Wohlwill, 1987; Spencer and Darvish, 1981). Possibly one reason for this poorer level of performance is the essentially abstract nature of tasks where affordances of places are represented symbolically, if at all.

Instead, research involving purposive navigational tasks in meaningful environments needs to be conducted. With these methodological concerns in mind, Heft and Blue (1990) conducted an exploratory experiment to determine whether the presence of salient place affordances along a route through a meaningful setting would facilitate the ease with which children learned that route. Children (ages 3 years 7 months to 7 years 11 months) were taken on a walk through a complex building interior. In one condition, the route led to a succession of places or sites which afforded specific activities (e.g., sharpening a pencil). Alternatively, children were led along the same route to these same places, but in this case, the places were merely employed as sites for an innocuous activity unrelated to their intrinsic features. Subsequently, children in both conditions were asked to lead the experimenter back along this same route.

Relatively few navigational errors were made by children who experienced the route as leading to distinctive place affordances as compared to a higher error rate when the route was less tied to functionally significant features along the way. These findings provide preliminary evidence that learning a path to functionally meaningful places, places with affordances – which is presumably the usual purpose of wayfinding – is more readily achieved than learning a path *per se* through the environment. Moreover, age was not correlated with successful wayfinding performance, suggesting that this type of navigational learning may be a basic perceptual skill. Future investigations which examine navigation as a purposive, goal-directed activity will likely enrich our understanding of navigational processes.

Concluding Comments

Perceptual theorists have historically taken the picture (e.g., the retinal image) as the starting place for an account of vision – as the basic visual phenomenon to be explained. And in doing so, they approach other issues, such as the 'perception of motion', as derivative, to be explained on the foundations established by an account of pictorially-based vision. James Gibson found this to be a peculiar strategy because, in his view, vision most fundamentally involves perceiving over time. In fact, movements by the perceiver help to reveal information, making perceiving easier rather than more difficult, as it certainly would do if visual perception started with a static picture on the eye. Instead, picture perception should be treated as a special case rather than a normative one (Gibson, 1971, 1979).

If it seems odd to claim that perceiving over time is the essential characteristic of vision, might this not be because our intellectual traditions have influenced us to think pictorially and spatially? At the very least, the claims of the ecological theory of perception are no more implausible than the assumptions on which the standard views are built.

> The notion of ambulatory vision is not more difficult, surely than the notion of successive snapshots of the flowing optic array taken by the eye and shown in the dark projection room of the skull (Gibson, 1979, p. 197).

The same state of affairs is present in the study of navigation. There has been an emphasis on static, configurational forms of environmental knowing, and a neglect of how environments are experienced over time. It has been suggested in the preceding pages that an essential and basic form of navigating involves following temporally structured information over time specifying a path of travel. From this perspective, a different approach to the acquisition of configurational knowing emerges: the configuration of the environmental layout is revealed with the detection of invariant stimulus information over paths of travel.

The primary intention of this chapter was to highlight the temporally-based character of navigation. Using this orientation as a framework, we may find the reexamination of our standard ways of thinking about navigation a worthwhile exercise.

Acknowledgements: My interest in this problem has its roots in discussions about navigating with James Gibson, as we traveled by car from Ithaca, NY to his weekly seminar in Binghampton, NY during the winter of 1975.

I would like to thank William Nichols, Edward Reed, Anne Pick, and Herb Pick for their helpful comments on an earlier draft of this chapter and Kerry Marsh for insightful discussions of these ideas in their early stages. I would also like to acknowledge the assistance of numerous students, who over a number of years, were instrumental in conducting this research. In particular, I would like to note the important contributions of Beth-Anne Blue, Hugh Campbell, Marion Kent, Claire Siegenthaler, and Micah Thompson. I am also very grateful to Frederick Prete who very generously helped in the construction of most of the figures. Thanks also to Graham Campbell who contributed in this vein as well.

Portions of this work were presented at the meetings of the American Collegiate Schools of Planning, (Columbus, Ohio, October, 1992) and the International Conference on Event Perception and Action (Vancouver, British Columbia, August, 1993). The empirical research reported here has been generously supported over the past decade by the Denison University Research Foundation. Work on this chapter was completed with the aid of an award from the Robert C. Good Fellowship Program, Denison University.

References

Allen, G.L. (1987). Cognitive influences on the acquisition of route knowledge in children and adults. In *Cognitive Processes and Spatial Orientation in Animal and Man*. Vol. II. (P. Ellen and C. Thinus-Blanc, eds.) pp. 274-283. Dordrecht: Martinus Nijhoff.

Allen, G.L., and Kirasic, K.C. (1985). Effects of the cognitive organization of route knowledge on judgments of macrospatial distance. *Memory and Cognition*, 13, 218-227.

Appleyard, D., Lynch, K, and Meyer, J.R. (1964). *The View From the Road*. Cambridge, MA: MIT Press.

Barker, R.G. (1963). On the nature of the environment. *Journal of Social Issues* 19, 17-38.

Barker, R.G. (1968). *Ecological Psychology*. Stanford, CA: Stanford University Press.

Beer, J.M.A. (1993). Perceived scene layout through an aperture during visually simulated self-motion. *Journal of Experimental Psychology: Human Perception and Performance* 19, 1066-1081.

Bergson, H. (1910). *Time and Free Will*. New York: Macmillan.

Bronfenbrenner, U. (1979) *The Ecology of Human Development*. Cambridge, MA: Harvard University Press.

Chatwin, B. (1987). *The Songlines*. New York: Penguin Books.

Chawla, L. (1992). Childhood place attachments. In *Place Attachment, Human Behavior and Environment* Vol. 12. (I. Altman and S.M. Low, eds.) pp. 63-86. New York: Plenum.

Cohen, R., Cohen, S., and Cohen, B. (1988). The role of functional activity for children's spatial representations of large-scale environments with barriers. *Merrill-Palmer Quarterly*, 34, 115-129.

Cohen, S., and Cohen, R. (1982). Distance estimates as a function of type of activity in the environment. *Child Development*, 53, 834-837.

Cooper-Marcus, C. (1992). Environmental memories. In *Place Attachment, Human Behavior and Environment* Vol. 12. (I. Altman and S.M. Low, eds.) pp. 87-112. New York: Plenum.

Cullen, G. (1971). *The Concise Townscape*. New York: Van Nostrand.

Dewey, J. (1896). The reflex arc concept in psychology. *Psychological Review* 3, 357-370.

Gallistel, C.R. (1990). *The Organization of Learning*. Cambridge, MA: MIT Press.

Gärling,, T., Böök, A., and Lindberg, E. (1984). Cognitive mapping of large-scale environments: The interrelationship of action plans, acquisition, and orientation. *Environment and Behavior* 16, 3-34.

Gibson, E.J. (1969). *Principles of Perceptual Learning and Development*. New York: Appleton-Century-Crofts.

Gibson, E.J. (1982). The concept of affordances in development: The renascence of functionalism. In *Minnesota Symposium on Child Psychology (Vol. 15): The Concept of Development* (W.A. Collins, ed.), pp. 55-81. Hillsdale, NJ: Lawrence Erlbaum.

Gibson, J.J. (1966). *The Senses Considered as Perceptual Systems*. Boston: Houghton-Mifflin.

Gibson, J.J. (1971). The information available in pictures. *Leonardo* 4, 27-35.

Gibson, J.J. (1979). *The Ecological Approach to Visual Perception*. Boston: Houghton-Mifflin.

Glotzbach, P.A. (1992). Determining the primary problem of visual perception: A Gibsonian response to the "correlation" problem. *Philosophical Psychology* 5, 69-94.

Glotzbach, P.A., and Heft, H. (1982). Ecological and phenomenological approaches to perception. *Nous* 16, 108-121.

Golledge, R.G. (1987). Environmental cognition. In *Handbook of Environmental Psychology* (D. Stokols and I. Altman, eds.) pp. 131-174. New York: John Wiley.

Hart, R. (1981). Children's spatial representation of the landscape: Lessons and questions from a field study. In *Spatial Representation and Behavior Across the Life Span* (L.S. Liben, A.H. Patterson, and N. Newcombe, eds.), pp. 195- 233. New York: Academic Press.

Hazen, N.L., Lockman, J.J., and Pick, H.L., Jr. (1978). The development of children's representations of large-scale environments. *Child Development* 49, 623-636.

Heft, H. (1979). The role of environmental features in route-learning: Two exploratory studies of wayfinding. *Environmental Psychology and Nonverbal Behavior* 3, 172-185.

Heft, H. (1980). What Heil is missing in Gibson: A reply. *Journal for the Theory of Social Behavior* 10, 187-193.

Heft, H. (1981). An examination of constructivist and Gibsonian approaches to environmental psychology. *Population and Environment: Behavioral and Social Issues* 4, 227-245.

Heft, H. (1983). wayfinding as the perception of information over time. *Population and Environment: Behavioral and Social Issues* 6, 133-150.

Heft, H. (1985) wayfinding and the flow of information along a path of locomotion. Unpublished manuscript.

Heft, H. (1988). Affordances of children's environments. A functional approach to environmental description. *Children's Environments Quarterly* 5, 29-37.

Heft, H. (1989). Affordances and the body: An intentional analysis of Gibson's ecological approach to visual perception. *Journal for the Theory of Social Behavior* 19, 1- 30.

Heft,H. (in press a). Toward a functional ecology of behavior and development: The legacy of Joachim F. Wohlwill. In *Children, Cities, and Psychological Theories: Developing Relationships*. (D. Görlitz, H. Harloff, J. Valsiner and G. May, eds.). Berlin: Walter de GruGrurter.

Heft, H. (in press b). Gibson's ecological approach and environment-behavior research and design. In *Advances in environment, behavior, and design* Vol. 4. (G.T. Moore and R.W. Marans, eds.), New York: Plenum.

Heft, H., and Blue, B. (1990). Affordances and children's way-finding. Unpublished manuscript.

Heft, H. and Kent, M. (1993). wayfinding as event perception: The structure of route information. *Paper presented at International Conference on Event Perception and Action*, Vancouver, BC, August, 1993.

Heft, H. and Wohlwill, J.F. (1987). Environmental cognition in children. In *Handbook of Environmental Psychology* (D. Stokols and I. Altman, eds.), pp. 175-204. New York: John Wiley.

Hochberg, J. (1986). Representation of motion and space in video and cinematic displays. In *Handbook of Perception and Human Performance* Vol. I. In K. Boff, J. Thomas, and L. Kaufman (eds.) pp. 22-1-22-64. New York: Wiley.

James, W. (1890). *The Principles of Psychology*. New York: Holt.

James, W. (1895). The knowing of things together. *Psychological Review* 2, 105-124. [Reprinted in *The writings of William James* (J.J. McDermott, ed.), pp. 152-168.. Chicago: University of Chicago Press.

Johansson, G., von Hofsten, C., and Jansson, G. (1980). Event perception. *Annual Review of Psychology*, 31, pp.27-63.

Jones, M.R. and Boltz, M. (1989). Dynamic attending and responses to time. *Psychological Review* 96, 459-491.

Kaplan, S., and Kaplan, R. (1982). *Cognition and Environment: Functioning in an Uncertain World.* New York: Praeger.

Lombardo, T. (1987). *The Reciprocity of Perceiver and Environment: The Evolution of James J. Gibson's Ecological Psychology*. Hillsdale, NJ: Lawrence Erlbaum.

Lukashok, A., and Lynch, K. (1956). Some childhood memories of the city. *Journal of the American Institute of Planners* 22, 142- 152.

Lynch, K. (1960). *The Image of the City*. Cambridge, MA: MIT Press.

Lynch, K. (1984). The immature arts of city design. *Places* 3, 10- 21. [Reprinted in City Sense and City Design: Writings and Projects of Kevin Lynch (T. Banerjee and M. Southworth, eds.), pp. 498-510, Cambridge, MA: MIT Press, 1990.]

Merleau-Ponty, M. (1963). *The Phenomenology of Perception*. (C. Smith, Trans.) London: Routledge and Kegan Paul.

Michaels, C.F., and Carello, C. (1981). *Direct Perception*. Englewood Cliffs, NJ: Prentice-Hall.

Moore, R.C. (1986). *Childhood's Domain: Play and Place in Child Development*. London: Croom Helm.

Neisser, U. (1976). *Cognition and Reality*. San Francisco: Freeman.

Newtson, D., and Enquist, G. (1976). The perceptual organization of ongoing behavior. *Journal of Experimental Social Psychology* 12, 436-450.

Newtson, D., Enquist, G., and Bois, J. (1977). The objective basis of behavior units. *Journal of Personality and Social Psychology* 35, 847-862.

Newtson, D., Hairfield, J., Bloomingdale, J., and Cutino, S. (1987). The structure of action and interaction. *Social Cognition* 5, 191-237.

Pick, H.L. Jr. (1993). Organization of spatial knowledge in children. In *Spatial Representation: Problems in Philosophy and Psychology* (N. Eilan, R. McCarthy, and B. Brewer, eds.), pp. 31-42. Oxford, UK: Basil Blackwell.

Reed, E.S. (1987). The ecological approach to cognition. In *Cognitive Psychology in Question* (A. Costall and A. Still, eds.), pp. 142-172. Brighton, UK: Harvester Press.

Reed, E.S. (1988). *James J. Gibson and the Psychology of Perception.* New Haven, CT: Yale University Press.

Rieser, J.J., Guth, D.A., and Hill, E.W. (1986). Sensitivity to perspective structure while walking without vision. *Perception* 15, 173-188.

Rieser, J.J., and Rider, R.A. (1991). Young chidren's spatial orientation with respect to multiple target when walking without vision. *Developmental Psychology* 27, 97-107.

Schoggen, P. (1989). *Behavior Settings: A Revision and Extension of Roger G. Barker's Ecological Psychology.* Stanford, CA: Stanford University Press.

Spencer, C., and Darvizeh, Z. (1981). The case for developing a cognitive environmental psychology that does not underestimate the abilities of young children. *Journal of Environmental Psychology* 1, 21-31.

Thiel, P. (1970). Notes on the description, scaling, notation, and scoring of some perceptual and cognitive attributes of the physical environment. In *Environmental psychology: Man and his physical setting* (H. M. Proshansky, W. H. Ittelson, and L.G. Rivlin, eds), pp. 593-619. New York: Holt, Rhinehart, and Winston.

Thiel, P. (in press). *People, Paths and Purposes.* Seattle, WA: University of Washington Press.

Thomson, J.A. (1987). Cognitive and motor representations of space and their use in human visually-guided locomotion. In *Cognitive Processes and Spatial Orientation in Animal and Man.* Vol. II. (P. Ellen and C. Thinus-Blanc, eds.) pp. 284-290. Dordrecht: Martinus Nijhoff.

Tolman, E.C. (1932). *Purposive Behavior in Animals and Men.* New York: Century.

Tolman, E.C. (1948). *Cognitive maps in rats and men. Psychological Review* 55, 189-208.

Turvey, M.T. (1977). Contrasting orientations to the theory of visual information processing. *Psychological Review*, 84, 67- 88.

Warren, W.H. (1984). Perceiving affordances: Visual guidance of stair climbing. *Journal of Experimental Psychology: Human Perception and Performance* 10, 683-703.

Waterman, T.H. (1989). *Animal Navigation.* New York: Scientific American Library.

Weisman, G. (1981). Evaluating architectural legibility: Way-finding in the built environment. *Environment and Behavior* 13, 189-203.

Wohlwill, J.F. (1974). The environment is not in the head! In *Environmental Design Research*, Vol. 1 (W.F.E. Preiser, ed.), pp. 166-181). Stroudsburg, PA: Dowden, Hutchinson, and Ross.

Harry Heft
Department of Psychology
Denison University
Granville Ohio 43023

VERBAL DIRECTIONS FOR WAY-FINDING: SPACE, COGNITION, AND LANGUAGE

Helen Couclelis

Abstract: This paper develops a tentative model of the cognitive mechanism underlying verbal direction-giving. It proposes the hypothesis that the cognitive map often assumed to be at the basis of that behavior, as well as the direction-giving discourse itself, are in fact generated by a common underlying mental model, which is itself structured by more primitive kinesthetic image-schemas and basic-level categories. Thus, theoretically, the paper synthesizes work on mental models from cognitive psychology with the perspective of experiential realism developed in cognitive linguistics. Empirically, it builds upon an analysis of transcripts from an informal experiment in direction-giving, rejoining similar studies by behavioral geographers and others. While distinct from image-based (in particular, cognitive-map based) as well as syntactic accounts of direction-giving, the proposed model offers an alternative that accommodates both these.

Introduction

Natural language provides direct access to how people perceive and understand space. The observation of spatial behavior, the analysis of sketch maps, and the vast array of psychological experiments devised to investigate particular aspects of spatial cognition, are all necessary parts of the study of cognitive maps and spatial decision making. But few other means of accessing "the world in the head" can match natural language in comprehensiveness and immediacy. For those who know how to listen, the way people talk about space reveals a lot about how they think of it, how they deal with it, and how they construct it in their minds.

This paper asks what may be learned about cognitive maps, and about spatial cognition more generally, through an investigation of how people give verbal route directions to one another. Route directions are readily available, natural protocols reflecting the direction givers' cognitive representations of certain critical aspects of their environment. Still, the relationship between the spoken words and the underlying cognitive structures is far from transparent. Responding effectively to a non-trivial request for route directions is a complex task during which different aspects of spatial cognition come into play at different stages. A number of questions can be raised about that process – difficult questions concerning fundamental aspects of cognition , that this study cannot hope to do

J. Portugali (ed.), The Construction of Cognitive Maps, 133–153.

more than raise. For example: How can a brief verbal description allow a representation of an unknown environment to be constructed in another person's mind? How is the structure of large-scale environments reflected in natural language? What is the connection between image-like and verbal representations of space? Are there structures more fundamental than cognitive maps that may be accessed through a linguistic perspective?

The paper describes a conceptual model of verbal direction giving. As it stands, the model is no more that a plausible account of how space, cognition and language may come together in the context of a practical everyday task. It is supported by a wide range of research results in behavioral geography, psychology, cognitive linguistics and discourse analysis, and illustrated with empirical evidence supplied by an informal experiment.

The model has two basic features: first, it involves the notion of an underlying *mental model* integrating specific and general knowledge about the environment with a range of intentional states and behaviors characterizing the direction-giving situation. Second, it suggests that if there is indeed a cognitive map involved in direction-giving, this is not more fundamental than the discourse itself: that is, direction-giving is not merely a question of some independently existing cognitive map being eventually translated into English for the purposes of communication. In this paper the term "cognitive map" will be used in the original, more literal sense of a map-like, imagery-based environmental representation, rather than as a synonym for cognitive representation in general, as is the case in some of the cognitive literature.

The paper is structured as follows: the next section puts this work in perspective, relating it to other work in the area, and also to some major debates in cognitive science. Prominent among these is the question of the relationship between discourse and mental imagery (in this case: between verbal direction-giving and cognitive maps). I argue that the two are connected by means of a common underlying mental model which in itself is neither linguistic nor map-like. That model is further discussed towards the end.

Following that background discussion I present a step-by-step description of direction-giving, drawing upon transcripts of actual interviews taken on the campus of the University of California at Los Angeles (UCLA). Five stages are distinguished in the direction-giving process: *initiation, representation, transformation, symbolization*, and *termination*, of which the middle three bear directly on the issue of spatial cognition. The process is also presented diagrammatically to illustrate the complex interrelationships between intentions, memories, beliefs, predispositions, and actions, all of which come into play in the course of giving route directions.

At the core of that direction-giving account is a mental model of the state of affairs as represented in the direction-giver's mind. The section that follows speculates as to what that mental model might be like, bringing in insights from cognitive linguistics as well as psychology. As with much other theoretical work of this kind that is not directly testable, that model should be judged primarily on the basis of the degree of clarification and understanding it might bring to the issue of spatial cognition.

The Cognitive Underpinnings of Direction-Giving

A source of the attraction of research on verbal direction giving, but also of its difficulty, is the large number of different aspects of cognition it involves. It is difficult to avoid the notion that a *cognitive map* somehow plays a key role in the process. A request for directions also calls up a conversational situation, to which the principles of *conversation analysis* apply. The route directions themselves form some kind of simple narrative, that can be approached from the perspective of *discourse analysis* for an elucidation of its structure. That narrative can be assumed to be based on a model, in the direction-giver's mind, of relevant aspects of the environment. Central to that mental model is a cognitive route-planning task, based on some combination of re-experiencing, remembering, and inferencing. The information necessary and sufficient to answer the query must be extracted from that model and put into *language*. Finally, as we all know from hard experience, the route directions people give often contain different kinds of *errors*. Unfortunate though these may be from the direction-seeker's viewpoint, they can provide the analyst with precious clues on the deeper structure and function of spatial cognition. Because of its multiple interesting dimensions, direction-giving has attracted researchers from many areas of cognitive science and artificial intelligence, as well as behavioral geography (Frank and Mark, 1991; Golledge, 1992; Gould, 1989; Klein, 1982; Ma, 1987; Mark, 1987, 1989a, 1989b; McGranaghan et al., 1987; Riesbeck 1980; Streeter et al., 1985; Wunderlich and Reinelt, 1982).

Underlying all these diverse aspects of direction-giving is one of the major controversies in cognitive science, that is, the question of the relationship between mental imagery and language (more particularly in this case, between cognitive maps and language). In much of the empirical work on direction giving, a common unstated assumption has been that the verbal directions are somehow "read off" a cognitive map, which is the more or less faithful image, in the direction-giver's mind, of an actual map that the subject has either memorized or mentally reconstructed. Thus, the image-like cognitive map is the protagonist, with the translation in language being a practically valuable but cognitively non-essential task. This is the view most commonly taken by geographers, who naturally tend to opt for the primacy of the map over the word.

To the extent that language is associated with propositional knowledge, the question of the relationship between verbal directions and spatial imagery is closely related to the continuing debate regarding the respective roles of propositional knowledge and imagery in cognition. At least four different positions have been taken on this issue:

a. The propositional form of knowledge is the primary or fundamental one, with imagery being a psychological epiphenomenon or manifestation of some aspects of that knowledge. This is the position taken over the years by Pylyshyn (1986) and his followers, but it has been losing ground following the experimental successes of the other, competing positions. (Figure 1a).
b. Whether or not some knowledge is primarily propositional, imagery is an independent and irreducible form of knowledge with its own set of mechanisms and rules. This school is best represented by Kosslyn (1980) and his seminal work on the mental manipulation of figures, which strongly supports the hypothesis that knowledge of concrete objets and their physical and geometrical properties is stored and used in a form closely analogous to their visual appearance (Figure 1b).
c. Knowledge of things is stored simultaneously in two parallel and equally fundamental formats, propositional and imagery, which are linked with each other. This is Paivio's Dual Coding theory, which has its own group of followers (Paivio 1991) (Figure 1c).
d. Both imagery and propositional knowledge are shaped by a finite set of pre-conceptual cognitive structures rooted in a person's bodily experience of growing up in, and interacting with a physical world modified and interpreted by culture. This is the "experiential" perspective proposed by a group of cognitive linguists, in particular Lakoff and Johnson (1980) and Lakoff (1987), based on their analyses of the metaphors of which all natural languages are replete. The pre-conceptual structures in question (basic categories, image schemas, action schemas...), are found over and over, according to these authors, underlying mental imagery as well as the most abstract forms of discourse (Figure 1d).

Adopting this latter experiential perspective, this paper argues that both the verbal route directions and the imagery-based cognitive map constructed in the mind of the direction-giver are rooted in a common underlying mental model of the direction-giving situation. "Mental model" should be understood here in the general sense of Johnson-Laird's (1983) work on the subject, Lakoff's (1987) "Idealized Cognitive Models" (ICM's), and Fauconnier's (1985) "mental spaces". A mental model in this sense is not necessarily an iconic representation of the reality it stands for (thus, here: it is not a map-like cognitive map of the part of the environment at issue in the direction-giving situation), although it normally will encompass imagery of different kinds.

Johnson-Laird (1988) begins his article entitled "How is meaning mentally represented?" with the question: "What do people construct when they understand discourse?" His answer is that they construct a mental model of the state of affairs described in the discourse. Both question and answer appear to fit the situation of the seeker of route directions, who presumably will need to construct an appropriate spatial representation from the verbal directions he or she receives. But the focus of this paper is on the production, rather than the understanding, of the route directions, thus Johnson-Laird's question must be reversed: what must people construct in order to produce (direction-giving) discourse? The answer will be, again, "a mental model", and the purpose of this paper is to explore just what kind of mental model this might be. This question is taken up again later in the final section, while the next section prepares the ground with some more concrete material.

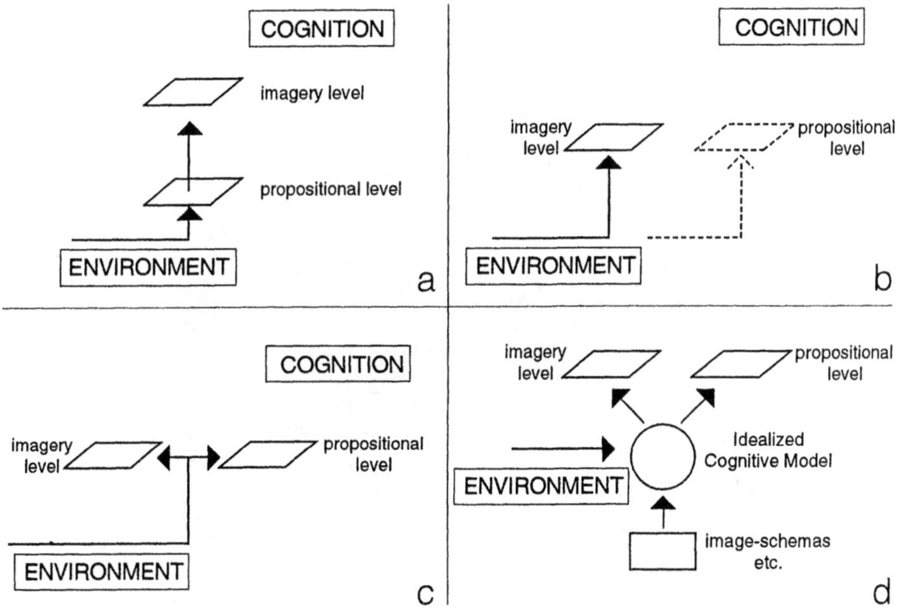

Figure 1: Various views on the relationship between the syntactic and imagery levels of cognition, and their main exponents: a. Primacy of syntactic level (Pylychyn); b. Autonomous status of imagery level (Kosslyn); c. Dual-coding theory (Paivio); d. Experiential realism (Lakoff)

A Sequential Model of Direction-Giving

This section describes the temporal organization of a model intended to provide a comprehensive account of direction-giving from the experiential perspective. The model represents a complex schema of direction giving assumed to govern the direction-giver 's responses, which integrates the goals, attitudes, and behaviors of the respondent with the spatial imaging and linguistic cognitive skills necessary for carrying out the task. The step-by-step unfolding of the model is presented here, while its logical organization is discussed in the next section.

The examples are taken from the transcripts of an informal experiment designed by S. Gopal and myself, involving 30 subjects, and run by S. Gopal on the UCLA campus. The chosen route, between the Research Library and the West Center, was of sufficient length and complexity to make direction-giving a challenging though not overwhelming task, while the exclusively pedestrian environment provided more degrees of freedom in route choice than would one geared towards vehicular circulation (Figure 2). The experimenter used an inconspicious tape recorder and the subjects were not told that they were participating in an experiment. The route was probed in both directions, so that half

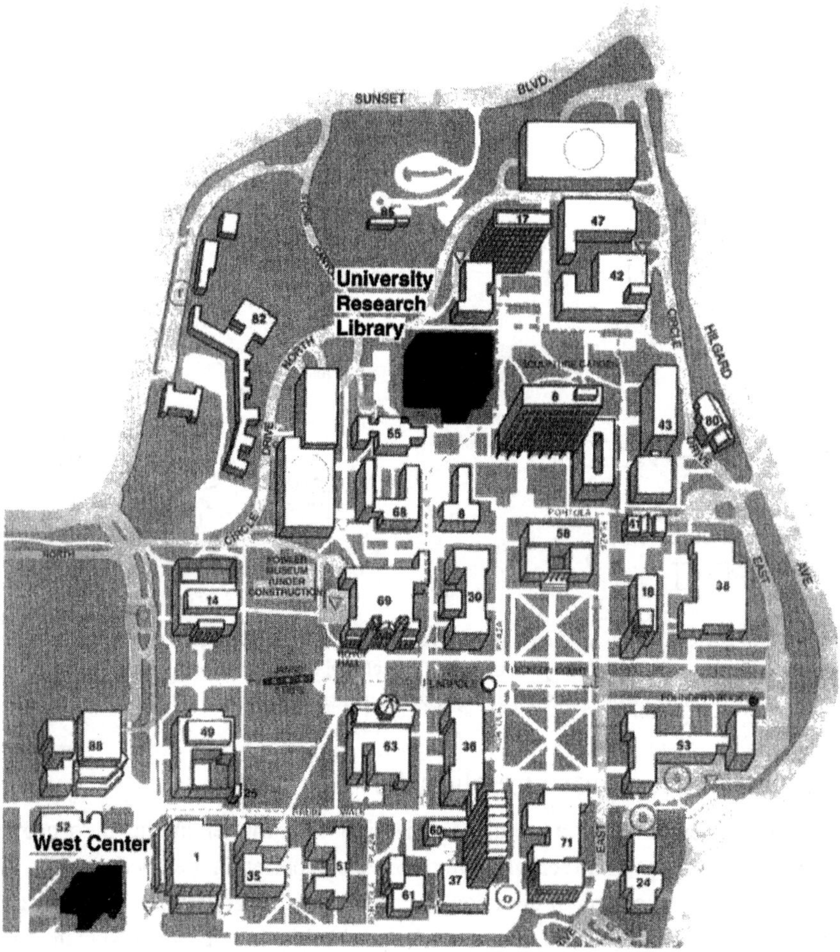

Figure 2: The environment of the way-finding experiment (UCLA campus)

of the subjects had to give directions to the West Center, with the experimenter at the University Research Library, and the other half to the University Research Library, with the experimenter at the West Center. Some results of that survey were reported in Freundschuh et al. (1990).

There are five major stages in the model (Figure 3): initiation, representation, transformation, symbolization, and termination. We assume two participants: the Seeker

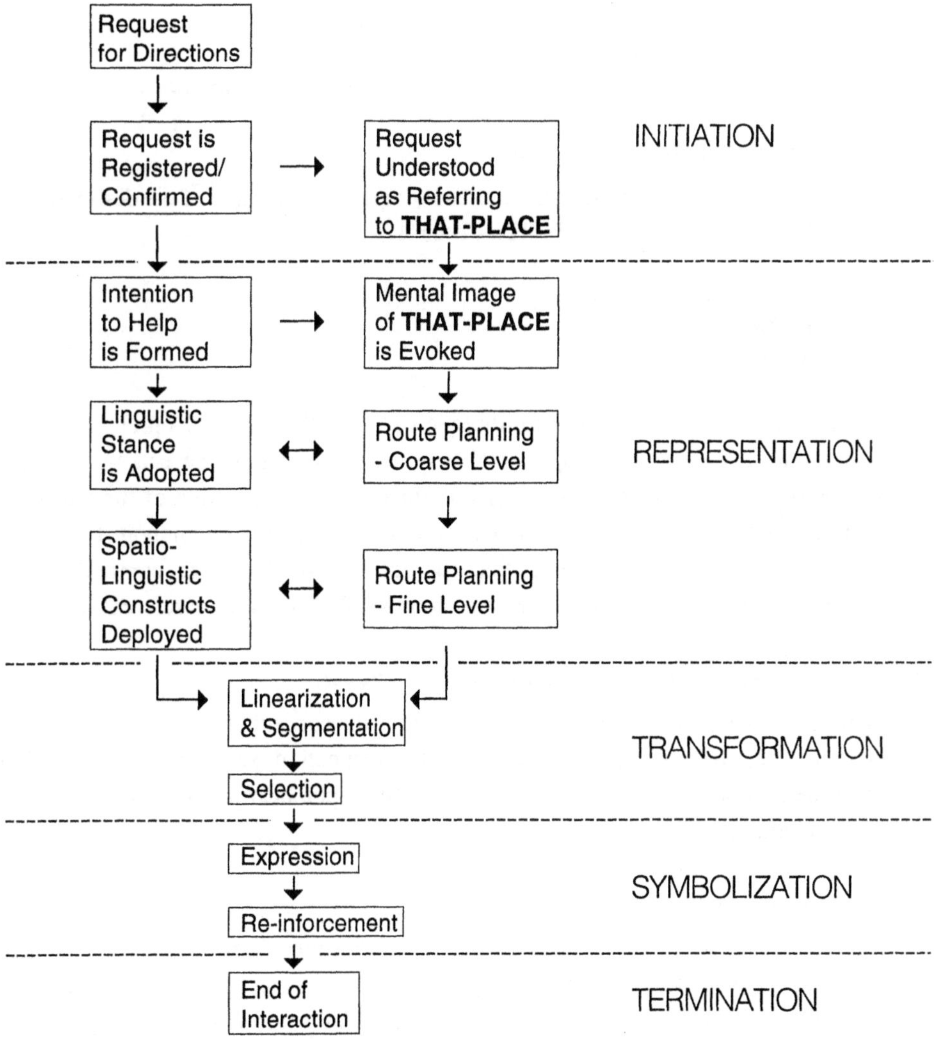

Figure 3: The logical organization of direction-giving discourse.

of directions (S) and the Respondent (R), though obviously more than one person may be involved on either side. The deictic terms THIS-PLACE and THAT-PLACE are used instead of "origin" and "destination" to stress the egocentric nature of much of the spatial thinking that goes on. All the pieces of conversation appearing in this section are taken verbatim from the transcripts.

A. Initiation

1. A request for directions is issued by S

We assume that the request is in a language comprehensible to R and that it takes the general format used for such requests in our society, for example

S *Do you know where the West Center is?*, or
S *Excuse me, do you know where the Univesity Research Library is?*

2. The request is registered and confirmed

R hears and acknowledges the question, often by repeating the name of the destination:

S *(Do you know where the West Center is?)*
R *The West Center?*
R *Ooh..The West Center...*
R *The West Center...*
R *West Center, oh yeh...*

R recognizes the query as a request for direction – rather than, say, as a probe for his or her local knowledge; clearly, a simple "yes" answer, even if literally correct, would have been inappropriate. This understanding by R of the underlying meaning of the question, and therefore of the kind of response called for, is a contextual phenomenon. Discourse analysis has a lot of insights to offer on that subject. Clearly language comprehension involves much more than a simple translation from symbols (words) to referents (things).

3. The request is understood as referring to THAT-PLACE

Recognizing the name of the destination triggers in R the concept of the place in question. Research on language and imagery, as well as introspection, strongly suggest that an image of the destination will be called up in R's mind upon hearing the name of a place she or he is familiar with (Paivio 1991).

4. The intention to help is formed

Human psychology as well as sociocultural conventions press R for a helpful response, unless an excuse is readily available (don't know, not from here, too much in a hurry...). This is expressed in the "principle of cooperation" well known in discourse analysis from

Grice's work (Grice 1975), stating that people tend to respond constructively to the conversational expectations of others. Obviously, willingness to help will vary in degree depending on a host of personal and interpersonal factors. Some work on direction-giving suggests a gender effect in the responses (that is, subjects tend to vary their response according to whether the experimenter is of the same sex or not: see Freundschuh et al. 1990). We did not find that effect in our UCLA survey, but even if we had, this would have been just another instance of the psychologicaly trivial truth that people's behavioral and affective responses to other people tend to be correlated. In any event, this is the stage where R will determine what amount of time and effort she or he is willing to invest in the response. The quality and detail of the directions that will be given will largely depend on this decision, made within the first few seconds of the interaction.

5. The direction-giving schema is activated

From now on the focus of R's intentions, thoughts, gestures, and verbal responses will be on the now salient goal of communicating to S the information needed to reach THAT-PLACE. A schema, in the sense developed by Arbib and Hesse (1986), is the cognitive structure capable of integrating attitudes, beliefs, representations, and actions in the required way. Two essential aspects of the direction-giving schema are described below, under (6) and (7).

B. Representation

6. The linguistic stance is adopted

By now R has a good grasp of the task at hand, and part of that understanding is that the directions will be given in language. Some respondents may ask for a map or attempt to draw one (several subjects in our survey asked for a map), but even in that case the primary response remains linguistic, with the map serving as a prop to the verbal account. Adopting the linguistic stance means that the spatial imagery to be deployed will be guided by the concepts and structures underlying language – rather than, say, being composed of fluid, dream-like flashes, or dominated by bodily memories of oneself getting increasingly weary over the distance traveled. There are indeed many different ways of knowing an environment or a route, but only some of these are relevant to the task of verbal direction giving.

7. Spatio-linguistic constructs are called up

A large part of the direction-giving schema develops around the kinds of cognitive primitives studied by Lakoff and Johnson (1980), Lakoff (1987) and others. These are the basic-level categories of things in the world (Rosch 1973), including, in this case,

Lynch's (1960) well-known pentad of landmarks, paths, nodes, districts and barriers; the image-schemas such as CONTAINER, CENTER-PERIPHERY, SOURCE-PATH-DESTINATION, based on fundamental spatial relationships known through our bodily interaction with the physical world; the basic-level actions (walking, turning, climbing, driving...) also discussed by Lakoff and Johnson (1980) (the notion of a basic set of cognitively primitive actions has also been explored by Schank and Abelson, 1977, and Jackendoff, 1983, 1987); and the geometrical idealizations underlying the use of locative expressions that Herskovits (1986), Talmy (1983) and others have studied. For example, the phrase "the tall building on your right" suggests, according to the analysis of Herskovits, that "the right" is conceptualized as either a surface on which the building rests (as in "the book on the table"), or as an area separated from the speaker by a boundary on which the building sits (as in "the house on the lake"). Similarly, "go straight" is a perfectly appropriate direction even if the route bends or winds somewhat, because routes (paths) are idealized as straight-line segments. According to the cognitive linguistics literature and the experiential perspective, these cognitive structures are inherently meaningful and not further decomposable, and together they form the preconceptual core of meaning underlying both language and imagery. The next section will attempt to set these experiential elements in the context of a more comprehensive mental model of direction-giving.

8. *The relative frame of reference is established*

While earlier in the interaction the mere mention of the name of the destination brought up the concept of THAT-PLACE in R's mind, it was the intrinsic (site rather than situational) characteristics of THAT-PLACE that were highlighted: its general appearance, function, or experience of R with it. Now, well into the direction-giving schema, THAT-PLACE is considered in its spatial relation to THIS-PLACE. Spatial concepts derived from the structure of physical space and R's egocentric frame of reference within it – concepts such as up, down, near, far, this way, that way – now come into play. The egocentric frame of reference is marked by the use of such deictic terms and is often underlined by pointing, gesturing, looking up, or turning one's body to face in the direction THAT-PLACE is believed to be. Typical for this stage of the interaction are responses of the following kind:

S *(Do you know where the West Center is?)*
R *We are far from it... uh... It's all the way down there and up*
R *OK, it's way on the other side of the campus*
R *Oh, gosh! It's that way – this way* (pointing)

A common first response is also to ask S back whether he or she knows some landmark in the vicinity of THAT-PLACE:

S *(Do you know where the West Center is?)*
R *Do you know where the Ackerman Center is?*
R *Do you know where the student bookstore is?*
R *Do you know where Royce Hall is?*
R *Do you know where the Powell Library is?*

Some responses were loaded with deictic terms throughout, suggesting that the respondent had difficulty moving past this stage of establishing an egocentric frame – a suspicion confirmed in the following case by the subject's admission that "I can't really tell you how". Here is an extended excerpt from one such transcript, with the egocentric spatial terms noted in bold italics:

S *(Excuse me, do you know where the Research Library is?)*
R *It's **up** and to the **left**. You've to keep going to the **left**...ask someone **over there*** (pointing)
S *(Down the walkway?)*
R *You go there and to the left. It's **way down there**. Not **too far**, but I can't really tell you how to get **there**. Turn to your **left**.*
S *(Thank you)*
R *Follow **all the way up** and then go **up** and ask...*

But normally, once the relative frame of reference is established, the problem-solving part of the task can begin.

9. Route planning – coarse level

This step appears to be contingent in the unfolding of the direction-giving schema. Studies of way-finding and direction-giving by Streeter and Vitello (1986) and others suggest that it may be used only, or primarily, by subjects with high spatial abilit y, who are capable of imaging the general layout of the area (configurational or survey knowledge). Coarse-level route planning requires R to take a detached mental view of the spatial structure of the entire area between THIS-PLACE and THAT-PLACE, and thus to transform the egocentric frame of reference into an objective one, as if looking at a map. In this case, R will mention cardinal directions and will point to major structural elements of the environment (for example, patterns such as a regular grid , or other features serving as axes in the basic spatial organization). The major channels between THIS-PLACE and THAT-PLACE will be highlighted before the route itself begins to be traced in detail. The following examples provide good indications of coarse-level planning:

S *(Do you know where the West Center is?)*

R *Do you know where the quad is? Do you know where the Ackerman is?... the main store... OK. There's a path that goes down the building straight down south. You will come to a spot with four big buildings... it's the center of the campus...*

R *We are on North Campus. The West Campus is right there* (pointing).

R *Yes. We are quite a distance from that. Which part of the campus are you familiar with?...*

S *(Excuse me, do you know where the Research Library is?)*

R *You gotta go north of the campus... go up there north of [inaudible] ...*

In our sample, there was more frequent mention of cardinal directions in the reverse-route half of the experiment (from West center to Research Libary). This may have to do with the "North Campus" area designation familiar to the subjects, with the center-periphery asymmetry between the two pairs of origins and destinations, or it may be fortuitous (Freundschuh et al. 1990).

10. Route planning – fine level

The next step for R is to trace the route mentally from beginning to end. In contrast with coarse-level planning, which requires taking a bird's eye view of the entire area, fine-level route planning involves imaging oneself actually navigating the route at street level (worm-eye view). The frame of reference becomes egocentric once again: the reconstruction is based largely on sensorimotor knowledge linking particular experienced states of the environment (in particular views) with particular motor activities, corresponding to basic action schemas (turn left, walk straight...). Very likely, this stage of direction-giving can be well approximated by computational models of navigation, such as those developed by Kuipers (1978) and Gopal et al. (1989). Views experienced at each point on the route are registered by their most striking or memorable physical elements (primarily landmarks, but any members of Lynch's pentad or basic urban categories may be appropriate). Together with the action schemas linked to them, these images constitute units of sensorimotor knowledge referred to as "VIEW-ACTION pairs" in Kuipers' (1978) work. Good examples of fine-level route planning:

S *(Do you know where the West Center is?)*

R *...walk between these buildings* (pointing)*... you see a little sidewalk. Just follow that... It will take you past through two more buildings and then it deposits you at the quad... a big grassy area.*

R *OK walk straight down.. You will come down through two flights of stairs like this. Go straight. You will pass this building, which is Royce. You will pass Royce, then you come to this big quad...like grassy. You follow the quad this building, which is Royce. You will pass Royce, then you come to this big quad like grass. You follow the quad and quad will go downhill by the grass and all that... then you get to the bottom and look straight ahead. You'll see the bear, and*

Ackerman Union, and turn left, go straight down. You will come to big buildings on your right, that's going to be John Wooden Center. And right across the John Wooden center is the West Center.

There is more to these extended excerpts than the mere illustration of fine-level route planning. In the first case, the phrases "it will take you" and then "it will deposit you", in reference to the sidewalk, suggest an underlying cognitive representation other than the prevalent one of the imaginary moving point. The last example is interesting also in the context of the next stage, linearization and segmentation.

C. Transformation

11. Linearization and segmentation

In behavioral studies, routes are typically conceptualized as one-dimensional elements. While the kinesthetic experience of traversing a route may indeed be linear, perceptually, routes are experienced and remembered as elungated 3-dimensional environments that can have considerable width or height whenever oblique views from different points on the route include distant or very tall elements. To conform with the linearity and modularity of speech, that 3-d continuum must be linearized and broken down into segments that can be fit into phrases. These segments are typically bounded by perceptually salient elements (landmarks) or kinesthetically salient decision points, and the phrases corresponding to them will be separated by pauses, linked together by the temporal proposition "then", or be otherwise marked as self-contained segments in an ordered linear sequence. Thus the multi-dimensional spatial experience of the route is transformed into a temporal flow isomorphic with the structure of language. Here are two classic examples of linearization and segmentation:

S *(Excuse me, do you know where the West Center is?)*

R *...what you've to do is just keep walking until you get to a clearing* ***and*** *it looks like a big lawn* ***and*** *there's the Library* ***and*** *you head off to the right* ***and*** *going down the hill* ***and*** *over to your left is the Student Union* ***and*** *if you passed the Student Union there's the Gym and Recreation Center* ***and*** *a little bit further down is the West center. It's right down there.*

R *West Center, oh yah...it's way down that way, you have to go through of the front door* ***and then*** *cross over to the library* ***and then*** *turn* ***and*** *there's a kinda walk that goes way down...you go way down* ***and then*** *turn* ***and*** *go way out to the street that's along there,* ***and*** *ask somebody where the West Center is.*

"Ask somebody else" is a common and significant end-of-segment marker. It may indicate a spatial decay threshold in the availability of spatial information away from THIS-PLACE, conversational factors such as fear of overloading the direction seeker with too much information, or sheer unwillingness to spend more time on the task.

12. Selection

The idealization of oneself as a moving point traversing a linked sequence of route segments is still too rich in detail, much of which is not relevant to the direction giving task. The conversational principle of quantity (make your contribution as informative as required; do not make your contribution more informative than required: see Grice, 1975), press R to select the information necessary for S to follow the route, but no more. If the route is judiciously segmented and the bounding elements well chosen, communicating the sequence of route segments should convey the necessary and sufficient information for the task. Less efficient respondents or subjects of lower spatial ability will often cling to superfluous detail in fear of 'getting lost' themselves in their mental travel. Contrast these two sets of directions, the second of which is both confused and confusing, suggesting a subject cognitively lost in the area around Powell Library (we have reason to believe that the respondent was familiar with the campus):

S *(Do you know where the Research Library is?)*

R *Just take this* [pointing to Bruin Walk] *and when you get to the top step where you can't go anymore make a left and you go all the way straight over and you're right there*

R *To the University Research Library? Go up the road. See where the bear is* [pointing]. *Go right, go that way, keep on walking, you will pass the Powell Library. Powell Library. Ummmm.... before you walk up to the Powell Library, go past the Ackerman and then you will be able to drift left, start going to the left and by that way you will be in front of Powell Library. If you go straight, you will be at the back of Powell Library. if you start drifting that way, you will be in front of Powell Library, when you get to that area, you are very close and then someone will give you more specific directions.*

D. Symbolization

13. Expression

The close interaction between the spatial and linguistic aspects of cognition culminates in the actual issuing of the appropriate verbal directions. Since the underlying model has already been structured around the linguistic stance (step 6), there is little additional effort involved in verbalizing the directions. Some decisions still have to be made: do you say "Royce Hall" or "the tall white building on the right"? There is also the more general question of the direction-giving style adopted: is it "command" ("go up the hill, turn left..."), or "advice" ("If I were you, I would ask again there.."), or "tour" ("...if you passed the Student Union there's the Gym and Recreation Center and a little further down is the West Center"), or "future" style (" You will come to the big buidings on your right...")? (see Freundschuh et al. 1990). These style variations very likely reveal corresponding differences in the underlying cognitive representation, but will not be further explored here.

14. Reinforcement

This phase has two characteristic expressions: the repetition of segments of the directions just given, and conversational query elements such as "OK?" "Allright?". Both kinds of reinforcement were found in our transcripts. Other non-verbal cues (especially, inquisitive eye contact) were noted by the experimenter at that stage, but were not recorded.

E. Termination

15. End of interaction

S *(OK, thank you!)*
R *OK, good luck*
R *Sure!*
R *Bye have a nice day!*

That's it!

From Mental Models to Cognitive Maps

The direction-giving discourse, as detailed in the previous section, reveals an underlying representation that has both map-like and non-map-like characteristics. On the one hand, the spatial structure and imagery implicit in the discourse suggests a cognitive map incorporating metric, topological, and attribute information on the particular environment of the study. On the other hand, the discourse itself has a striking regularity across respondents in both its large-scale and detailed organization that can neither be reduced to the cognitive map nor dismissed as irrelevant to the question of spatial cognition.

The hypothesis proposed in this paper is that the task of giving directions for way-finding suggests the existence of an underlying mental model in the sense expounded by Johnson-Laird (1983). Furthermore, at the root of that mental model is a set of more primitive cognitive structures, image-schemas and basic-level categories and actions as discussed in Lakoff (1987). This section builds an argument for such a model of direction-giving, which differs from most imagery-based and syntactic accounts of cognition by positing a common pre-conceptual root for both.

According to Johnson-Laird (1989), a mental model is a representation of a body of knowledge that has the following elements: (a) its structure corresponds to the structure of the situation it represents; (b) it may be realized as a mental image and/or may contain abstract elements; and (c), it does not contain variables (as in propositional representations) but tokens, i.e. concrete instantiations of possible elements. The first condition establishes the mental model as a representation, albeit a structural rather than an iconic or an analog one. The second condition distinguishes mental models from imagery-based representations (in particular, cognitive maps) by allowing for elements that are not perceptual (e.g., logical conditions, abstract properties, or action schemas). The third one contradicts the theory of propositional (syntactic) representations as expressed in formal linguistic and computational models (Pylyshyn 1986). Thus a mental model is not a frame representing all possible situations compatible with the known information, but functions as a representative sample from that universe, which can be revised and updated on the basis of subsequent information.

Translated into the domain of the direction-giving task, these conditions may be interpreted as follows:

a. The direction-giver's mental model represents the entire direction-giving situation, that is, the fact that another person is asking for help in finding a route, as well as the spatial context and knowledge necessary to answer the query. As suggested by work in discourse analysis (Grice, 1975), part of that model is the assumption that the other person can understand the situation in a similar way, i.e. construct a similar mental model. Thus the direction giver's task is to produce the kind of directions she/he would need if the roles were reversed.
b. The direction-giving mental model incorporates a cognitive map of the relevant section of the campus as well as more fundamental schemas having to do with how people experience and interact with urban environments, as will be further discussed below.
c. It is natural to assume that the direction giver's mental model incorporates imagery pertaining to specific features of the environment he or she is trying to describe in words. A critical point in Johnson-Laird's theory is that the reverse is also true, that is, the representation constructed from the direction-giving discourse is a mental model that incorporates imagery pertaining to specific elements (tokens) in the situation at hand, rather than representations of generic types.

This account does raise further questions: How can the direction-seeker construct a good enough model of an environment never before experienced on the basis of only a brief verbal description? And how does the direction-giver know what information to give for such a construction to be possible? Clearly both sides of this process often go wrong, and errors made in both giving and following directions can be in themselves valuable clues in the study of spatial cognition. Still, that this complex communication between strangers works most of the time is reason enough to suspect solid common roots underlying the mental models of both the seeker and giver of directions.

The empirical evidence supports the view that the spatial knowledge the direction-giver uses to describe the requested route is made up of two parts. The hypothesis put forward here is that the second kind of knowledge is common to the mental models of both the giver and seeker of directions, and this makes it possible for the latter to construct a cognitive representation of the situation sufficiently similar to that of the former. The asymmetry between the two mental models lies in the richer knowledge of the first kind normally available to the direction-giver. The two types of knowledge in question are:

a. knowledge specific to the *environment in question*; and
b. a store of *general schemas* regarding urban environments in general and the ways people normally interact with them.

These schemas serve to organize and interpret the items of concrete knowledge people have of specific places, and help draw inferences about places not well remembered or not known. The relationship between b and a is akin to the well-known *type/token* distinction in cognitive psychology (Jackendoff, 1987), the distinction between recognizing things in the world as unique individuals, and recognizing them as members of particular categories with defining properties. For example, "Royce Hall", a concrete landmark on the specific route, is a token of the type "building", which suggests any number of things about how big it might be, what it may look like, what kind of material it may be made of, and whether or not it may be relied on to be still found at its last known location. Similarly, the direction "walk down a hill" in our interview transcripts refers to a specific hill on the UCLA campus that will be encountered by the direction seeker after "you bear to your right" at that particular point of the specific route, but it is uttered following the activation of a "walking-down-an-incline" action schema that could be substantiated in any number of different ways and places for any typical member of the human species.

Neither kind of knowledge is in itself sufficient to find a route, so the direction-seeker attempts to make up for the missing place-specific knowledge by combining the verbal directions with the appropriate general schemas to generate tentative mental images of the relevant part of the environment. Informal evidence suggests that some of the problems people encounter in following directions are due to the fact that these fairly specific hypotheses about what the route may look like are often greatly at odds with reality: a case of frustrated expectations that can be very literally disorienting, even though the directions obtained may be both accurate and correctly remembered.

The perspective of experiential realism in cognitive linguistics, best represented by the work of Lakoff and Johnson (1980) and Lakoff (1987), can take us one step further in explaining the cognitive mechanism of the direction-giving exchange. Central to that work is the notion that a small number of pre-conceptual kinesthetic *image schemas* universally structure human cognition: the container schema, the part-whole schema, the link schema, the center-periphery schema, the source-path-goal schema, and a handful more (Lakoff 1987). These are generated through a developing individual's interaction with the physical world, are expressed in a myriad of metaphors and other non-literal forms of expression found in every human language, underlie both concrete and abstract thinking, and, being shared by all members of the human race, are at the basis of interpersonal communication. Moreover, most of these constructs are inherently spatial and can be seen to be directly relevant to spatial imagery. More specifically, the source-path-goal (or -destination) schema dominates both the production and the decoding of way-finding directions, and helps explain why the direction-giving exchange works as well as it does. The structural elements of the source-path-destination image-schema (a starting position, an endpoint, a direction, a sequence of contiguous locations: see Lakoff 1987, p. 275) appear to correspond directly to the structure of not just the cognitive map but also of the discourse of direction-giving (THIS- PLACE, THAT-PLACE, coarse-level planning, fine-level planning and segmentation).

According to the experiential perspective, the other fundamental preconceptual building blocks of cognition are the *basic-level categories* as established through the work of Rosch (1973) and others. Among the properties of basic-level categories is that they are characterized by gestalt perception and evoke the kind of strong imagery constituting the "tokens" in Johnson-Laird's mental models. The buildings, paths, and hills mentioned in direction-giving are names of such categories. Basic-level concepts exist not just for objects but also for actions and properties (the former are sometimes also referred to as action schemas): going, walking, turning, seeing; tall, white, big. The direction-giving discourse is replete with basic-level concepts that help impart some representational specificity on the structure provided by the SOURCE-PATH-GOAL image schema.

Thus the cognitive mechanism of direction-giving described in this section integrates the psychological robustness of the theory of mental models with the explanatory power of Lakoff's notion of Idealized Cognitive Models (ICM). The former contributes the considerable experimental evidence detailed in Johnson-Laird's work, the latter suggests a plausible causal foundation for cognitive primitives in ontogenesis and phylogenesis. In this way, the infinite regress plaguing many attempts to explain the root s of cognition can be avoided.

Conclusion

The purpose of this paper was to put forward a tentative hypothesis of direction-giving, compatible with the empirical evidence and much of the relevant literature, and with potential wider implications for the theory of spatial cognition and its applications. The hypothesis, as summarized in Figure 4, is that both the discourse and the cognitive map often assumed to be at the basis of direction-giving behavior are in fact generated by a common underlying mental model, which is itself structured by more primitive kinesthetic image-schemas (in particular, the source-path-goal schema), and basic-level categories

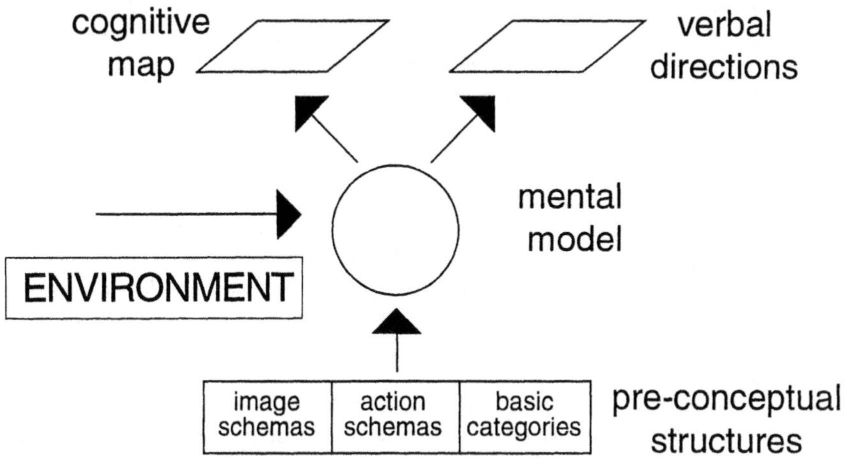

Figure 4: The cognitive structure of direction-giving: a hypothesis.

and action schemas. The model proposed is thus a close adaptation of the experiential view represented schematically in Figure 1d. Just like the literature it builds on, such an account cannot be tested directly. Rather, it must by judged by the plausibility of the story it tells, the breadth and power of the literature it synthesizes, the resilience of the hypotheses it is at variance with, and the implications that may be drawn from its further development.

The hypothesis proposes a particular relationship among the elements: cognitive map, direction-giving discourse, mental model, kinesthetic image-schemas, and prototype categories, all of which (with the exception of the discourse) are themselves hypothetical entities. In doing so, it contradicts other popular hypotheses that set different priorities within different collections of hypothetical entities. For instance, if cognitive maps are not more basic that discourse, then it is not the case that spatial imagery is the privileged form of encoding environmental knowledge; or, if the imagery format is convenient for storing such knowledge after it is generated, it may not be the proper means for arriving at it in the first place. Debate on these issues is not about to come to an end, and (to use the source-path-destination metaphor), this paper, which is slightly off the beaten path, hopes to contribute a vision of another possible direction. It is not a new direction, since Kant (quoted in Johnson-Laird, 1988) already wrote:

- In truth, it is not images of objects, but schemata, which lie at the foundation of our pure sensuous conceptions.

Acknowledgements: Sucharita Gopal carried out the experiment described in this study and substantially contributed to its design. Partial funding for this research was provided by the National Center for Geographic Information and Analysis (NCGIA). Ideas in this paper were stimulated by various research-related activities in the context of Initiative 2: Spatial Languages and Spatial Relations of the NCGIA.

References

Arbib, M. and Hesse, M. (1986) *The Construction of Reality*, Cambridge: Cambridge University Press.

Fauconnier, G. (1985). *Mental Spaces*, Cambridge, MA: MIT Press.

Frank, A. and Mark, D. (1991). Language issues for GIS. In *Geographical Information Systems: Principles and Applications*, (D. J. Maguire, M. F. Goodchild and D. W. Rhind, eds), London: Longman Books, 147-163.

Freundschuh, S. M., Mark, D. M., Gopal, S., and H. Couclelis (1990). Verbal directions for way-finding: implications for navigation and geographic information and analysis systems. In *Proceedings, Fourth International Symposium on Spatial Data Handling*, Zurich, Switzerlannd, v.1, 478-487.

Golledge, R. G. (1992). Place recognition and wayfinding: making sense of space, *Geoforum* 21:2, 199-214.

Gopal, S., Klatzky, R., and Smith, T. R. (1989). NAVIGATOR: a psychologically based model of environmental learning through navigation. *Journal of Environmental Psychology* 9, 309-331.

Gould, M. D. (1989). Considering individual cognitive ability in the provision of usable navigation assistance. In *Proceedings, First Vehicle Navigation and Information Systems Conference* (VNIS '89), IEEE Vehicular Technology Section, Toronto, Canada, pp 443-447..

Grice, H. P. (1975). Logic and conversation. In *Syntax and Semantics 3: Speech Acts* (P. Cole and and J. Morgan, eds.), New York: Academic Press.

Herskovits, A. (1986). *Language and Spatial Cognition: an Interdisciplinary Study of the Prepositions in English*, Cambridge: Cambridge University Press.

Jackendoff, R. (1983). *Semantics and Cognition*, Cambridge, MA: MIT Press.

Jackendoff, R. (1987). On beyond zebra: the relation of linguistic and visual information. *Cognition* 26, 89-114.

Johnson-Laird, P. N. (1983). *Mental Models:Towards a Cognitive Science of Language, Inference, and Consciousness*, Cambridge, MA: Harvard University Press.

Johnson-Laird, P. N. (1988). How is meaning mentally represented? In *Meaning and Mental Representations* (U. Eco, M. Santambrogio and P. Voli, eds.), pp. 99-118, Bloomington: Indiana University Press.

Jonhson-Laird, P.N. (1989). Mental models. In *Foundations of Cognitive Science* (M. I. Posner, ed.) pp. 469-499. Boston, MA: Bradford Books, MIT Press.

Klein, W. (1982). Local deixis in route directions. In *Speech, Place, and Action: Studies in Deixis and Related Topics*, (R. J. Jarvella and W. Klein, eds.), pp. 161-182, Chichester:John Wiley & Sons.

Kosslyn, S. M. (1980). *Image and Mind*, Cambridge, MA: Harvard University Press.

Kuipers, B. (1978). Modelling spatial knowledge. *Cognitive Science* 2, 129-153.

Lakoff, G. (1987). *Women, Fire, and Dangerous Things: What Categories Reveal About the Mind*, Chicago: University of Chicago Press.

Lakoff, G., and Johnson, M. (1980). *Metaphors We Live By*, Chicago: University of Chicago Press.

Lynch, K. (1960). *The Image of the City*, Cambridge, MA: MIT Press.

Ma, P. (1987). An algorithm to generate verbal instructions for vehicle navigation using a geographic database, *The East Lake Geographer* 22, 44-60.

Mark, D. M. (1987). On giving and receiving directions: cartographic and cognitive issues. In *Proceedings, 8th International Symposium on Computer-Assisted Cartography*, Baltimore, Maryland, pp. 562-571.

Mark, D. M. (1989a). A conceptual model for vehicle navigation systems. In *Proceedings, First Vehicle Navigation and Information Systems Conference* (VNIS '89), IEEE Vehicular Technology Section, Toronto, Canada, pp 448-453.

Mark, D. M. (1989b). Languages of spatial relations: researchable questions and NCGIA research agenda. *Report 89-2*, National Center for Geographic Information and Analysis, Santa Barbara, CA.

McGranaghan, M., Mark, D. M., and Gould, M. D. (1987). Automated provision of navigation assistance to drivers. *The American Cartographer* 14, 121-138.

Paivio, A. (1991). Dual coding theory: retrospect and current status, *Canadian Journal of Psychology* 45:3, 255-287.

Pylyshyn, Z. (1986). *Computation and Cognition: Toward a Foundation for Cognitive Science*, Boston, MA: MIT Press.

Riesbeck, C. K. (1980). You can't miss it: judging the clarity of directions, *Cognitive Science* 4, 285-303.

Rosch, E. H. (1973) . Natural categories. *Cognitive Psychology* 4, 328-50.

Schank, R. and Abelson, R., (1977) *Scripts, Plans, Goals, and Understanding: an Inquiry Into Human Knowledge Structures*, Hillsdale, N.J.: Lawrence Erlbaum assoc.

Streeter, L.A., and Vitello, D. (1986). A profile of drivers' map reading abilities. *Human Factors* 28, 223-239.

Streeter, L.A., Vitello, D., and Wonsiewicz, S.A. (1985). How to tell people where to go: comparing navigational aids. *International Journal of Man-Machine Studies* 22, 549-562.

Talmy, L. (1983). How language structures space. In *Spatial Orientation: Theory, Research and Application* (H. Pick and L. Acredolo, eds.), New York: Plenum Press.

Wunderlich, D. and Reinelt, R. (1982). How to get there from here. In *Speech, Place, and Action: Studies in Deixis and Related Topics*, (R. J. Jarvella and W. Klein, eds.), pp. 183-202, Chichester:John Wiley & Sons.

Helen Couclelis
Department of Geography
University of California
Santa Barbara, California, USA

Part Two: Transformations

FROM VISUAL INFORMATION TO COGNITIVE MAPS

M. Jeanne Sholl

Abstract: It has been suggested that much of animal navigation takes place without reference to visual information in the environment (Gallistel 1990). A geocentric dead reckoning process, which tracks the travel trajectory of animals internally with only periodic reference to external visual cues, is thought to be a major component underlying animal navigation. Following Gibson (1979),and in contrast to the non-visual emphesis for animal navigation, two types of structure in visual flow are thought to be important for human navigation: perspective structure, which specifies self-motion and self-to-object relations, and invariant structure, which specifies object-to-object relations. A review of the cognitive mapping literature for sighted and blind human adults suggests that the invariant structure in visual flow is important to the formation of cognitive maps, or survey knowledge, of environments. In contrast, a geocentric dead reckoning process which uses the self-movement information provided by perspective structure may be importantly involved in the formation of route knowledge. Findings reviewed from studies of the development of navigational skills in children are consistent with the idea that cognitive mapping skills develop in concert with the ability to perceptually differentiate perspective and invariant structure in visual flow.

Introduction

Cognitive maps represent mentally the geometric layout of the differentiated topography of a space. By definition, a cognitive map or survey representation of a spatial layout codes the Euclidean relations (straight line distances and directions) among behaviorally relevant landmarks within a coordinate reference system centered on the environment. Cognitive maps function to support navigation, and, in turn, are created by navigation and exploration of space (e.g., Gallistel, 1990; Thinus-Blanc, Save, Buhot and Poucet, 1991). During navigation and exploratory spatial behavior, landmarks are experienced sequentially in space and time. The process of constructing a cognitive map can be thought of as a process that places a mental "copy" of each sequentially experienced landmark into a simultaneous system that preserves metric information about the straight-line distance and direction of landmarks relative to one another (Levine, Jankovic and Palij, 1982). An important, emergent property of a simultaneous system is that the spatial relations between all landmarks entered in the system are equally available, even those relations not directly experienced.

J. Portugali (ed.), The Construction of Cognitive Maps, 157–186.

How are cognitive maps formed and what role does vision play in the construction of cognitive maps in humans? The role of vision in navigation and cognitive mapping has been studied extensively in animals, so one approach to addressing this question would be to summarize the wealth of findings from animal studies. Studies conducted on rodents have examined how the hippocampus transforms visual input into a cognitive map of space (e.g., Poucet, 1993), and primate studies have explored the role of the visual "where" system in governing spatial behavior (e.g. Mishkin, Ungerleider and Macko, 1983). More is known about the role of vision in rodent than in primate cognitive mapping; however, it is unclear at this time to what extent the rodent findings can fully inform an account of the role vision plays in human navigation and cognitive mapping. This is not to say that the literature on animal cognitive mapping has little relevance for understanding human cognitive mapping. Spatial orientation and cognitive mapping are functions fundamental to the survival of all navigating animals, and they are probably governed by fairly primitive systems (that is, systems that arose early in evolutionary development), with overlapping neurobiological and neurocomputational instantiations across species. However, vision may have a more complex role to play in human than rodent navigation, at least, in part because human visual association cortex is more differentiated and much larger than rodent visual association cortex. Moreover, for rats, hippocampal function appears to be exclusively dedicated to spatial memory; whereas, for primates, spatial memory is only one type of episodic or declarative memory controlled by the hippocampus (e.g., Rolls, 1991).

In the next section, an animal model of navigation developed by Gallistel (1990) will be described briefly, and then human data will be reviewed that suggest visual information may be more important for human than for animal navigation. The animal model has not been extensively tested on humans, and while empirical studies may find that the model correctly describes human navigation in broad strokes, some already existing findings suggest that the model does not fully explain the particulars of human navigation. Subsequent sections of the chapter will review the literature on the role of vision in human cognitive mapping and navigation. The review will focus on behavioral studies that manipulate either the visual properties of a space or visual experience with a space.

An Animal Model of Navigation and Cognitive Mapping

There are two types of spatial representation created by navigation, which refers to travel between places in the environment that cannot be simultaneously seen. One type is, of course, a cognitive map or survey representation of space, and the other is a route representation (e.g., O'Keefe and Nadel, 1978; Thorndyke and Hayes-Roth, 1982).

Route knowledge is a sequentially organized series of stimulus-response associations or procedures for how to get from one location in space to another. Gallistel's (1990) animal model explains cognitive map formation, and it is based on an exhaustive review of the spatial behavior of animals, observed in both laboratory and field settings. A geocentric dead reckoning process, which uses nonvisual cues to monitor the animal's orientation and trajectory within a coordinate system anchored on the physical environment, plays a major role in animal navigation and cognitive map formation. In addition to the dead reckoning process, the animal model contains subsystems which function to represent the spatial relations among behaviorally relevant landmarks (i.e., a cognitive mapping system) and to compute trajectories from one represented location to another in order to set a course, plan a route, take a shortcut, etc.

The animal uses its cognitive map and its orientation with respect to geocentric coordinates in order to set a course to a goal. Orientation is established in relation to geocentric directions by the dead reckoning system. Geocentric directions are computed from various globally available, external cues, including the sun's azimuth, magnetic north, and large-scale features of the environment. Once the animal's course is set, movement toward the goal is monitored by the dead reckoning system from signals generated internally as a result of body movement. These signals include: the motor efferent commands that control locomotion; output from the vestibular apparatus signaling angular and linear acceleration of the body; and the proprioceptive signals from the muscles, tendons, and joints generated by the act of locomotion (vestibular and proprioceptive cues signaling movement will subsequently be referred to jointly as mechanical kinesthesis). The internally generated angular and linear velocity signals are integrated with internally generated estimates of time to compute the distance of each segment of the animal's route and the magnitude of the turns connecting the segments. The resultant moment-to-moment, non-visual representation of the travel trajectory is plotted in relation to the animal's cognitive map of the space, and periodic visual fixes of the environment correct any accumulated error. As the animal approaches the goal, navigational progress is monitored with respect to local visual cues, not dead reckoning.

In an unfamiliar environment, the dead reckoning process functions to construct a cognitive map of the environment. As an animal engages in exploratory activity, its sensori-motor system codes the location of visible landmarks in body-centered coordinates, while the dead reckoning process concurrently updates the position of the body in geocentric coordinates. A cognitive map is formed by a process which uses the known position of the body within the geocentric reference system to transform the body-centered coordinates of behaviorally relevant landmarks into geocentric coordinates.

A geocentric dead reckoning process can account for the uncanny ability of many animals to navigate successfully without relying on local visual cues. An example is reported by Carr and Watson (1908) and cited by Gallistel (1990). Rats were trained in a "Hampton Court" maze until they were sufficiently familiar with the layout of the maze to run a route from start to goal quickly and without error. The maze was then physically altered by cutting a wide slice out of it (like taking a big slice out of the middle of a loaf of bread and putting the two ends back together), thus shortening some of the pathways. When first introduced into the altered maze, the rats, running full speed along the route to the goal, ran smack into the walls at the end of the shortened pathways. Clearly, the act of navigation was not controlled by the rapidly expanding visual angle projected by the approaching wall at the end of the shortened alleys; if it were, the rat would have slowed to a stop to avoid collision. Instead the rat appeared to be relying on dead reckoning and a mental map of the original shape of the maze.

While some type of geocentric dead reckoning may very well form a component of human navigation, it is likely that local visual cues have a more central role to play in human than animal cognitive mapping. Existing evidence suggests that in the absence of information about the location of nearby objects in the environment, humans have difficulty monitoring their travel trajectories, even over short paths. Studies show both that vestibular signals by themselves do not provide reliable information for dead reckoning and that motor-kinesthetic representations of travel trajectories decrease in precision with an increase in the complexity of short paths.

I have tested the dead reckoning ability of human adults over simple paths consisting of two to four linear segments (Sholl, 1989). In a passive transport task, subjects were pushed along a path in a wheel chair while wearing a blindfold and with auditory cues masked, thus requiring them to rely on vestibular cues to plot their course. At the end of the path, subjects pointed in the direction of the starting point. Paths were composed of 2, 3, or 4 straight-line segments connected by 135° or 90° turning angles. Subjects performed at chance on simple paths composed of more than two linear segments and one turning angle. When the same task was repeated, but blindfolded subjects now walked along the path under the guidance of the experimenter, performance was considerably better. Mean pointing responses were reliably accurate for most paths, although the variability in pointing responses increased for paths composed of oblique turning angles and three or more linear segments. Klatzky and colleagues have conducted a similar experiment, but instead of pointing to start, subjects walked the straight-line trajectory from the end of the path to its start point (Klatzky, Loomis, Golledge, Cicinelli, Doherty and Pellegrino, 1990). Using paths containing oblique angles and with the additional requirement of executing the computed trajectory, they found that error increased as the number of path segments increased from one to three.

Summary

For humans, the research on unsighted navigation under blindfold shows that vestibular cues alone cannot support a dead reckoning process, but that the motor efferent and proprioceptive signals generated by the act of walking support effective dead reckoning over short trajectories. The human evidence suggests a dead reckoning process that retains a motor-kinesthetic representation of a simple travel trajectory, coding the length of the path segments and the magnitude and direction of the turning angles, and updating the position of the body relative to the starting point of the path. The shape of the navigated path can be recovered from the motor-kinesthetic trace, and new trajectories can be computed from the representation.[1]

A process that retains a motor-kinesthetic trace of the shape of the travel trajectory, without relating the trjectory to environmental reference points, will be defined functionally as an egocentric dead reckoning process. The experiments on human subjects reviewed in this section showed that egocentric dead reckoning was most effective if paths contained right angle turns, but lost its effectiveness in fairly short order when the paths contained oblique angle turns. Because egocentric dead reckoning is unable to monitor with precision travel trajectories that contain as few as one or two oblique angled turns, it is unlikely that a dead reckoning process plays the same role in human navigation that it plays in animal navigation.

The studies of navigation that will be reviewed latter in the chapter suggest that it may be possible to functionally differentiate geocentric from an egocentric dead reckoning process for humans. The behavioral data suggest that geocentric dead reckoning differs from egocentric dead reckoning by using a self-to-object updating process to relate the travel trajectory to a stable configuration of objects in the environment. Thus, unlike animals, who in Gallistel's model use geocentric dead reckoning to navigate in relation to a geocentric or cardinal directions, humans may use a geocentric dead reckoning process to monitor navigation in relation to a configuration of landmarks in the immediate environment. Moreover, for humans, the geocentric dead reckoning process may use self-to-object information to integrate the location of landmarks in a motor-kinesthetic representation of the path, forming route knowledge. Thus, in contrast to animals who use a geocentric dead reckoning process to form a cognitive map of space, it is hypothesized that humans use geocentric dead reckoning to form route knowledge of a space.

For animals, Gallistel describes a dead reckoning process that sets a course with respect to a geocentric reference axis, and then monitors the animal's trajectory on the cognitive map. To set a course, animals appear to be able to use locally available sensory cues (azimuth of the sun, configurations of stars, magnetic force fields) to ascertain compass directions. However, humans do not appear to orient themselves on the basis of magnetic signals (Gould & Able, 1981) or the sun's azimuth. Instead, as alluded to above, the metric interrelations among immovable objects or landmarks in the environment are likely to form an environment-centered coordinate system for human navigation.

Human Adult Navigation and Cognitive Mapping

Vision provides a rich source of spatial information to the human navigator. As a person moves through a cluttered environment, his or her continually changing perspective with respect to stationary objects and surfaces produces a dynamically changing mosaic of images on the retina. The solid visual angles projected by the objects and surfaces in the person's field of view flow continuously across the retina, as new vistas open up in front and old vistas are closed off behind the person. The flow of solid visual angles contains structural regularities that specify directly both the interrelations among objects and the relation of the self to the field of objects. Gibson (1979) calls these regularities in visual flow *invariant structure* and *perspective structure*, respectively.

Spatial Information in Visual Flow

perspective structure

Gibson uses the term perspective structure to refer to the constant patterns in the changing mosaic of the solid visual angles produced by body movement. The act of walking straight ahead creates one continuous pattern in the central visual field, i.e., radial flow, and another in the peripheral field, i.e., lamellar flow.[2] Radial flow is the centrifugal outflow or radial expansion of the solid visual angles projected by objects in front of the person. The rate of radial expansion provides partial information for estimating the length of time before the person comes into contact with an object, and the origin of the radial expansion corresponds to the person's heading in the environment. (Whether the origin of radial expansion actually serves a steering function in human navigation is a subject of debate, e.g., Warren, Morris and Kalish, 1988). Lamellar flow is the parallel streaming in the direction opposite body movement of the solid visual angles projected by objects to either side of the body. Their relative rate of movement is a continuous function of their distance from the person, with more distant objects displaced at a slower rate than nearer objects. Another geometric pattern, the horizontal movement of solid angles in a plane perpendicular to the line of sight, is produced by whole body turns.

The hierarchically nested array of solid visual angles projected by objects and surfaces in the forward field of view can be described at a more fine-grained level of analysis in units of texture elements. Texture elements are patches of uniform texture, differentiable from neighboring elements in terms of their intensity and/or spectral composition (Lee,

1980). Analysis of optical flow in terms of the movement of texture elements provides a level of description that is indifferent to the actual shape of objects and surfaces in the environment. The patterns of texture flow produced by body movement serve as stimuli for the perception of body movement, producing illusory linear or rotary movement in a stationary observer (e.g., Dichgans & Brandt, 1974; Telford & Frost, 1993). Because perspective structure stimulates the perception of body movement, it has been referred to as visual kinesthesis by some (Lishman & Lee, 1973). Visual kinesthesis along with mechanical kinesthesis and motor efferent signals may be used by a human dead reckoning process to compute the angular and linear displacements of the body during travel.

In the perception of body movement, visual kinesthesis dominates mechanical kinesthesis and motor efferent signals (Lishman & Lee, 1973; Lee, 1980). Rieser and colleagues (Rieser, Pick, Ashmead and Garling, in press) have shown that when the rate of optical flow is decoupled from the act of walking, signals from the motor-joint system are calibrated to the rate of optical flow. For example, if the rate of optical flow is greater than it should be (as happens when a person walks along a moving conveyer belt like those found in large airports), the motor-joint system is recalibrated so that each step is expected to bring about a greater displacement of texture elements than it actually does. As a consequence, if a person whose visual-locomotor system has been recalibrated in this way is asked to view a target, and, then with eyes closed, walks straight ahead until the target is reached, the person will not walk far enough. Presumably, the velocity signals generated by the motor-joint system have been modulated by the prior visual kinesthesis, and the dead reckoning process integrates an overestimate of linear body velocity with an accurate estimate of time to produce an underestimate of the linear displacement required to reach the target.

When the level of description shifts from texture elements to the solid visual angles projected by objects and surfaces in the environment, perspective structure provides information about how individual objects in the person's field of view are displaced relative to the body as the person moves through the environment. At this level of analysis perspective structure provides information for tracking body movement in relation to individual objects in the environment. The angular velocity of the solid visual angle projected by a stationary object varies predictably as a function of its direction and distance from the person. Thus, perspective structure provides the information needed to perceptually update the position of the body relative to all of the individual objects in the field of view.

Invariant structure

In contrast to perspective structure, the invariant structure in the optical flow provides information about the stable features of the environment that do not change with changes in the viewer's perspective, information about object identity and about the spatial relations among the textured surfaces and objects in the environment. Invariant structure is the geometric pattern in the optical flow that remains constant under transformation. Gibson (1979) describes reversible occlusion as one source of invariant information about object-to-object or surface-to-surface relations. Reversible occlusion is the reversible movement of one textured surface at the boundary of another surface. When one immovable object is hidden behind another, its surface will be revealed at the edge of the occluding object as the person moves at certain angles of regard in the direction of the hidden object. If the observer reverses direction by walking backward along the same trajectory, the surface that had been revealed now disappears at the occluding edge.

The pattern of deletion and accretion of textured surfaces at occluding edges along various movement trajectories specifies the metric interrelations between large-scale, immovable objects in the environment. During navigation, people typically steer toward openings in cluttered environments, and as a consequence, the deletion and accretion of textured surfaces occurs primarily in the periphery of the visual field. Invariant structure provides the information needed to form a survey representation or cognitive map of the metric interrelations among an ever expanding network of spatially contiguous objects. The formation of a cognitive map or survey representation can be thought of a process whereby the occluding surfaces of large-scale objects in the environment become functionally transparent over time (Gibson, 1979; Thorndyke & Hayes-Roth, 1982).

In geocentric navigation, a course is set and movement is tracked in relation to coordinates centered on the environment. The position taken here is that for humans, the metric configuration of immovable objects in the environment forms the geocentric system for orientation. Object-to-object relations are stored in a cognitive map and locally available information specifying self-to-object relations both locates the body in the object-to-object representation and geocentrically orients the person in relation to the network of objects that extend beyond the person's field of view (see Easton & Sholl, in press, for a further discussion of the interaction between an object-to-object and self-to-object system in the retrieval of spatial information). In geocentric navigation, movement of the body with respect to a configuration of landmarks may be monitored periodically by a dead reckoning process that uses the flow rate of texture elements to compute body displacement. However, for humans it is thought that self-to-object information is of primary importance for tracking the travel trajectory through object differentiated space (e.g., Rieser, 1990; Rieser et al., 1986; Rieser et al., 1992).

Navigation by Sighted Adults

Lindberg and Garling (1983) have studied the type of spatial knowledge created by sighted adults under guided navigation. They found that if the path is traversed while counting backwards, route knowledge of the path is formed; whereas, if the path is traversed without backward counting, survey knowledge is created. The results are interpreted to indicate that survey knowledge is not formed automatically but requires attentional resources, and backward counting uses up the central resources needed to form a survey representation. It is Lindberg and Garling's view, that when walking from one landmark to the next, people must continuously update where they are in relation to the first landmark, so that when they arrive at the next landmark they are able to code the straight-line distance and direction between the two. These processes are proposed to require cognitive effort.

I conducted a study (Sholl, 1993) designed to test the alternative explanation that route knowledge is produced by backward counting because the load it puts on cognitive resources causes the functional or useful field of view to constrict. That is, route knowledge is produced not because the acquisition of survey knowledge requires cognitive effort, but because visual information from the peripheral visual field is not fully processed. The idea that backward counting causes a functional constriction of the visual field is suggested by findings showing that in visual detection tasks peripheral field performance deteriorates when people are under a cognitive load (Miura, 1990; Williams, 1982). It has also been reported that people who traverse a path while counting backwards are less likely than controls to report peripheral details about the path, such as items on the walls or carpet on the floor (Smyth & Kennedy, 1982). Interestingly, Smyth and Kennedy reported that backward counters were just as likely as controls to know the shape of the path traversed, but were less likely to maintain their geocentric orientation with respect to the outside environment.

One reason why restriction of the functional visual field might inhibit the establishment of survey knowledge is that a full field of view is needed in order to process fully the invariant structure in optical flow. Others have discussed the importance of a broad visual field for abstracting self-to-object relations from lamellar flow (Rieser, Hill, Taylor, Bradfield and Rosen, 1992). Using low vision subjects, Rieser et al. tested whether the more important factor underlying people's ability to form cognitive maps of navigated environments was the breadth of the visual field or visual acuity. Their findings suggested that a broad visual field, irrespective of visual acuity, is needed from early childhood in order to precisely represent the Euclidean relations among landmarks experienced when navigating a space.

I tested the hypothesis that counting backwards produces route knowledge because it causes a constriction of the functional visual field by replicating Lindberg and Garling's backward counting and control conditions and adding a condition in which subjects wore goggles which physically restricted vision to the central 5° of the visual field. It was expected that if route knowledge is produced because backward counting reduces the processing of information projected to the peripheral retina, then the restricted-vision group should perform the same as the count-backwards group and show route knowledge of the path. In contrast, if route knowledge is produced because counting backwards places a load on cognitive resources, then the restricted-vision group should acquire survey knowledge, because subjects in this condition have full cognitive resources available for navigation.

Subjects were guided along a path laid out in the interior of a building and were instructed that they were to learn the interrelations among landmarks encountered along the path. The landmarks were separated from one another by a varying number of turns. After traversing the path seven times, the subject's knowledge of the landmark interrelations was tested with a point-to-target task. Subjects imagined themselves at one landmark on the path, the test-site, and, facing in the direction of travel, pointed in the straight-line direction they imagined a second, or target, landmark to be.

Route and survey knowledge can be differentiated behaviorally by examining pointing error and latency as a function of the number of turns in the path between the test site and target. With route knowledge, the location of the test site relative to the target is not directly known and must be calculated. The computations become more complex as the number of path segments or turns between the test and the target sites increase. The more complex the computations, the greater the expected pointing error and latency. With survey knowledge, the location of one landmark relative to another is directly available in the mental representation; therefore, pointing error and latency are independent of the number of path segments connecting the test site and target landmark.

The findings showed that the count backwards and restricted vision groups formed route knowledge of the path, while the control group, who navigated under normal walking conditions, formed survey knowledge. People in the restricted vision condition had full cognitive resources available to process spatial relations during navigation, so their failure to form a cognitive map of the space cannot be attributed to depleted central resources. Moreover, all subjects had the same visual experience at each of the landmarks along the path. At each landmark there was a 10 second pause, and during this interval all subjects thought about where landmarks were located in relation to one another, with no restrictions on attention or vision. So the findings suggest that broad visual fields are needed while walking in order to create a cognitive map of space.

Summary

If visual attention is restricted to the central visual field, either mechanically or because a person is under a cognitive load, the creation of a cognitive map of the space is delayed or inhibited, and route knowledge is formed instead. The hypothesis advanced here is that a broad visual field is needed while walking so that the visual system can process the invariant structure in the optical flow, which specifies the metric relations between spatially contiguous surfaces along the path. In my experiment the path consisted of interior hallways, which do not provide a broad field of view under normal circumstances. So visual field restriction did not restrict people's ability to see a panorama of large-scale objects in spatial relation to one another. Instead, it primarily had the effect of restricting the person's ability to see the accretion of previously hidden surfaces at the occluding edges of the walls and doorways located along the path. The accretion of surfaces in the visual periphery occurred as a result of turning from one hallway into another or moving through a doorway into a room.

Some have proposed that the failure to form survey maps as a consequence of a restricted visual field is attributable to a failure to process self-to-object relations embedded in perspective structure (Rieser, et al., 1992). While self-to-object information is provided by lamellar flow in the visual periphery, existing evidence suggests that self-to-object information is also communicated nonvisually by motor-kinesthetic signals, and that motor-kinesthetic signals can substitute effectively for visual self-to-object information (to be discussed fully in the next section; see also, Rieser, 1990; Rieser et al., 1986; Rieser et al., 1992). If so, visual restriction did not reduce available self-to-object information.

Route knowledge is established under visually restrictive conditions, and it is of interest to consider how it is formed. The information that was available to a person in the restricted field condition primarily consisted of motor-efferent and mechanical kinesthetic signals produced by the act of walking. Full visual information was available only when subjects stopped at the landmarks. Thus, information about the length of the linear segments of the path and the magnitude of the turning angles (which were all right angled) connecting segments to one another was computed largely from motor-kinesthetic cues, which presumably were integrated to form a motor-kinesthetic representation of the route. The formation of a representation of path trajectory from motor-kinesthetic signals is a dead reckoning function.

In the previous section, research was described showing that humans can use an egocentric dead reckoning process to generate a representation of a simple travel trajectory when the trajectory's position in relation to the surrounding environment is irrelevant or unknown. Egocentric dead reckoning relies on motor efferent signals and mechanical kinesthesis, and on visual kinesthesis, when available. In the path-learning experiment described in this section, people learned a path through a sequence of hallways and rooms, and landmarks were introduced at various points along the path. In this situation, a dead reckoning process computed a representation of the trajectory and incorporated information about landmarks/objects encountered along the path to form route knowledge. As described earlier, a dead reckoning process that incorporates landmark information is a geocentric process.

As mentioned earlier, geocentric dead reckoning differs from egocentric dead reckoning by its reliance on a self-to-object updating process. In the restricted vision conditions of the path learning experiment, people can relate motor-kinesthetic cues to external landmarks in order to estimate object displacement relative to their moving bodies (e.g., Rieser, 1990; Rieser et al., 1986; Rieser et al., 1992). Self-to-object updating may facilitate non-visual navigation by providing a frame of reference for tracking travel trajectories. The self-to-object reference frame may coordinate path segments and turning angles into a more coherent representation of the travel trajectory, than that produced by motor-kinesthetic signals alone. The introduction of landmarks also breaks the total path down into subpaths, each connecting one landmark to the next, so that the landmarks may serve as anchor points for organizing route information. To summarize, the finding that visual restriction produces route knowledge suggests that a geocentric dead reckoning process computes and stores the location of behaviorally relevant landmarks in a motor-kinesthetic representation of the path traveled.

In sighted navigation of an environment, route knowledge emerges first, and then survey knowledge. The length of time it takes for a cognitive map to develop varies with the complexity of the space to be learned. In the experiment just reviewed, I found evidence that people had formed a cognitive map of an 11-segment route laid out in an area spanning about 50 by 100 feet after seven path traversals. When testing knowledge of the spatial layout of the first floor of an office building, which covered an area spanning a little under 650 by 900 feet, Thorndyke and Hayes-Roth (1982) found that people employed in the building for less than a year showed route knowledge of its layout and people employed for more than a year showed survey knowledge. These researchers hypothesized that over time route knowledge is reorganized into survey knowledge. However, another possibility is that the two types of knowledge are formed by separate subsystems, with survey knowledge taking longer to develop than route knowledge. According to this account, the route subsystem generates a motor-kinesthetic representation that incorporates object/landmark information using a self-to-object updating process. The survey subsystem generates an object-to-object representation from invariant structure in visual flow.

Navigation by the Blind

The navigation performance of the blind has been compared to performance of the blindfolded sighted in order to study (a) the role of vision in the development of competent navigation skills and (b) the type of spatial representation formed by unsighted as opposed to sighted navigation. Some studies divide the blind into groups of early or congenitally blind and late or adventitiously blind so that the effect of early visual experience can be evaluated. One approach to assessing the role of vision in cognitive mapping ability has been to compare the nature of the errors blind and sighted people make when asked to solve problems involving familiar spaces (e.g., Byre & Salter, 1983; Rieser, Lockman and Pick, 1980; Rieser, Hill, Talor and Bradford and Rosen, 1992). Generally, the findings from studies of both adults and children have shown that the congenitally or early blind are less likely to represent the straight-line distances and directions between pairs of familiar landmarks than the sighted (except for Byrne and Salter, 1983, who show similar accuracy between the two groups in straight-line distance estimates). In other words, a person with no visual experience is less likely than a sighted person to form a cognitive map of a navigated environment.

An obvious conclusion is that visual information is an important factor in the cognitive mapping process. However, there are problems inherent in studying people's pre-experimental representations of an environment, especially if the purpose of the research is to draw conclusions about the type of spatial representation created by navigation. When testing knowledge of familiar spaces, pre-experimental exploration and navigation of the space is uncontrolled, and experienced trajectories (either on foot or by looking at a map) are unknown. In order to infer that a person has formed a survey representation or cognitive map of a space, it needs to be shown that judgments of the straight-line distance and direction between two landmarks that have never been experienced along a straight-line trajectory (either directly or by map) are as accessible (both in terms of retrieval speed and/or accuracy) as judgments about two landmarks directly experienced on a straight-line trajectory (e.g., Levine, et al., 1982). Thus, an alternative explanation for the conclusion that vision is necessary for cognitive map formation is that the sighted use more varied routes when traveling in an environment, and that they are more likely than the blind to have traveled straight-line trajectories between pairs of landmarks or to have studied maps of the space. The advantage of experiments testing people's knowledge of unfamiliar environments is that the travel trajectories are controlled.

Because the blind must rely on mechanical kinesthetic and motor efferent signals for monitoring their movement through space, the blind are required to use some form of dead reckoning for navigation. Experiments that study blind people's ability to navigate in unfamiliar environments can be divided functionally into navigation tasks that require egocentric dead reckoning and those that require geocentric dead reckoning. The results of the study reviewed in the previous section, which examined the kind of spatial knowledge formed from navigating a path, suggested that a geocentric dead reckoning process creates route knowledge under visual restriction (Sholl, 1993). Of interest, is whether geocentric dead reckoning also generates survey knowledge as contended in Gallistel's (1990) animal model.

Egocentric dead reckoning

In egocentric dead reckoning tasks, people navigate a path nonvisually and their knowledge of the shape of the path is tested. Typical tests include: retracing the path from end-point to start-point, estimating the distance of its linear segments and/or the magnitude of the angles connecting segments, and computing new trajectories from one point on the path to another (e.g., often the straight-line trajectory from the end-point to the start-point). In tasks of this type, the blind have been found to perform at least as well as, if not better than, the blindfolded sighted (Juurmaa, 1965; Loomis et. al., 1993; Passini, Proulx and Rainville, 1990), unless the travel trajectories form a shape familiar to the sighted (Worchel, 1951). It is noteworthy that when retracing or completing multi-segment paths, performance is not very accurate for either the blind or the sighted (Loomis et. al., 1993). However, on simple navigation tasks in which subjects estimated the length of a single straight-line segment or the magnitude of a single turning angle, Loomis et al. (1993) reported that performance of both groups was fairly accurate and similarly biased. In a distance estimation task, both the blind and the sighted overestimated a 2 m distance and underestimated distances between 4 and 10 m. When estimating turning angles, both groups overestimated turns between 60° and 120°, underestimated turns between 240° and 300°, and showed relatively greater accuracy in estimating turns which were multiples of 90° In sum, the findings suggest similar nonvisual mechanisms support egocentric dead reckoning in the sighted and the blind.

Geocentric navigation

The act of navigating in relation to stationary objects in the environment is generally more difficult for the blind than the sighted (see also Passini & Proulx, 1988). In a typical geocentric navigation task, the blind and the blindfolded sighted are guided along a path that contains stationary objects at various locations. At each object location, subjects stop in order to tactually identify the object. After traversing the path one or more times in order to become familiar with its layout, subjects' knowledge of straight-line (i.e., Euclidean) self-to-object and/or object-to-object relations is tested. A typical finding is that the early blind perform worse than the sighted both when their path knowledge is tested while occupying a test site on the path and when it is tested from a location remote from the path (Herman, Chatman and Roth, 1983; Passini & Proulx, 1988; Rieser et al., 1986; Veraart & Wanet-Defalgue, 1987, but see Loomis, et al., 1993 for an exception). Interestingly, the late blind show a different pattern of performance than the early blind. The late blind do not differ from the sighted in the accuracy of their Euclidean judgments when the judgments were made from an occupied test site on the navigated path (Rieser et al., 1986; Veraart & Wanet-Defalgue, 1987). However, like the early blind they performed worse than the sighted when they retrieve spatial relations from a remote location by imagining themselves at the test site (Rieser et al., 1986).

Why do the blind perform the same as the sighted on egocentric dead reckoning but differ from the sighted on various aspects of geocentric navigation?[3] Findings from egocentric tasks suggest that the blind may be somewhat better than the sighted at forming and using motor-kinesthetic representations of path trajectories. The geocentric tasks required that subjects use dead reckoning to navigate with respect to stationary objects. As described earlier, the visual flow created by the act of walking provides important visual information about how stationary objects are displaced relative to one another and relative to the moving body. The early blind have never had this visual experience, while the late blind have an early history of processing the geometric structure in optical flow. It was argued earlier that self-to-object updating plays a major role in geocentric dead reckoning; therefore, the difficulty experienced by the early blind on all geocentric navigation tasks may be attributable to the difficulty they experience extracting self-to-object information from nonvisual sources. The late blind experienced difficulty on some geocentric tasks and not others. Their performance can be characterized as good on the tasks that could be solved by processing self-to-object relations (this hypothesis was first advanced by Rieser, 1990; Rieser, et al., 1986), but poorer on tasks that rely on processing object-to-object relations.

Rieser, et al. (1986) have hypothesized that both the late blind and the sighted are able to update self-to-object relations during unsighted navigation because of their history of sighted navigation. This hypothesis is based on the finding that the two groups performed comparably when pointing to targets from "home base" in a baseline condition (all target objects had been directly experienced in relation to home base) and when pointing to targets from an occupied test site in a "locomotion" condition (no target object had been directly experienced in relation to the test site). Rieser et al. assumed that both the late blind and the blindfolded sighted used motor-kinesthetic cues to update self-to-object relations as they were guided from the home base to the test site, and responded on the basis of updated self-to-object relations once they reached the test site, accounting for the ease with which they made their pointing responses in the locomotion condition. Because the early blind had never experienced perspective structure in visual flow, they had no sense of how the act of walking caused the systematic displacement of stationary objects relative to the body, and therefore they performed poorly on this task. According to Rieser (1990; Rieser, et al., 1986; Rieser, et al., 1992), a history of sighted locomotion causes the motor-joint signals produced by walking to become synchronized with the visual signals registering self-to-object changes in optical flow. Thus, the sighted and the late blind know that every forward step of the body changes the location of stationary objects relative to the body. Because the covariation between visual and mechanical kinesthesis is known, motor-joint signals can substitute for visual signals and function as a source of information for perceptually updating self-to-object relations during unsighted navigation.

As described in the earlier section on sighted navigation, a self-to-object updating process may contribute importantly to the creation of route knowledge. The finding that the sighted and the late blind, who appear to have similar self-to-object updating ability (Rieser et al., 1986), have equally accurate route knowledge of a path (Veraart & Wanet-Defalgue, 1987) is consistent with the idea that self-to-object information is important to the formation of route representations. The discrepant performance of the early blind compared to the late blind and the blindfolded sighted in the formation and use of route knowledge (Veraart & Wanet-Defalgue, 1987) is consistent with the idea that the early blind use a less reliable, egocentric dead reckoning process for geocentric tasks. Egocentric dead reckoning would presumably allow the early blind to track the travel trajectory using motor-kinesthetic cues to encode and store the shape of the trajectory. However, the imprecision of this process and the apparent inability of the early blind to use biomechanical signals to update consciously self-to-object location could account for their less accurate route knowledge. In addition, a cognitive self-to-object updating process may have a role to play in the retrieval of object-to-object relations from route representations, by providing a mechanism for simulating movement between objects along the path.

Summary

Although explicitly tested in only one study reviewed for this section (Herman et al., 1983), there is little evidence that unsighted navigation produces survey knowledge of a space. It is possible that with sufficient exposure, a cognitive map of spatial layout would eventually develop. It could also be argued that the ability to update multiple self-to-object relations simultaneously, an ability which was demonstrated by the blindfolded sighted and the late blind in the Rieser et al. (1986) locomotion condition, implies that these groups had access to an object-to-object mental representation. However, unsighted self-to-object updating does not necessarily require access to a cognitive map of the space

(Pick, 1993). Subjects may perceptually update self-to-object relations in working memory during the act of unsighted walking. Alternatively, they may update body location in either a cognitive map or a route representation of the test space. However, if the late blind and the sighted had formed a cognitive map of object interrelations in the Rieser et al. experiment, they should not have experienced the difficulty that they did when retrieving object-to-object relations from an imagined position in space. Nevertheless, until explicitly tested, failure to find convincing evidence that unsighted navigation produces survey knowledge does not permit the conclusion that it is unable to produce survey knowledge.

Navigation Performance of Infants and Children

In this section the role of visual information in the development of infants' and children's sighted navigational ability will be reviewed. A critical developmental milestone in navigation ability occurs when the child becomes self-mobile, and research conducted by Acredolo (1987) suggests that self-produced versus passively produced movement has functional consequences both for navigational ability and for how spatial information is encoded in memory. Accordingly, the review of the developmental literature will be divided into two subsections: one on passive navigation and the other on active navigation.

Passive Navigation

Passive navigation performance has been studied in infants and children by researchers interested in the development of spatial frames of reference for coding object location. The research in this area explores the infant's ability to monitor his/her movement within a space that can be seen in its entirety from a single observation point. Although movement within such small spaces is not typically thought of as requiring navigation ability, the size of the space is appropriate in scale to the size of the infant, and it is large enough to challenge the infant's underdeveloped navigational system. One experimental paradigm typical of research in this area has been adapted from studies of place versus response strategies in animals.

In the "place-vs.-response" paradigm, the infant is instrumentally conditioned to make a motor response in the direction of an anticipated target event (e.g., a 90° head turn to the right). The target event (e.g., a toy display, an experimenter playing peek a boo) always occurs in the same location and is only visible for a brief period of time following an alerting signal. Once the child has reached a criterion level of performance on the training trials, a transfer test is given. Immediately prior to the transfer test the child is moved relative to the target event, which is not visible to the infant during the movement.

Typically the travel trajectory is a 180° rotation in place or a 180° rotation combined with a translation. Of interest is the direction of the orienting response made by the child in anticipation of the target event after movement to a new position in the test space. If on the transfer trial, the child responses as if his or her body hadn't been moved and makes the same anticipatory response that s/he made during the training trials (e.g., 90° head turn to the right), the child's response is described as egocentric. But if the child compensates for his/her new position within the space and correctly turns his/her head in the direction of the event's location in the environment (e.g., 90° head turn to the left), the child is said to have made an objective response.

Acredolo (1978; Acredolo & Evans, 1980) has shown that there is a systematic progression in development from egocentric to objective responses during the first year and a half of life. At 6 months, the infant's predominant mode of response is egocentric; whereas, at 16 months of age, infants respond objectively. At these two ends of the developmental continuum, the preferred response mode is largely indifferent to the visual characteristics of the experimental space. That is, in general it doesn't seem to matter if the test space contains distinctive visual features or not, 6 month old infants respond egocentrically and at 16 months, they respond objectively. Eleven months is a transitional age, with egocentric responses if the visual surroundings are non distinctive and a partial shift to objective responses in the presence of a distinctive visual cue.

Unfortunately, neither an egocentric nor an objective response allows a direct inference about the type of spatial knowledge acquired during training. An egocentric response indicates either that a particular motor response has been conditioned (e.g., "turning my head to the right produces the target event") or that the location of the target event has been coded in coordinates centered on the body (e.g., "the target event occurs directly to my right"). Although both types of learning rely on egocentric information, egocentric spatial codes, but not conditioned motor responses, require a realization that the target event actually occupies a location in extrapersonal space.

Objective responses also have two interpretations, either they may reflect that the child has perceptually encoded the interrelations among all of the salient cues in the environment and has a true idea of place (which can be thought of as the smallest unit of knowledge in a cognitive map of a space), or they may reflect cue learning, which involves learning a spatial proximity relation between a visual cue and the target event. In cue learning, when the target event is not visible, the visual cue serves as a "beacon" signaling the target event's location. Cue learning eliminates the child's need to update self-to-object relations when moving through space. At the end of the movement trajectory, the child simply has to search for the visual cue in order to find the location of the target event.

One other variable that needs to be considered when interpreting performance on transfer trials is whether the experimental space contains one or more distinctive visual cues. If the experimental space contains no distinctive cues and the child makes an objective response, then the response cannot reflect either cue or place learning because there are no visual landmarks to associate with the target event or with one another. In this situation, the response may be governed by dead reckoning.

Research conducted by McKenzie and colleagues (Keating, McKenzie and Day, 1986; McKenzie, Day and Ihsen, 1984; Tyler & McKenzie, 1990) suggests that the progression from egocentric to objective responses observed by Acredolo and others is most likely a progression from instrumentally conditioned motor responses to instrumentally conditioned cue responses. McKenzie and colleagues (e.g., Tyler & McKenzie, 1990) demonstrated that egocentric responses were most likely conditioned motor responses, by training infants to respond to a single target event from multiple facing directions. In the multiple-heading paradigm, infants sat at the center of a circular enclosure (i.e., a low wall), centered within a visually homogeneous, circular room. Immediately after fixating a jiggling ball at the center of his or her field of view, a target event briefly emerged in the peripheral visual field from behind the wall (the experimenter popped up waving a musical toy and saying "peek-a-boo"), and the infant made a head turn in order to fixate the target. During instrumental training, the target event always occurred in the same fixed location, and the infant was rotated in place to face in three different directions. Thus, instead of turning the head in a single direction during training, as was typical of single-heading paradigms, the infant sometimes made a head turn to the right and sometimes to the left in order to fixate the target. On the transfer trial, the infant was rotated to yet another orientation, and the direction and magnitude of the infant's head turn after fixating a jiggling ball was recorded.

In the multiple-heading paradigm, infants either responded objectively by turning the head to fixate the target or they emitted a conditioned motor response along a gradient of stimulus generalization. In the latter case, infants made the head-turning response that they had learned during training from the facing direction that was most similar to the facing direction at transfer. Note that knowledge of the similarity in facing direction is not based on visual information, because the test space is visually homogeneous. Therefore, it must depend on a kinesthetic record of the sequence of rotational displacements, probably created by an egocentric dead reckoning process, as described below.

In a visually homogeneous environment, 8 month old infants were most likely to make conditioned motor responses. By comparing instrumental training to "associative" training, Tyler and McKenzie found that 8 month old infants who made conditioned motor responses under instrumental training made objective responses under associative training. Associative training differed from instrumental training in that reinforcement (i.e., the target event) was not contingent upon a head-turning response, and there were fewer training trials at each facing direction (2 trials in the associative paradigm and an average of 3.3 to 6.0 trials to criteria in the instrumental paradigm). Tyler and McKenzie also found that associative training produced objective responses at a much younger age than typically observed with instrumental training. More than half of the 6 month old infants trained associatively responded objectively, which contrasts sharply with Acredolo's finding (1978; Acredolo & Evans, 1980) that single-heading, instrumental training almost never produced objective responses in 6 month old infants.

Objective responses in a visually homogeneous environment require dead reckoning to track body movement relative to the hidden target event, and the Tyler and McKenzie (1990) findings suggest that infants as young as 6 months can use dead reckoning to monitor body rotation. Other experiments suggest that eight-month, but not six-month old infants, have the ability to navigate by dead reckoning during combined rotations and translations of the body (McKenzie, Day, Colussa and Connell, 1988). In principle, dead reckoning could be used to update body location in terms of the remembered location of the target event. Remembering location requires coding the place where the target event occurred and perceptually updating the target location relative to the body during movement. However, there is little evidence that infants under 9 months of age process either the visual object-to-object relations that specify place (e.g., Acredolo, 1978; Acredolo & Evans, 1980; Millar & Schaffer, 1972) or the visual self-to-object relations needed to update body position (e.g., Bertenthal & Bai, 1989). If 6 month old infants are not sensitive to the visual information relevant to geocentric navigation, their ability to navigate by dead reckoning in the associative training paradigm (Tyler & McKenzie, 1990) is most likely mediated egocentrically.

Memory of simple travel trajectories can be governed by fairly low-level sensory-motor routines, as long as the target event's distal location remains in the infant's forward field of view. By fixating target-event location prior to movement and maintaining fixation during movement, visuo-ocular motor and visuo-cephalo motor routines will specify the location's new eye and head coordinates at the end of the movement trajectory[4]. Presumably, over time these sensory-motor tracking routines are internalized and integrated, perhaps into something like a linear associative mapping function (e.g., McNaughton, Chen and Markus, 1991), so that the infant does not have to actually fixate the target location during movement in order to compute its new location in body-centered space. The egocentric dead reckoning system may come to "know," for example, that in order to bring an object which is straight ahead into view after a 60° body turn to the right, the head must turn 60° to the left. Consistent with the idea of visuo-ocular motor based dead reckoning are findings showing that infants old enough to navigate egocentrically under associative training conditions (Tyler & McKenzie, 1990) are also able to visually track a target event (Lepecq & Lafaite, 1989),[5] and that visual tracking predicts objective responses (Bai and Bertenthal, 1992).

If, as hypothesized, infants use kinesthetic signals to track rotary displacements and if sensory-motor routines recompute motor coordinates during body movement, it suggests that a primitive form of egocentric dead reckoning develops within the first 8 months of life. The findings also suggest that egocentric dead reckoning is dominated by response conditioning during the first year of life. That is, infants who use dead reckoning to correctly orient toward a target event after passive displacement will, if trained instrumentally, make incorrect conditioned motor responses on the same task. Instrumentally conditioned objective responses are observed in a visually distinctive environment at about the time the child becomes self mobile (e.g., Acredolo, 1987; 1988; Bai & Bertenthal, 1992). Increasing the salience of the visual cue increases the likelihood of objective responses at all ages (Acedolo & Evans, 1980), and, as alluded to earlier, instrumental conditioning across multiple headings appears to free up the "habit" system to analyze distal visual cues (Tyler and McKenzie, 1990).[6]

It is quite likely that the instrumentally conditioned objective responses that are observed around the time the child becomes self mobile reflect cue rather than place learning. A true idea of place requires encoding the spatial relations among distal objects/ cues. If infants were learning the place, they should be just as likely to respond objectively when cues are spatially remote from the target event as when cues are located in close proximity to the target event. However, 11 month old infants, who are 80% likely to respond objectively when cues are in close proximity to the target event, are only 37% likely to respond objectively when cues are spatially remote from the target event (Acredolo & Evans, 1980). Although 8 month old infants show some sensitivity to the geometry of the distal experimental space (44% make place responses when the space is square versus 12.5% when the space is circular), more than half of the 8 month old infants tested in the differentiated square space respond incorrectly (Keating, et al., 1986). One consequence of self mobility is that increased attention is paid to the visual surround in order to navigate around obstacles, approach goals, avoid barriers, etc. (e.g., Acredolo, 1987; Bai & Bertenthal, 1992). Increased visual attention to objects beyond reaching space may encourage attention to the spatial relation between the target event and a nearby distinctive object, thereby promoting cue learning.

Summary

Spatial behavior in pre-mobile infants appears to be particularly susceptible to response conditioning. When infants are instrumentally conditioned to make a single response directed toward a fixed distal location, the infant is likely to continue to make the same response after a passive rotation and/or translation of the body, even though it is then incorrect. However, if instrumental conditioning is not used to familiarize the infant with the location of the target event, infants as young as 6 months demonstrate dead reckoning ability and will correctly orient to the location of the hidden target event after body displacement. Cue learning, which requires making an association between two stimuli based on spatial proximity, is most likely to occur in self-mobile infants. Presumably, self mobility sets the stage for the formation of associations between spatially proximate objects by directing visual attention outward to objects located beyond reaching space. However, even self-mobile infants at 11 months of age are likely to emit a conditioned motor response on a transfer test if instrumentally conditioned in a visually undifferentiated space.

Active Navigation

Once children are capable of self-produced movement, they begin to show evidence that they can navigate in relation to coordinates anchored to the physical environment. Instrumental conditioning no longer inhibits their ability to navigate passively in small experimental spaces that contain no distinctive cues. In Acredolo's (1978) instrumental conditioning paradigm, 16 month old infants responded objectively on a transfer test regardless of the visual distinctiveness of the experimental space. The finding of objective responses in both visually distinctive and non-distinctive spaces suggests that 16 month old infants navigate geocentrically. That is, they were perceptually able to update their location relative to the interrelations between the four walls of the room in the visually undifferentiated condition and relative to the interrelations between the visual cues/objects in the differentiated condition.

It is my hypothesis that geocentric navigation does not develop until the child is able to differentiate perspective from invariant structure in optical flow and separately process self-to-object and object-to-object relations. To my knowledge there are no studies that examine directly the newly self-mobile child's ability to functionally differentiate perspective and invariant structure in optical flow. However, there are studies focusing on the processing of perspective structure, which show that the radial and lamellar flow comprising perspective structure are gradually differentiated over time (e.g., Bertenthal & Bai, 1989; Schmuckler & E. J. Gibson, 1989; Stoffregen, Schmuckler,. & E. J. Gibson, 1987). The developmental trends evidenced in these studies are likely to generalize to the higher-order perceptual and functional differentiation of perspective from invariant structure.

The development of functionally differentiated sensitivity to radial and lamellar flow has been studied in the moving room paradigm. In the moving-room paradigm, a three-walled, floorless "room" is moved back and forth parallel to the stationary child's line of sight. The walls are patterned to provide optical texture and the "room" is mounted on wheels so that it moves a few inches above a stationary floor. Sensitivity to visual self-motion information is measured by the extent to which the discrepancy between visual flow and stationary proprioceptive feedback disrupts body posture. For adults (Stoffregen, 1985) and for children over two (Stoffregen, et al., 1987), lamellar flow (side walls move but not the front wall) elicits strong compensatory postural responses, causing the children over two to stagger or fall down and adults to sway, while radial flow (front wall moves but not the side walls) elicits minimal postural compensation. The sensory-motor routines for the control of standing and walking are less well developed in children under two than in children over two and are more easily disrupted by the moving room. However, unlike adults, children under two are not selectively sensitive to lamellar flow.

Generally for children under two, there is a progression from a global sensitivity to full field, undifferentiated structure in premobile infants (7 month old; Bertenthal & Bai, 1989), to separate levels of sensitivity for radial flow in the central field and lamellar flow in the peripheral field in infants who have recently become self-mobile (9 months to 2 years; Bertenthal & Bai, 1989; Stoffregen, et al., 1987). For children between 9 months and 2 years, radial flow elicits weaker compensatory postural responses than lamellar flow. The finding that small postural responses are elicited by radial flow indicates that, unlike adults, postural adjustments have not yet been brought under the almost exclusive control of lamellar flow (Bertenthal & Bai, 1989; Stoffregen, et al., 1987). In summary, evidence for rudimentary separation of radial and lamellar flow emerges around nine months, which is the age at which most infants have begun to crawl, and children need to have been walking for about a year before they show the same level of functional differentiation in perspective structure as adults.

Studying the effect of a moving room on locomotor control, Schmuckler and E. J. Gibson (1989) found that children who have been walking less than a year are more disrupted by the moving room when walking around obstacles than when walking in an uncluttered room, whereas children who have been walking more than a year do not show this effect. They hypothesize that novice walkers have not yet learned that the different geometric patterns in optic flow each provide functionally distinct information, that is, that radial flow provides information largely relevant to steering and that lamellar flow specifies self motion. As children become more experienced at walking, their visual systems become selectively tuned to the separate functions of lamellar and radial flow, and these functions become better integrated in the control of sighted navigation. The position taken here is that, like lamellar and radial flow, perspective and invariant structure also become progressively differentiated, both perceptually and functionally, as the child becomes more adept at walking. As the child learns to discriminate invariant from perspective structure, s/he will begin to dissociate object-to-object relations from self-to-object relations, setting the foundation for geocentric navigation.

What kind of navigation ability is demonstrated in a place-vs.-response paradigm by children who walk along the movement trajectory instead of being passively transported? As with passive transport, navigation ability appears to be, in part, a function of how the space was learned, and instrumental conditioning can mask the child's underlying navigation ability. When an instrumental conditioning paradigm is used in a visually undifferentiated, 12' by 12' square experimental space (a space very similar to the one used by Acredolo, 1978, in which 16 month old infants showed objective responses after passive displacement), between 40 and 50% of three and four year old children made egocentric locomotor responses on a transfer test (Acredolo, 1977). Only five year olds consistently made objective, locomotor responses. A self-produced rotational displacement following instrumental training in a circular experimental space produced essentially at random responses by 18 month olds, although they did perform as if they remembered the direction and trajectory of the turn and had some knowledge of the geometry of a circle (Rieser and Heiman, 1982). However, if experimental procedures do not involve instrumental conditioning, there is clear evidence for accurate active navigation with respect to hidden target sites in children as young as 16 months of age (Pick, 1993).

It is difficult to draw any firm conclusions about the role of visual information in the development of cognitive mapping ability in the self-mobile child, because the experiments conducted in this area have often not been designed in a way that directly addresses the question. One experiment (Acredolo, 1976) that does explicitly manipulate the visual properties of the experimental space shows that 3 and 4 year old children navigate in relation to the surface-to-surface relations between distinctively different walls of a room. However, in a slightly larger, less clearly differentiated space, Acredolo (1976) reports evidence suggesting that 3 year olds navigated on the basis of egocentric dead reckoning (they retraced their travel trajectory); while, 4 year olds navigated in relation to a single object. Only 5 year old children were able to navigate in relation to the spatial interrelations among the walls of the room.

Other studies show that self-mobile children can accurately retrace a path traversed within a non distinctive environment (Hazen, Lockman and Pick, 1978; Pick, 1993), but that the ability to compute new trajectories from a representation of the path varies with the size and complexity of the experimental space. Two year olds can take shortcuts if the path trajectory is in the shape of a square (Pick, 1993), but for more complex U or Z-shaped paths, it is not until the children reach the age of 5 or 6 that they are able to compute new trajectories between non-directly experienced points on the paths (Hazen, et al., 1978). It is not clear from these experiments whether the child is using route or survey knowledge to compute new trajectories, or for that matter, whether trajectories were directly available in perceptually updated self-to-object relations. Another general finding is that the more visually differentiated the experimental space the more accurately children are able to code the spatial location of an event (Acredolo, Pick and Olsen, 1975).

Summary

I have hypothesized that the geometric structure embedded within optical flow becomes progressively differentiated in the self-mobile child, and that the ability to navigate geocentrically develops as the self-to-object and object-to-object information in visual flow becomes functionally separated by the visual perceptual system. Not many studies address the development of geocentric navigation and its constituent subsystems in a systematic way. However, existing evidence does suggest that instrumental conditioning can dominate geocentric navigation in children under five, and that the ability to navigate geocentrically is sensitive to the size, complexity, and visual differentiation of the experimental space in ways which are not yet fully understood.

Conclusions

Behavioral studies of human adults show that route knowledge of environments navigated by foot is established fairly quickly, and that survey knowledge takes longer to develop (Thorndyke and Hayes-Roth, 1982). I have hypothesized that route knowledge is a motor-kinesthetic representation of the travel trajectory, that allows flexible traversal of the path trajectory (e.g., by walking, running, skipping, etc.) and incorporates information about the location of behaviorally relevant objects. Two forms of information provided by visual perspective structure contribute to the creation of route knowledge. One is the information provided by the rate of texture element flow, which, when coordinated with motor outflow and mechanical-kinesthetic signals and integrated with estimates of time, forms a representation of the shape of the travel trajectory. The other is self-to-object information which is used to monitor body location in relation to fixed objects in the environment. When navigating through an environment, it is hypothesized that route knowledge is created by integrating visual self-to-object information with the motor-kinesthetic representation of the route. As shown by the unsighted navigation of the sighted and the late blind, signals from the motor-joint system can substitute for visual self-to-object information (e.g., Rieser, 1990) in the formation of route knowledge.

A full field of view during the act of walking is an important factor in the formation of a cognitive map of the environment. At a moving point of observation, invariant structure unfolds across the entire visual field, specifying surface-to-surface relations at occluding edges and object-to-object relations between bounded surfaces. Cognitive maps emerge over time in a process which, at a functional level of analysis, is comparable to the occluding surfaces of the environment becoming gradually transparent, so that all objects represented in the cognitive map can be "seen," and their straight-line distances and directions known, from any observation point within the environment. The importance of a full field of view in the cognitive mapping process is indicated by findings suggesting that environments are not cognitively mapped by either sighted people who have their vision artificially restricted, low-vision people with narrow fields of view, or adventiously and congenitally blind people. I hypothesize that the same self-to-object updating process that contributes to route knowledge formation monitors location of the self in relation to the configuration of objects that form a geocentric system of spatial reference. But this process alone is not sufficient for cognitive map formation. Invariant object-to-object relations in visual flow must be processed in order to form a cognitive map of a space.

Developmental studies suggest that instrumental conditioning dominates early spatial learning. Prior to becoming self-mobile, the spatial behavior of infants is very susceptible to instrumental conditioning, regardless of the visual characteristics of the environment. However, under other learning conditions, infants as young as six months show evidence of an egocentric dead reckoning process that retains a record of a simple travel trajectory. The onset of self-mobility is associated with the ability to form spatial relations between visual stimuli located close together in space. The child later develops the ability to associate visual stimuli separated in space, and once the child can form associations between spatially separate objects, s/he can form an idea of place. The processing of object-to-object relations may require self mobility in order that visual attention be directed outward beyond reaching space. The distribution of visual attention across the entire visual field while walking, as opposed to the concentration of attention to one focal area while reaching or grasping (Trevarthen, 1968), may be a necessary precondition for perceptual differentiation of the various forms of geometric structure in optical flow. The developmental findings are not inconsistent with the idea that cognitive mapping skills develop in concert with the ability to perceptually differentiate perspective and invariant structure in visual flow.

In summary, geocentric navigation in humans appears to be heavily reliant on the self-to-object and object-to-object information in the visual flow generated by the act of walking. Before they are self-mobile, infants show little evidence that they are capable of geocentric navigation. Although there is little direct evidence to this effect, I have suggested that with increased walking experience, the child's developing visual system becomes increasingly adept at differentiating and processing the invariant and perspective structure in optical flow important to geocentric navigation ability. Based on Gibson's

(1976) analysis of optical flow fields, I have proposed that the invariant structure in optical flow is an important source of information for the formation of cognitive maps of navigated space. When access to visual information in the peripheral field is mechanically restricted during sighted navigation, human adults form route knowledge but not survey knowledge of spatial layout (Sholl, 1993). It could be argued that restricted access to the self-to-object information provided by lamellar flow in the peripheral field is the factor important to cognitive map formation (e.g., Rieser, 1990; Rieser, et al., 1986; Rieser, et al., 1992). However, the findings suggesting that the sighted and the late blind are both able to update self-to-object relations when navigating through space (Rieser et al., 1986; Veraart & Wanet-Defalgue, 1987), but that only the sighted form survey representations of space argues against this interpretation. To date, very little is known about the recovery of invariant structure from optical flow and future research on the role of vision in cognitive mapping should explore this problem.

Notes

1. Here and elsewhere, a motor-kinesthetic representation is defined as a representation that provides a level of description that functions to support travel along a trajectory by using a variety of motor behaviors, e.g., running, walking, crawling, etc. A motor-kinesthetic representation is not a stored sequence of specific motor programs, permitting only a rigid reenactment of the original sequence of motor responses.
2. To simplify discussion, the eyes and head are assumed to be fixed in the straight ahead position when describing opitical flow.
3. Loomis et al. (1993) using a task similiar to Rieser et al. reported no important differences in geocentric navigation between the congenitally blind and blindfolded sighted. Lommis et al. attribute the discrepency in results to their having sampled a more highly mobile group of congenitally blind people than Rieser et al. There are clearly individual differences in the blind population, and some exceptional congenitally blind people have good geocentric nagivation skills (see also Landau, Gleitman & Spelke, 1981), but they are the exception rather than the rule, as the studies reviewed here indicate.
4. The role of sustained visual tracking in facilitating the infant's ability to make objective responses was proposed originally by Acredolo (1987), who hypothesized that sustained tracking during self-locomotion was responsible for the increased liklihood of objective responses observed in self-mobile infants.
5. Although 7 month old infants tracked a target event when it was turned on, they did not track it under continuous rotation when it was turned off (Lepecq & Lafaite, 1989). Infants may not have the motor control needed to visually track location under a continuous rotation in place, in the absence of a visual cue to "grab" attention. The experiments conducted by McKenzie and colleagues did not provide an attention grabbing cue, but displacements were small and discrete.
6. Mishkin makes a neuroanatomical distinction between a cortical memory system that controls declarative memory and a habit system that controls instrumental conditioning in primates, and O'Keefe and Nadel (1978) distinguish between a local system that creates a cognitive map of the environment and a taxon system that controls habit learning in rodents. The developmental findings suggest the habit system dominates spatial learning in the first year of life, with later development of a cognitive mapping system for coding object-to-object relations.

References

Acredolo L.P. (1987). Early development of spatial orientation in humans. In *Cognitive processes and spatial orientation in animal and man* (P. Ellen and C. Thinus-Blanc, eds.), pp. 185-201. Dordrecht: Marinus Nijhoff.

Acredolo, L. P. (1978). Development of spatial orientation in infancy, *Developmental Psychology*, 14, 224-234.

Acredolo, L. P. (1977). Developmental changes in the ability to coordinate perspectives of a large-scale space, *Developmental Psychology* , 13, 1-8.

Acredolo L.P. (1976). Frames of reference used by children for orientation in unfamiliar spaces. In *Environmental knowing* (G. Moore and R. Golledge, eds.), pp. 165-172. Stroudsburg, PA: Dowden, Hutchinson & Ross.

Acredolo L.P. & Evans D. (1980). Developmental changes in the effects of landmarks on infant spatial behavior, \i *Developmental Psychology*, 16, 312-318.

Acredolo L.P., Pick H. L. & Olsen M.G. (1975). Environmental differentiation and familiarity as determinants of children's memory for spatial location. *Developmental Psychology*, 11, 495-501.

Bai D.L. & Bertenthal B.I. (1992). Locomotor status and the development of spatial search skills. *Child Development*, 62, 215-226.

Bertenthal B.I. & Bai D.L. (1989). Infant's sensitivity to optical flow for controlling posture. *Developmental Psychology*, 25, 936-945.

Byrne R.W. & Salter E. (1983). Distances and directions in the cognitive maps of the blind. *Canadian Journal of Psychology*, 37, 293-299.

Dichgans J. & Brandt Th. (1974). The psychophysics of visually induced perception of self-motion and tilt, In J. Dichgans & Th. Brandt, *The Neurosciences: Third Study Program*, Cambridge, MA: MIT Press.

Easton R.D. & Sholl M.J. Frames of reference and accessibility of spatial knowledge. *Journal of Experimental Psychology: Learning, Memory and Cognition*, In press.

Gallistel C.R. (1990). *The organization of learning*, Cambridge MA: MIT press.

Gibson J.J. (1979). *The Ecological Approach to Visual Perception*, Boston: Houghton Mifflin.

Gould J.L. & Able K. P. (1981). Human homing: An elusive phenomena. *Science*, 212, 1061-1063.

Hazen N.L., Lockman J.J. & Pick H. L. (1978). The development of children's representations of large-scale environments. *Child Development*, 49, 623-636.

Herman J.F., Chatman, S.P. and Roth, S. F. (1983). Cognitive mapping in blind people: Acquisition of spatial relationships in a large-scale environment, *Journal of Visual Impairment & Blindness*, 77, 161-166.

Juurmaa J. (1973). Transposition in mental spatial manipulation: A theoretical analysis, *AFB Research Bulletin*, 26, 87-134.

Juurmaa J. (1965). *An analysis of the components of orientation ability and mental manipulation of spatial relationships*, Helinski: Reports from the Institute of Occupational Health (No. 28).

Keating M.B., McKenzie, B.E. and Day, R.H. (1986). Spatial localization in infancy: Position constancy in a square and circular room with and without a landmark, *Child Development*, 57, 115-124.

Klatzky R.L., Loomis J.M., Golledge R.G., Cicinelli J. G., Doherty S. & Pellegrino J.W. (1990). Acquisition of route and survey knowledge in the absence of vision, *Journal of Motor Behavior*, 22, 19-43.

Landau B., Gleitman H. & Spelke E. (1981). Spatial knowledge and geometric representation in a child blind from birth, *Science*, 213, 1275-1278.

Lee D.N. (1980). The optical flow field, *Philosophical Transactions of the Royal Society* (B), 290, 169-178.

Lee D.N. (1974). Visual information during locomotion, In *Perception: Essays in honor of James J. Gibson* (R. B. MacLeod & H. L. Pick, eds.), pp. 250-269. Ithica: Cornell University Press.

Lepecq J.C. & Lafaite M. (1984). The early development of position constancy in a no-landmark environment, *British Journal of Developmental Psychology*, 7, 289-306.

Levine M., Jankovic I.N. & Palij M. (1982). Principles of spatial problem solving, *Journal of Experimental Psychology: General*, 111, 157-175.

Lindberg E. & Garling T. (1983). Acquisition of different types of locational information in cognitive maps: Automatic or effortful processing? *Psychological Research*, 45, 19-38.

Lishman J R. & Lee D.N. (1973). The autonomy of visual kinesthesis, *Perception*, 2, 287-294.

Loomis J.M., Klatzky R.L., Golledge R.G., Cicinelli J.G., Pellegrino J.W. & Pry P.A. (1993). Nonvisual navigation by blind and sighted: Assessment of path integration ability, *Journal of Experimental Psychology: General*, 122, 73-91.

McKenzie B.E., Day R.H., Colussa, S. & Connell S. (1988). Spatial localization by infants after rotational and translational shifts, *Australian Journal of Psychology*, 40, 165-178.

McKenzie B.E., Day R.H. & Ihsen E. (1984). Localization of events in space: Young infants are not always egocentric, *British Journal of Developmental Psychology*, 2, 1-9.

McNaughton B.L., Chen L.L. & Markus E.J. (1991). "Dead reckoning," landmark learning, and sense of direction: A neurophysiological and computational hypothesis, *Journal of Cognitive Neuroscience*, 3, 190-202.

Millar W.S. & Shaffer H.R. (1972). The influence of spatially displaced feedback on infant operant conditioning, *Journal of Experimental Child Psychology*, 14, 442-452.

Mishkin M., Ungerleider L.G. & Macko K.A. (1983). Object vision and spatial vision: two cortical visual pathways, *Trends in Neuroscience*, 6, 414-417.

Miura T. (1990). Active function of eye movement and useful field of view in a realistic setting. In From *eye to mind: Information acquisition in perception, search, and reading*/

(R. Groner, G. d'Ydewaile, and R. Parham, eds.), pp. , North-Holland: Elsevier Science.

O'Keefe J. & Nadel L. (1978). *The Hippocampus as a Cognitive Map*, Oxford: Clarendon Press.

Passini R. & Proulx G. (1988). Wayfinding without vision an experiment with congenitally totally blind people. *Environment and Behavior 20*, 227-252.

Passini R., Proulx G. & Rainville C. (1990). The spatio-cognitive abilities of the visually impaired population, *Environment and Behavior*, 22, 91-118.

Pick H.L. (1993). Organization of spatial knowledge in children. In *Spatial Representation* (N. Eilan, R. McCarthy, B. Brewer, eds.), pp. 31-42. Oxford: Blackwell.

Poucet B. (1993). Spatial cognitive maps in animals: New hypotheses on their structure and neural mechanisms, *Psychological Review*, 100, 163-182.

Rieser J.J. (1990). Development of perceptual-motor control while walking without vision: The calibration of perception and action. In *Sensory-Motor Organizations and Development in Infancy and Early Childhood* (H. Bloch & B. I. Bertenthal, eds.), pp. 379-408. Netherlands: Kluwer Academic.

Rieser J.J. & Heiman M.L. (1982). Spatial self-reference systems and shortest route behavior in toddlers, *Child Development*, 53, 524-533.

Rieser J.J., Guth D.A. & Hill E. W. (1986). Sensitivity to perspective structure while walking without vision, *Perception*, 15, 173-188.

Rieser J.J., Lockman J.J. & Pick H. L. (1980). The role of visual experience in knowledge of spatial layout, *Perception & Psychophysics*, 28, 185-190.

Rieser J.J., Pick H.L., Ashmead D.H. & Garling A. E. (in press). The calibration of human locomotion and models of perceptual-motor organization, *Journal of Experimental Psychology: Human Perception and Performance.*

Rieser J.J., Hill E.W., Taylor C.R., Bradfield A. & Rosen S. (1992). Visual experience, visual field size, and the development of nonvisual sensitivity to the spatial structure of outdoor neighborhoods explored by walking, *Journal of Experimental Psychology: General*, 121, 210-221.

Rolls E.T. (1991). Functions of primate hippocampus in spatial processing and memory, In *Brain and Space* (J. Paillard, ed.), pp. 334-352. Oxford: Oxford University Press.

Schmuckler M.A. & Gibson E.J. (1989). The effect of imposed optical flow on guided locomotion in young walkers, *British Journal of Developmental Psychology* 7, 193-206.

Sholl M.J. (1987). Cognitive maps as orienting schemata, *Journal of Experimental Psychology: Learning, Memory, and Cognition*, {\i 13}, 615-628.

Sholl M.J. (1993, November). *The effect of restricting vision to the central field on spatial knowledge acquisition*, Paper presented at the Annual Meeting of the Psychonomic Society, Washington, DC.

Smyth M.M. & Kennedy J.E. (1982). Orientation and spatial representation within multiple frames of reference, *British Journal of Psychology*, 73, 527-535.

Stoffregen T.A. (1985). Flow structure versus retinal location in the optical control of stance, \i *Journal of Experimental Psychology: Human Perception and Performance 11*, 554-665.

Stoffregen T.A., Schmuckler M.A. & Gibson E.J. (1987). Use of central and peripheral optical flow in stance and locomotion in young walkers, *Perception 16*, 113-119.

Telford L. & Frost B.J. (1993). Factors affecting the onset and magnitude of linear vection, *Perception and Psychophysics*, 53, 682-692.

Thinus-Blanc C., Save E., Buhot M.C., Poucet B. (1991). The hippocampus, exploratory activity, and spatial memory, \i *Brain and Space* (J. Paillard, ed.), pp. 334-352. Oxford: Oxford University Press.

Thorndyke P.W. & Hayes-Roth B. (1982). Difference in spatial knowledge acquired from maps and navigation, *Cognitive Psychology 14*, 560-589.

Trevarthen C. B. (1968). Two mechanisms of vision in primates, *Psychologische Forschung* {\i 31}, 299-337.

Tyler D. & McKenzie B.E. (1990). Spatial updating and training effects in the first year of life, *Journal of Experimental Child Psychology 50*, 445-461.

Ungerleider L.G. & Mishkin M. (1982). Two cortical visual systems. In D.J. Ingle, M.A. Goodale & R.J.W. Mansfield, *Analysis of Visual Behavior*, Cambridge, MA: MIT Press.

Veraart C. & Wanet-Defalque M.C. (1987). Representation of locomotor space by the blind, *Perception & Psychophysics*, 42, 132-139.

Warren W.H., Morris M.W. & Kalish M. (1988). Perception of translational heading from optical flow, *Journal of Experimental Psychology: Human Perception and Performance*, 14, 646-660.

Williams L.J. (1982). Cognitive load and the functional visual field, *Human Factors*, 24, 683-692.

Worchel P. (1951). Space perception and orientation in the blind, *Psychological Monographs*, 65, 1-28.

M. Jeanne Sholl
Department of Psychology
Boston College Chestnut Hill, MA 02167

CONSTRUCTING COGNITIVE MAPS WITH ORIENTATION BIASES

Robert Lloyd and Rex Cammack

Abstract: The purpose of this research is to investigate cognitive maps constructed using different encoding processes. Different learning processes have been shown to produce cognitive maps with different characteristics. Two critical research issues are the fixed-orientation bias and the equiavailability principle. Previous research has indicated that studying a north-at-the-top cartographic map encodes a cognitive map biased in the orientation of the cartographic map. Such cognitive maps are images that have information in all parts of the map equally available. Other research has shown cognitive maps encoded by environmental navigation produced cognitive maps with no orientation bias. Subjects, however, had faster access to information in front of them than information behind them. These results suggested that exposure to a single versus multiple orientations of the spatial information explained the biases. Others have argued the two situations coincide with encoding the spatial information from secondary and primary sources. The current study considered five different learning experiences that were used to encode information about the same seven landmarks in a space. Encoding the information from three-dimensional spaces resulted in longer reaction times for an identification task. Although all learning experiences were secondary, some produced cognitive maps with orientation biases and some without. Learning experiences that provided multiple orientations eliminated an orientation bias. A single perspective oblique view learning experience appeared to produce a bias for front-back over left-right. Orientation-free higher-order cognitive maps, as described by Taylor and Tversky, could account for all these results.

Introduction

Cognitive maps express the essential structure of spatial information encoded in our memories through learning processes. Like cartographic maps, cognitive maps can be constructed using many different sources of information and encoding processes (Lloyd, 1993). Some cognitive maps may be stored as permanent structures in long-term memory, e.g., a cognitive map of a familiar city, while others may be temporary structures of the current state of a dynamic environment, e.g. a parent keeping track of the locations of children as they play in a park. In either case the characteristics of objects are thought to be stored along with their spatial locations (Kahneman, Treisman and

J. Portugali (ed.), The Construction of Cognitive Maps, 187–213.

Gibbs, 1992; Lloyd and Hooper, 1991). Expressed in the simplest terms possible, cognitive mapping is a recording in memory of the existence of an object and its known location in space. A cognitive map is, in these simplest terms, the encoding of a structure in our memory of *what* is *where* (Kosslyn and Koenig, 1992; Sagi and Julesz, 1985). Although this seems simple enough, the processes used to acquire spatial knowledge appear to have a fundamental impact on the character of a cognitive map.

The purpose of this study is to investigate the nature of cognitive maps produced by different encoding processes. Specifically the study focuses on understanding the circumstances that produce cognitive maps with fixed orientations and those that produce cognitive maps that are orientation free. The literature has discussed both types of cognitive maps and theoretical interpretation have been offered to explain their differences (Carpenter and Just, 1986; Evans and Pezdek, 1980; Lloyd, 1993; MacEachren, 1992; Lowe, 1987; Presson and Hazelrigg, 1984; Sholl, 1987). The current research used an experimental design to compare five different processes used to encode a simple cognitive map. In all cases the locations of seven objects were learned and subjects confirmed if triads of objects selected from the space were being presented in their correct positions or in a mirror representation of their correct positions. Triads were presented for identification rotated away from a north-at-the- top orientation. The efficiency of the subject's decision-making process revealed if the spatial information they had acquired using a specific learning process was encoded in a fixed or free orientation structure.

Review of Concepts

Spaces and Viewers: What's up?

The conventional map has a spatial structure that relates it directly to the earth. Traditional cartographic maps have a north-at-the-top orientation and a perspective which positions the viewer 90° above the plane depicting the map. Orientation, used in this sense, relates the map to vertical and horizontal axes as the map might hang on a wall or appear on a monitor. Shepard and Hurwitz (1984) argued that we persistently map spaces onto vertical and horizontal axes. They reason that we always have immediate knowledge of the vertical axis because up versus down is defined by gravity, This allows a viewer who is considering a space defined by vertical and horizontal axes to easily distinguish top from bottom and left from right. This means relative location on cartographic maps can be expressed with north to south along the vertical axis and west to east along the horizontal axis. These terms have a consistent meaning relative to the plane even if the viewer's perspective is changed.

Navigation experiences provide multiple perspectives for a viewer. A person standing on the plane has a view just above the surface. A specific view is from a point on the plane facing in a particular direction. During navigation, orientation is expressed in terms of an egocentric spatial structure, i.e., ahead, behind, left or right. When we look at important objects ahead of us they naturally have a top and bottom and a front and back. Shepard and Hurwitz (1984) argued we also map navigation experiences onto vertical and horizontal axes. They argued straight ahead is naturally considered as "up" within an egocentric frame. Our awareness of gravity along the vertical axis associates away from us and the backs of objects with the top of the scene. The relationship of environmental objects to vertical and horizontal axes, however, is not consistent because the relationship is dependent on the viewer's perspective within the space. If you are in the center of a space looking at an object in an extreme northern location in the space, the object will be ahead of you if you are facing north, behind you if you are facing south, to your left if you are facing east, and to your right if you are facing west.

Types of Knowledge and Processes

Some authors have argued for categories of spatial knowledge based on the processes used to encode and decode the information. For example, Thorndyke and Hayes-Roth (1982) argued that cognitive maps can be encoded with two different types of information. They argued that *procedural knowledge* is encoded by navigating through an environment. This type of knowledge documents the procedures one would use to go from one location in the environment to another. It is stored as verbal information and is decoded using a serial process. They argued that a distinctly different type of knowledge, *survey knowledge*, is encoded in our memories as mental images and can easily be obtained through reading maps. This type of knowledge gives a more holistic impression of the environment and can be decoded using a parallel process. Although *procedural knowledge* may be transformed into *survey knowledge* over time, some evidence has suggested this transformation may not automatically occur at the urban scale (Lloyd, 1989; Thorndyke and Hayes-Roth, 1982).

Other authors (Presson and Hazelrigg, 1984; Presson, DeLang and Hazelrigg, 1989) have argued for a different categorization of spatial knowledge. In their studies *primary knowledge* is any spatial knowledge acquired directly from the environment by a viewer. It may be from a single perspective, e.g., looking from the roof of a tall building, or from multiple perspectives as when navigating through the environment. They define *secondary knowledge* as spatial knowledge acquired indirectly as when reading a map that represents the environment. Although the distinction between direct and indirect contact between the viewer and the environment may be important, these studies have only considered very simple cartographic maps defined in traditional ways, i.e., line maps with a fixed orientation and vertical perspective. It is now possible, through computer generated map displays, to easily provide viewers with *secondary knowledge* on maps with multiple orientations and viewer perspectives or even provide an indirect navigation experience through a map animation. MacEachren (1992) provided map readers with multiple versions of the same map so they could encode a cognitive map and not be biased by a single north-at-the-top orientation. His results suggested that having access to multiple orientations of the same map eliminated an orientation bias, but at a cost of decreased accuracy and slower reaction times when using the map.

Orientation Biases

The form of the spatial information available for processing can affect how a cognitive map is encoded (Lloyd, 1982; Lloyd, 1989). One also might expect the intended use of the cognitive representation of space to affect the selection of an encoding process (Taylor and Tversky, 1992A). Consider two examples. In the first example you have a map of a country that has the cities, rivers, and highways presented in the usual way. Your goal is to learn about the cities in the country; their names, locations, and connections are important information. You could encode the map into your memory as an image (Kosslyn, 1980). If you later wished to recall information related to a particular city in the country, you could recall the image of the map and "look" at it like you would the original map. Assuming the original map was created with the usual north-at-the-top orientation, the image of the map would be encoded in your memory in the same way. The cartographic map and the map image in your memory would be functionally equivalent (Kosslyn, 1980; Lloyd, 1982). Your effective use of the cognitive map would, therefore, be biased by this fixed orientation.

Evans and Pezdek (1980) assumed that college student subjects had learned about cities in the United States from maps oriented with north at the top. Triads of cities were presented to the subjects as either correct or mirror presentations. Mirror presentations reflected the displays around the vertical axis. Each display was also rotated a specific number of degrees away from north at the top (Figure 1). For example, rotating clockwise 90° would present the display with west at the top (Figure 1B) and rotating 180° would present it with south at the top (Figure 1C). Subjects decided as quickly as possible whether each presentation was a correct or mirror display. This research design followed the procedures initially developed by Shepard (1978) to study the rotation of mental images of two and three-dimensional objects. The results indicated map images have properties similar to images of other objects. The amount of rotation from north at the top was directly related to the subjects' reaction times. Subjects answered fastest for displays with north at the top and slowest with displays with south at the top. This suggested that experiences with maps in a fixed orientation creates functionally equivalent

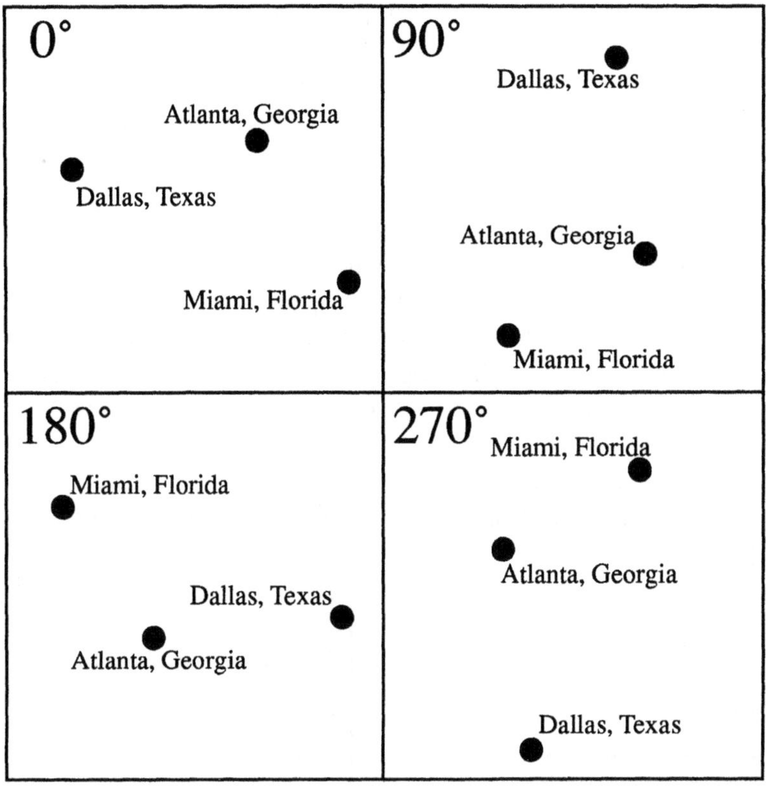

Figure 1: An example of a triad of cities rotated 0°, 90°, 180°, and 270° from north at the top.

map images with the same fixed orientation. One's ability to use such a cognitive map to make decisions is biased by its fixed orientation so that decisions related to other orientations require additional processing. For example, identifying a map presented in a south-at-the-top orientation would require the map to be rotated 180° so that it is congruent with the cognitive representation before it can be identified. This, of course, takes additional processing time (Figure 2A).

For a second example, suppose you have just moved to a new city and need to learn your new environment. You have a set of particular landmarks in the environment that you have to interact with and set about learning where they are by navigating within the environment from one landmark to another. You continue this type of interaction with the environment until you can comfortably go from any origin landmark to any destination landmark.

Evans and Pezdek (1980) assumed that college students learned about the landmarks on their campus by interacting with them. Subjects determined if triads of landmarks were presented as correct or mirror images for displays that were rotated away from north at the top. Unlike the city triads discussed above, no rotation bias was evident in the results, i.e., reaction times for south-at-the-top displays were as fast as for north-at-the-top displays (Figure 2B). This suggested a fundamental difference between the two example situations with map reading experiences producing fixed orientation cognitive maps and navigation experiences producing orientation free cognitive maps. This difference was confirmed with a third experiment that had subjects not familiar with the college campus learn its landmarks by studying a map. They then identified the same correct and mirror displays of landmark triads. Learning from the map produced reaction times with a fixed orientation bias, i.e., north-at-the-top displays were fastest and south-at-the-top displays were slowest (Figure 2A). The authors suggested that the multiple perspectives provided by navigation experiences is an important factor in producing orientation free cognitive maps (Figure 2B).

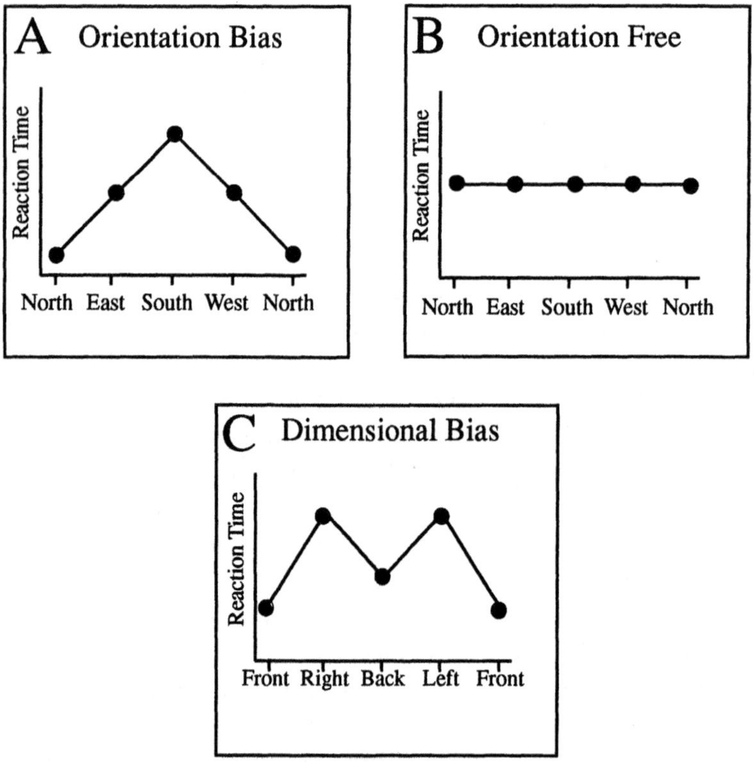

Figure 2: Theoretical representations of reaction times that should result when a cognitive map A. has a north-at-the-top orientation bias; B. has no orientation bias; and C. has a dimensional bias for front-back over left-right. The observer is viewing from outside the space for A and B and inside the space for C.

The above examples used a research paradigm that considers the relationship between the time needed to decide if a stimuli is a correct or mirror version of a memorized object and the degree of rotation of the object from its original learned orientation. The paradigm has been used successfully by psychologists in many different contexts (Shepard and Cooper, 1983). Geographers have also used it to consider if imagery is used to identify outline maps (Steinke and Lloyd, 1983). Another study has shown that the type of information displayed on maps affects reaction times with more complex distributions requiring more processing time at all orientations (Lloyd and Steinke, 1984).

Conerway (1991) presented correct and mirror outlines of states that varied in boundary complexity to subjects who were to identify the state. The orientation bias was found for all states, but the complexity of the state boundaries also affected reaction time. States with complex boundaries were identified faster than states with simple boundaries. The slope of the relationship between reaction time and degrees of rotation from north at the top, however, was not significantly different for the states. This suggested that subjects rotated images of all states to north at the top at the same rate, but that clues provided by the characteristics of the boundaries made identification easier for states with complex boundaries. Rice (1990) used this same paradigm to study the identification of three-dimensional prism maps. He also suggested that subjects used prominent outline features as part of the identification process.

The Equiavailability Principle

Cognitive maps represented as images are stored in a fixed orientation and this can bias some decision-making processes. Another important characteristic of cognitive maps encoded as map images, however, is desirable. These types of cognitive maps can be parallel processed which allows information at all locations on the maps to be equally available (Levine, Jankovic, and Palij, 1982). This is not the case for maps encoded through navigation. Some researchers have argued that navigation experiences encode cognitive maps that are *orienting schemata* (Neisser, 1976; Shepard, 1978). It has been demonstrated that cognitive maps of campus landmarks provide information that is biased in a way consistent with the *orienting schemata* argument (Sholl, 1987). For a viewer at a given location within the space facing in a particular direction, information in front of the viewer is accessed faster than information behind the viewer. Cognitive maps of cities learned from cartographic maps did not have this type of bias, i.e., information ahead and behind a viewer was equally accessible (Sholl, 1987).

Hintzman, O'Dell, and Arndt (1981) reported on 14 different experiments that had subjects "point to" targets that surrounded them in an environment while imagining themselves in a particular spot facing in a particular direction. For a variety of research designs, e.g., visual maps, cognitive maps, tactile maps, they reported subjects were faster answering along the vertical axis (front and back) than the horizontal axis (left and right). Generally, subjects could point to objects in front of them faster than behind them, but pointing in other directions was slower than pointing behind. The configuration of mean reaction times plotted for quadrants around a circle (front, left, back, right, front) looked like the letter "M" (Figure 2C). They argued that mental rotation was used with visual tasks, but not with cognitive maps. It was suggested that "cognitive maps are not strictly holistic, but consist of orientation-specific representations, and-at least in part-of relational propositions specific to objects" (Hintzman, O'Dell, and Arndt, 1981: 149).

Two and Three-dimensional Representations

Traditional cartographic maps present the environment as two-dimensional representations because the viewer has a parallel perspective planimetric view of the plane depicting the map. This does not reflect actual human experiences because the viewer has a vantage point orthogonal to all points on the plane simultaneously (Muehrche, 1986). Objects on the plane appear as two-dimensional because only the tops are rendered as visible. Changing the view to a single perspective at some angle less than 90° above the plane allows the third dimension of objects on the plane to be rendered as visible to the viewer. Studies that have considered the "mirror" versus "correct" decision for three-dimensional map displays report that some subjects use a non-rotational strategy to solve the problem (Holmes, 1984; Goldberg, MacEachren and Kotval, 1992). One study presented pairs of three-dimensional maps that were judged to be either the "same" or "different". The maps could be either two "correct" maps, two "mirror" maps or one of each. Some subjects used the usual rotational strategy while others used a verbal strategy that determined if objects in the space were organized in a clockwise progression or a counterclockwise progression (Holmes, 1984). Oblique views of spaces may present information in a way that encourages or at least makes it possible for viewers to encode a more complex cognitive map. Such cognitive maps might be able to produce map images for rotational solutions or, if needed, generate verbal information about relationships among objects in the space.

Although Rice (1990) argued that his subjects rotated three-dimensional prism maps of Wisconsin to identify them as "correct" or "mirror" he indicated that south-at- the-top maps did not always produce the longest reaction times. He argued that some views of the three-dimensional space had foreground objects blocking critical information from the viewer and, thus, causing longer reaction times.

Mental Models

Bryant, Tversky, and Franklin (1992) made a distinction between having an *internal* (viewer is in the space) and *external* (viewer is outside the space) perspective of three-dimensional spaces they called scenes. Subjects did not actually view the scenes, but read descriptions that put the viewer in one of the two perspectives. The researchers were interesting in the mental models that were formed of the scenes. They argued that the *internal* and *external* spatial frameworks represented different perceptual experiences of observers. Reaction times for an object identification task indicated subjects reading descriptions with an *internal* spatial framework identified objects in front of the viewer faster than object in back of the viewer. Subjects reading descriptions with an *external* perspective had all objects in front of the viewer and responded equally fast to objects in the front and back of a space.

In another study that considered mental models formed from described scenes, it was concluded that subjects formed separate mental models for viewers in different locations in the scenes (Franklin, Tversky and Coon, 1992). The authors also argued that subjects were able to switch from one perspective to another as necessary to answer specific questions.

Taylor and Tversky (1992B) had subjects encode information about naturalistic environments by reading route descriptions, survey descriptions, or studying maps. Route descriptions used terms such as front, back, left, and right. Survey descriptions used terms such as north, south, east, and west. One important research question was whether or not a perspective is encoded into a mental model of the environment. Subjects appeared to be able to answer inference questions, i.e., questions whose answers could not be explicitly derived from the descriptions, equally well regardless of the type of description they had read. It was argued that people form situation models that include perspective and update them as new perspectives are experienced. Readers were thought to use these individual perspective views to construct abstract comprehensive mental models. It was concluded that:

> Subjects' spatial mental models, whether acquired by map, route description, or survey description, contained information about the spatial relations among landmarks in a way that did not favor one perspective over another. What might such a spatial mental model look like? We would like to suggest that it may not look like anything that can be visualized. Rather, it may be like an architect's 3D model of a town; it can be viewed or visualized from many different perspectives, but it cannot be viewed or visualized as a whole. Particular spatial perspectives can be derived from a more abstract spatial mental model that is perspective free (Taylor and Tversky, 1992B: 289).

One might consider the abstract representation, referred to above as a mental model, as being a higher-order perspective-free cognitive map. A higher-order perspective-free cognitive map might be used to create a particular view required to answer a particular question. This derived view could have an orientation-bias or a bias for front over back even if the higher-order cognitive map is free of these biases.

Research Design

Subjects

Subjects were volunteers and students at the University of South Carolina. They were recruited with signs posted in campus buildings and with advertisements in the school newspaper. They were paid $5.00 for participating an experiment which took less than one hour to complete. Five different experiments were run over a 6 month period. This resulted in data for a total of 147 subjects being available for analysis.

Lamdmarks

The same seven landmarks were represented in each experiment. The landmarks were created as uniquely colored three-dimensional shapes. Five landmarks were familiar two-dimensional shapes extruded up from the surface, i.e., red circle, yellow triangle, cyan hexagon, blue parallelogram, and white cross. The remaining two were familiar three-dimensional shapes, i.e., green cone and magenta pyramid. The location of each landmark was the same in all spaces. The first three experiments presented the landmarks in two-dimensional displays and the last two in three-dimensional displays (Figure 3). The landmarks were referred to using color as an identifying label in all experiments.

Experimental Spaces

Although the basic landmarks were the same for all five experiments, the spatial information was presented in a different way for each experiment. The space for the first experiment was a two-dimensional display. It had an outline map of the state of South Carolina with the seven symbols on it (Figure 3A). This first case was to represent a traditional map display. It partially replicated the design used in previous studies that used outline maps (Conerway, 1991; Lloyd and Steinke, 1984; Steinke and Lloyd, 1983). Subjects in the first experiment learned from a South Carolina map with north at the top and the map reader viewed the map from 90° above the plane. It was expected that subjects in the first experiment would use the state outline as well as the distribution of the landmarks to make their decisions. As in previous studies using outline maps, it was expected that an orientation bias for the subjects' cognitive maps would be revealed by the relationship between angle of rotation and reaction times. This relationship was expected to indicate that subjects rotated the trial maps back to north at the top to make their decisions. Evidence from the first experiment represented a baseline result for the configuration of seven landmarks.

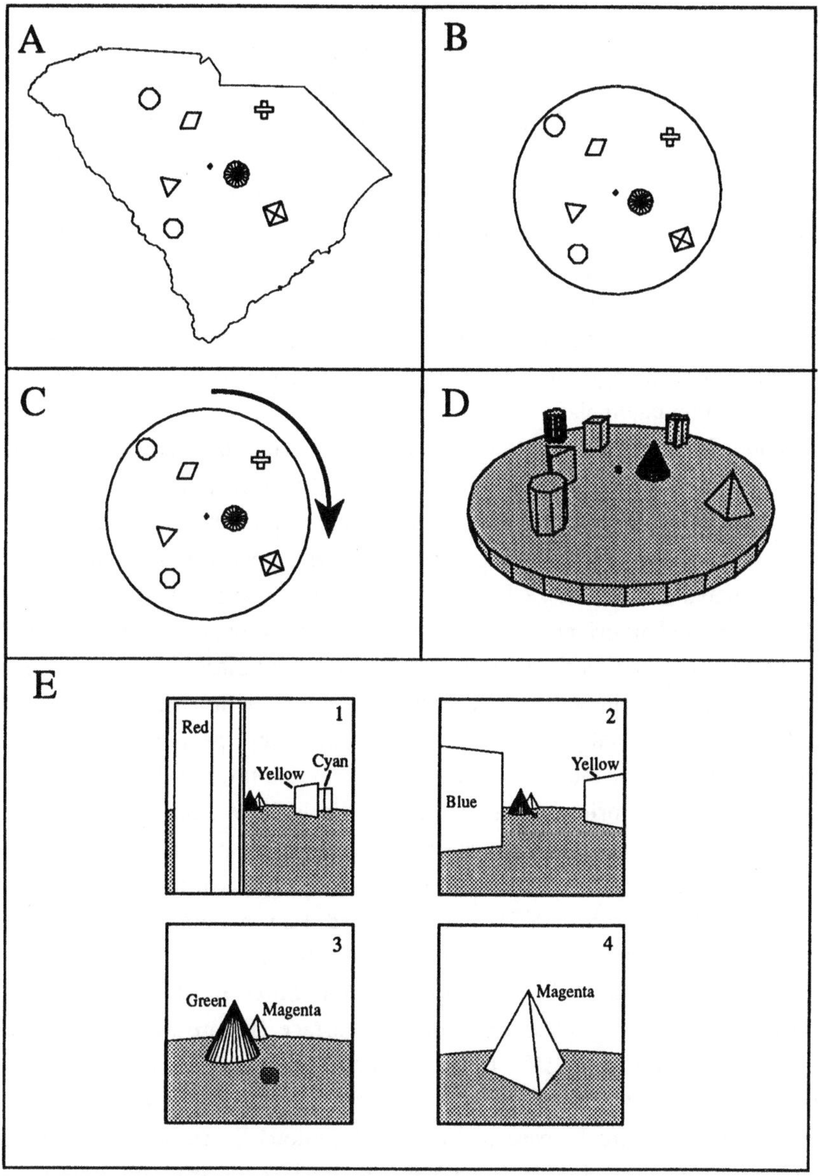

Figure 3: Example of the spaces used for learning in Experiments I through V. The seven landmarks are represented as planimeteric views for A, B, and C and as a single perspective oblique view on D and a multiple perspective simulated navigation on E. The landmarks are on a map with a complex boundary in A, on static circular maps on B and D, and on a spinning circular map on C. The four frames in E represent a navigation from the Red landmark to the Magenta landmark.

The second experiment presented the same seven landmarks in a fashion similar to the first experiment, except the South Carolina outline map was removed and replaced by a circular outline (Figure 3B). Subjects in the second experiment learned from a circular map with north at the top and the map reader viewed the map from 90° above the plane. Subjects in the second experiment could not use prominent features of the outline to help them decide if the current rotated trial was a correct or mirror version because the circle looked the same in all orientations. They would be forced to use only the distributions of the landmarks to make their decisions. It was expected that Experiment II subjects would have a more difficult task than Experiment I subjects. If Experiment II subjects also rotated the space to north at the top before making their decisions, however, the relationship between angle of rotation and reaction time should reveal cognitive maps with the same orientation bias.

Experiment III tested the single versus multiple orientation theory. It used the same seven objects on the circular base that was used in Experiment II. The basic difference was that Experiment II presented the map as a static display that always had north at the top. Experiment III presented the map during the learning phase as a dynamic display that constantly rotated in a clockwise direction as it was being studied by the subjects (Figure 3C). This presented all orientations at the top as the map rotated a full 360°. The single versus multiple orientation theory argues that orientation biases naturally occur during map reading because the convention is to make maps with a single north-at-the-top orientation. Since virtually all maps are made this way, people seldom experience multiple orientations of the same map or any orientation other than north at the top. Exposure to multiple orientations of the same map have been shown to eliminate orientation biases in cognitive maps encoded during map reading just as multiple orientations seem to eliminate orientation biases in cognitive maps encoded through navigation (MacEachren, 1992). We reasoned that a computer animation that continuously rotated a map should produce similar results.

Experiment IV presented the test map as a three-dimensional space with the seven landmarks displayed as objects on a circular surface (Figure 3D). The viewer was positioned due south and 25° above the plane during the learning phase of the experiment. This display is like the one used in Experiment II in that it has the same seven landmarks on a circular base, has a north-at-the-top orientation, and is viewed from a single perspective. The oblique perspective (Figure 3D), however, provided a different looking display than the planimetric perspective (Figure 3B). Presson and Hazelrigg (1984) found that *primary* knowledge provided by a single oblique view of the environment produced the same equiavailability biases found when environments were learned through navigation, i.e., information in front of a subject could be accessed faster than information behind a subject. The oblique view used in Experiment IV provided *secondary* knowledge and was used to test for an orientation bias.

Experiment V provided a completely different learning experience. This experiment had subjects learn the space by watching multiple computer simulated navigations through the space (Figure 3E). Simulations were created that allowed subjects to learn the space by starting at any origin landmark and traveling to any destination landmark. The viewer traveled through the space parallel to the plane at an elevation equal to one half the height of the objects. Figure 3E illustrates four frames of the simulation traveling from the red landmark to the magenta landmark. These types of navigation experiences presented views of the space to subjects that had multiple orientations. This was an egocentric learning experience because viewers appeared to be inside the space moving in various directions among the landmarks. Navigation experiences in previous studies have been in real environments and have provided primary knowledge (Evans and Pezdek, 1980; Lloyd, 1989; Presson and Hazelrigg, 1984; Sholl, 1987; Thorndyke and Hayes-Roth, 1982). Experiment V provided a navigation experience that enabled subjects to encode their cognitive maps of the space as *secondary* knowledge. Experiment V addressed a number of important questions. What type of cognitive maps will be encoded from these *secondary* egocentric learning experiences (Figure 3E)? How will these cognitive maps compare with cognitive maps encoded from maps with multiple orientations, but a planimetric perspective (Figure 3C) or from maps with a single orientation, but with an oblique perspective (Figure 3D)?

For all experiments the seven landmarks were used to create 35 correct displays and 35 mirror displays. This represents all the possible correct and mirror triads that can be produced from seven objects. For Experiments I, II, and III the mirror triads were created by rotating the plane depicting the map 180° around the vertical axis, i.e., north-south axis. This is consistent with previous studies that have used standard outline maps (Lloyd and Steinke, 1984, Steinke and Lloyd, 1983). For Experiments IV and V the mirror images were for three-dimensional views of the space. They were created by rotating the plane for the correct view 180° around the vertical axis, i.e., ahead-behind axis.

Procedures

The same basic procedure was used for all subjects in the five experiments. In the first phase of each experiment subjects learned the locations of the seven landmarks by viewing the landmarks represented in a space (Figure 3). The subjects were instructed to study the space until they believed they knew the location of each symbol. For subjects in Experiments I through IV this meant looking at the space displayed on a monitor for as long as they wished. For Experiment V subjects this meant running as many navigation simulations as they thought necessary to learn the space. In the next phase subjects were given a test that presented the space with a grey oval shape marking six of the locations and a white oval shape marking one of the locations. Subjects were asked to name the color of the landmark that occupied the white oval's location in the original space. Subjects who could not successfully match the colors with the locations on the first try repeated the study and test phase until they were successful. This verified that each subject had learned the space before attempting the experimental trials.

In all these trials subjects were presented a rotated view of three isolated symbols and asked to indicate if the trial was a correct representation of the locations of the symbols in the space they had learned or a mirror representation (Figure 4). Visual examples of correct and mirror displays were provided and the meanings of the terms were explained to the subjects before practice trials were attempted. For the two-dimensional displays used in the first three experiments, the rotation away from north at the top was determined by generating a random number between 0 and 360 for each trial to determine its degrees of rotation. This provided a unique set of trials for each subject. This procedure was not possible for the three-dimensional displays used in Experiments IV and V for two reasons. First, unlike the two-dimensional displays used in the first three experiments, the three-dimensional displays could not be created during the experiment as needed. They had to be rendered and stored before the experiments with separate software. It was not practical to create and store every possible display that could be generated by a random number process. Second, previous studies that used three-dimensional displays had indicated specific map effects were caused by views that had foreground objects blocking background objects (Rice, 1990). To limit such effects we created 35 correct and 35 mirror displays for Experiments IV and V that presented the triads so that all three landmarks could easily be seen and recognized. All subjects in these two experiments responded to the same 70 displays.

All subjects were given at least 10 practice trials and exactly 70 experimental trials. Additional practice trials were repeated as needed for individual subjects until they could perform the task accurately. The trials were presented in a different random order for each subject. In all cases the experimental trials were represented in a manner consistent with the subjects' learning experiences (Figures 1 and 2). The motion of the map in Experiment III and the viewer in Experiment V was not part of the experimental trials. Subjects in these two cases considered views of triads that were like "snapshots" of their dynamic learning experiences.

For each experiment the reaction times for the 70 experimental trials were aggregated for each subject by considering what was "up" for a given trial. Consider the map of South Carolina (Figure 3A). It shows the space with north at the top. Random numbers were used to rotate the trial maps between 0° and 360° before being presented to the

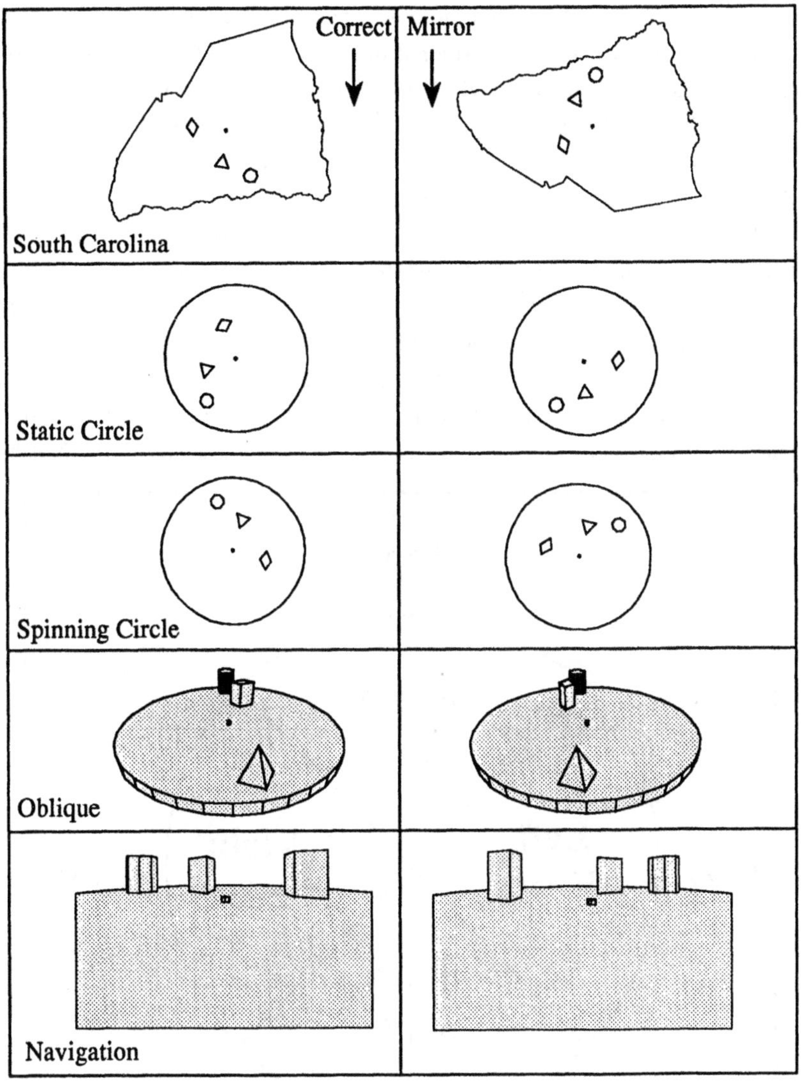

Figure 4: Examples of correct and mirror trial shown at various rotations for the five experiments.

subjects. We considered the trial to be a TOP trial if that random number positioned the map so that north stayed within 45° clockwise or counterclockwise of its original vertical position (Figure 5). If the rotation positioned due East within 45° clockwise or counterclockwise of vertical, we considered this a RIGHT trial because East was in the "up" position. If the rotation positioned due South within 45° clockwise or counterclockwise of vertical, we considered this a BOTTOM trial because South was in the "up" position. Finally, if the rotation positioned due West within 45° clockwise or counterclockwise of vertical, we considered this a LEFT trial because West was in the "up" position. Reaction times for TOP, RIGHT, BOTTOM, and LEFT quadrants were averaged for each subject as geometric means using only those trials that were responded to correctly for both "correct" and "mirror" responses. The percentage of the trials in the TOP, RIGHT, BOTTOM, and LEFT quadrants responded to correctly was also computed for each subject for both "correct" and "mirror" responses. Similar aggregation processes were used for the last two experiments except that the TOP of the vertical axis corresponded to the line-of-sight notion of what was in front of the viewer (Figure 5). For Experiment IV learning was done with the viewer external to the space. On the circular map in front of the viewer South was in the foreground and North was in the background at the top of the display (Figure 3D). Subjects in Experiment V had learned by navigating among the landmarks and had no way of being aware of the North-South

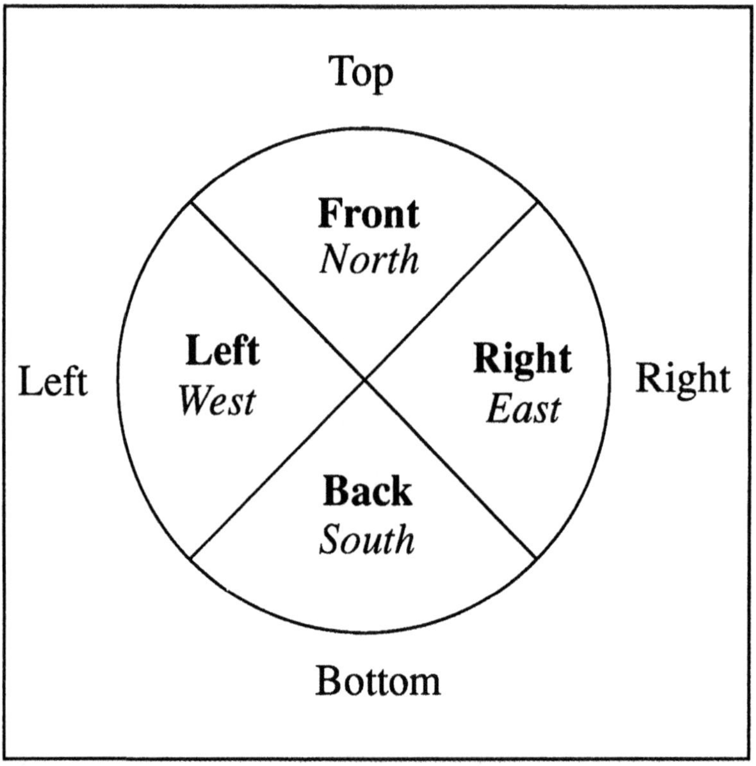

Figure 5: The quadrants used to aggregate the trials. For the two-dimensional spaces the Top, Right, Bottom, and Left quadrants correspond to North, East, South, and West at the top of the display. At the top of three-dimensional displays Top, Right, Bottom, and Left quadrants correspond to Front (ahead), Right, Back (behind), and Left of a viewer looking north from the center of the space.

and East-West reference axes that were explicit for subjects in other experiments (Figure 3E). The data for Experiment V was aggregated using notions of TOP and BOTTOM consistent with Experiment IV to make the results comparable with the other experiments.

Results

Both reaction time and percentage correct were averaged over the subjects for each experiment (Figures 6 and 7). Two initial analyses of variance were computed using both reaction time and percentage correct as dependent variables and EXPERIMENT (I,II,III,IV and V) and QUADRANT (TOP,RIGHT,BOTTOM, and LEFT) as main effects. This provided a general look at differences among the experiments. For the analysis of reaction times EXPERIMENT was significant (F(48.52), P>F=0.0001), but QUADRANT was only marginally significant (F(2.86), P>F=0.06). The differences in response times among the experiments are indicated by the separation of lines in Figure 6. The shapes of the lines indicate differences in the response times by quadrants (Figure 6). If the subjects in a given experiment were rotating images of the displays to compare them to an orientation-biased cognitive map, a signature pattern should be apparent. The BOTTOM reaction time should be significantly slower than the TOP reaction time with the LEFT and RIGHT times being intermediate (Figure 2A). If the learning process for a particular experiment produced an orientation-free cognitive map one would expect the mean reaction times for the BOTTOM and TOP quadrants not to be significantly different from each other or from the LEFT or RIGHT quadrants (Figure 2B).

For the analysis of percentage correct neither EXPERIMENT (F(1.92), P>F=0.14) nor QUADRANT (F(1.22), P>F=0.32) were significant. A visual expression of the percentage correct means indicates that accuracy levels for the experiments were about the same and, within experiments, quadrants were about the same (Figure 7). Since the five experiments produced accuracy results that were not significantly different, the comparisons of individual experiments that follows will focus on the reaction time results.

Another consideration is whether a pattern revealed for the average subject is reflected by most individual subjects. If individual subjects consistently responded in the same way to an identification task, this would strengthen any arguments based on aggregate data. The means for individual subjects were used to compute frequency counts by quadrant for four different variables. For each of the four quadrants (Figure 5), the percentage of subjects that had their highest mean reaction time, lowest mean reaction time, highest percentage correct, and lowest percentage correct were computed and displayed as pie charts (Figure 8). Specific pie charts can be compared with representations of aggregate data (Figures 6 and 7) to assess how consistently individual

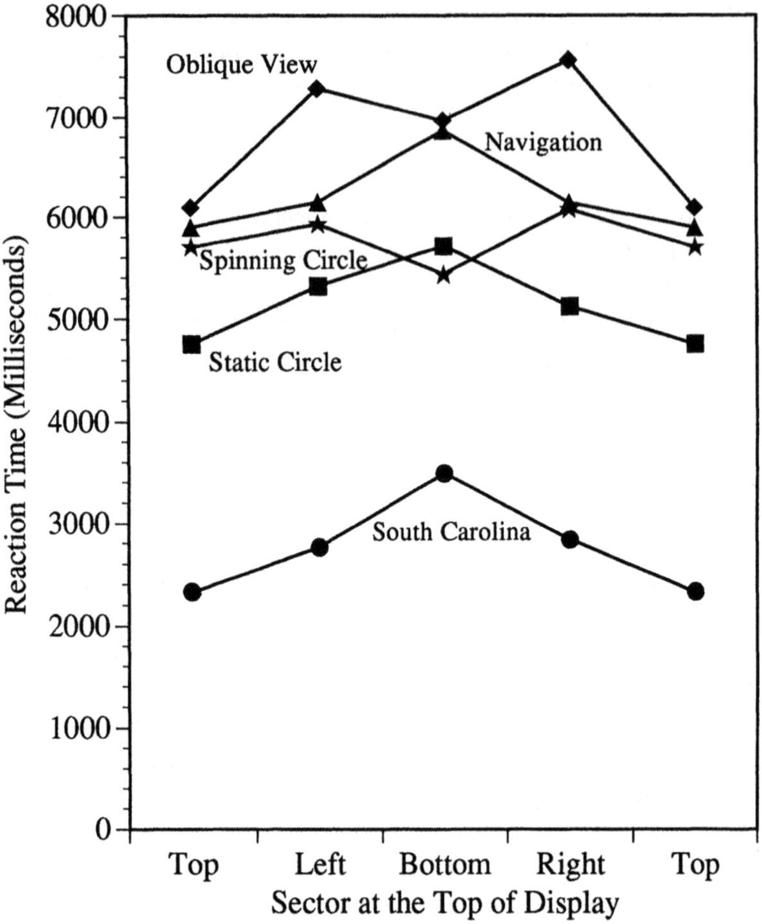

Figure 6: Mean reaction times for each experiment plotted for each quadrant and repeated for the top quadrant for closure.

subjects reflect aggregate patterns.

The Map versus the Static Circle

Comparing reaction times for individual experiments (Figure 5) one can see that Experiment I subjects responded much faster than any other group. A difference in means test comparing the two fastest mean reaction times, Experiment I (2855 msec) and Experiments II (5230 msec), indicated a significant difference (t(-9.86), P>t=0.0001). The faster response time for the subjects learning the landmarks on a map of South Carolina is undoubtedly explained by the additional information in the state outline that is

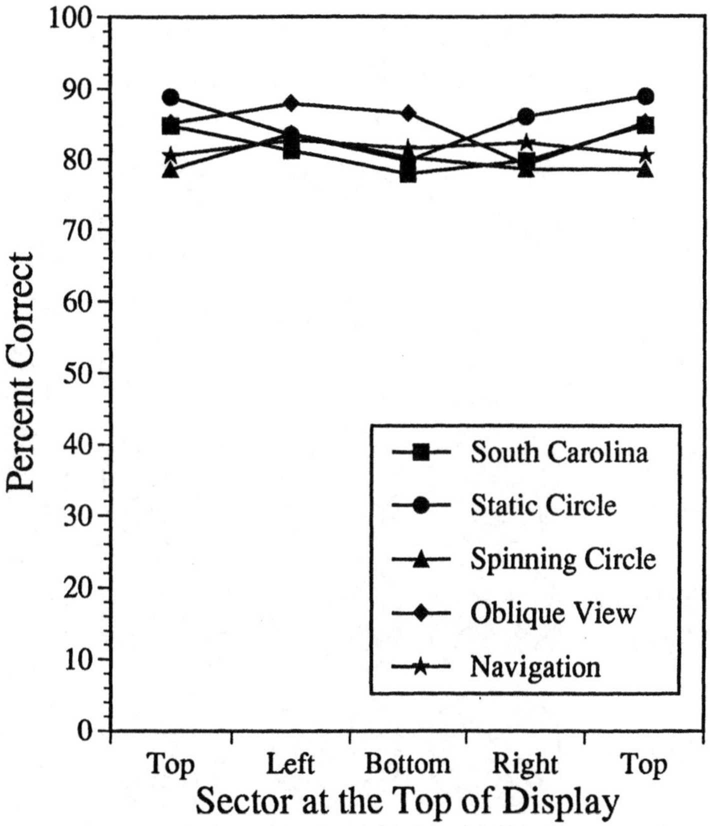

Figure 7: The percentage of the trial correct plotted for each quadrant and repeated for the top quadrant for closure.

not in the circle outline. In fact, Experiment I subjects could have answered the correct versus mirror questions without ever considering the landmarks. For Experiment II subjects, the circular outline was no help. They had to rely totally on the distribution of landmarks to determine if a trial was "mirror" or "correct". The general shape of the reaction time plots for Experiment I and II is similar and appears to indicate subjects were rotating images of the displays back to north at the top to make their decisions (Figure 6 and Figure 2A). The fastest reaction times for both experiments were for the quadrant having north on the map at the top of the display and the slowest were for south on the map at the top of the display. A statistical comparison of the difference in means for the TOP and BOTTOM categories gives additional support for the rotation argument. A significant difference was indicated for Experiment I (t(-2.41), P>t=0.04) and a marginally significant difference for Experiment II (t(-1.98), P>t=0.08).

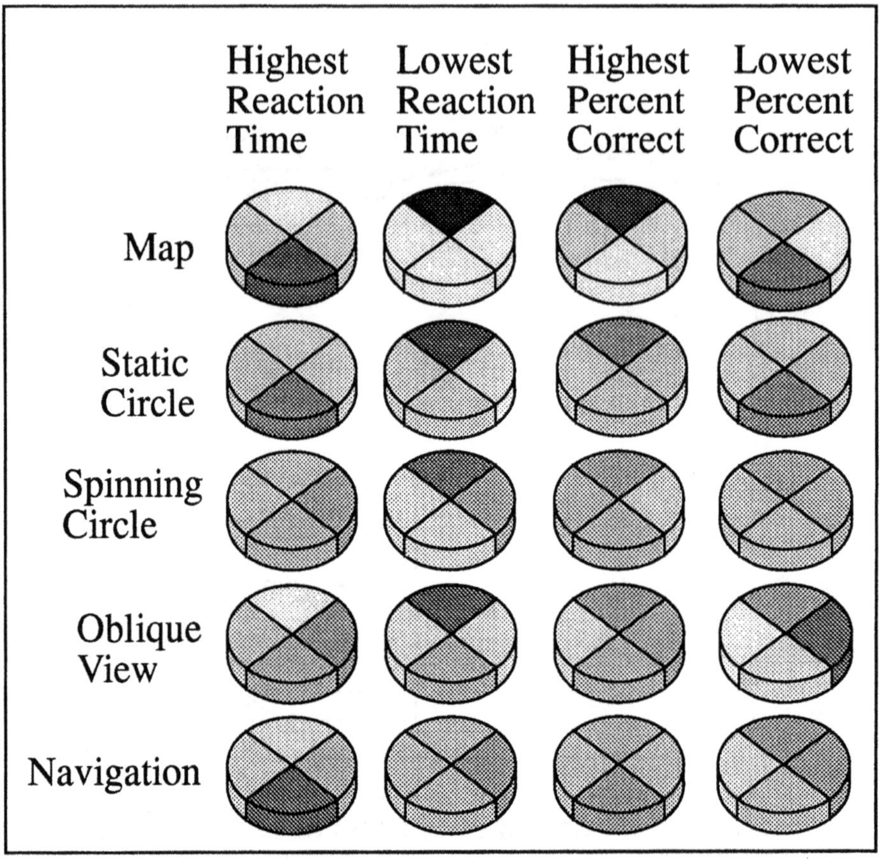

Figure 8: Individual subject responses aggregated by quadrant. Each of the five rows represent one of the experiment. Each column collects either high or low value for reaction time or percentage correct. The grey values in the quadrants correspond to the proportion of the subjects having the column value in that quadrant. For example, most subjects in the first experiment (top row) had their highest reaction times with the BOTTOM quadrant, lowest reaction times with the TOP quadrant, highest percentage correct with the TOP quadrant, and lowest percentage correct with the BOTTOM quadrant.

Data for individual subjects in both experiments also support the rotational argument (Figure 8). Individual subjects tended to have their highest mean reaction time associated with the BOTTOM quadrant and their lowest mean reaction time associated with the TOP quadrant. This same pattern is evident in the percentage correct data. Individual subjects tended to have their highest accuracy with north-at- the-top maps and lowest accuracy with south-at-the-top maps.

The Static Circle Versus the Spinning Circle

Both Experiments II and III had the landmarks presented within a circle. Experiment II subjects learned the space in a fixed north-at-the-top position while Experiment III subjects learned as the space rotated and, thus, were provided multiple orientations. The mean reaction time for subjects who learned with the static circle (5320 msec) compared to those who learned with the spinning circle (5743 msec) was significantly different (t(-5.32), P>t=0.0002) (Figure 6). The two groups of subjects appeared to use different encoding strategies for their cognitive maps. The subjects learning from the spinning circle did not have one dominant orientation that was used consistently by subjects to form an image (Figure 6). The BOTTOM quadrant mean reaction time was actually faster than the TOP quadrant's time. This is the opposite of what would be expected if images were being rotated. The difference, however, was not significant for spinning circle subjects (t(0.19),P>t=0.85), but was significant for static circle subjects (t(-2.32),P>t= 0.05). This suggests the spinning circle subjects were probably not rotating images of the displays back to a standard position to compare them with a cognitive map with a fixed orientation. It is possible that individual subjects may have encoded cognitive maps as images, but used different orientations for their maps, i,e, some had north at the top, but others had east, west or south at the top. Averaging such subjects would produce flat reaction time plots. The individual subject data suggests this might be a possible explanation (Figure 8). The distinct pattern of higher values associated the TOP and BOTTOM quadrants evident for static circle subjects is not present for spinning circle subjects. Another and possibly better explanation came from the subjects themselves.

In informal interviews conducted after the experiments most Experiment I and II subjects had described using imagery to make their decisions. Experiment III subjects described a different process that amounts to verbal coding. They indicated that they used the center of the circle and the edge of the circle as reference locations. The "green symbol was closest to the center" and the "red symbol was closest to the edge." They also considered the order around the circle. The "magenta symbol is clockwise from the white symbol." This type of encoding uses a polar coordinate system of distances and angle to fix locations of landmarks in the space. Some subjects described grouping the symbols into linear categories and remembering the order along the line. Red, blue, green, and magenta make one linear group and cyan, yellow, blue, and white make another group (Figure 3). Correct versus mirror decisions then could be made by considering the correctness of the order around the space or the correctness of the order in one of the linear categories. These verbal procedures produce equal reaction times and percentages correct within the quadrants because there is no orientation bias in the information being encoded (Figure 2B). The cognitive map in this case is not an image in memory, but an encoding of the relationships among landmarks and between the center or edge of the space and the landmarks.

The Oblique View Versus Planimetric Views

The Experiment IV subjects learned the same landmarks as subjects in Experiment II, and III and all three groups were presented the landmarks in a circular space. The subjects who learned with the static circular space (Figure 3B) and the spinning circular space (Figure 3C) both had views from 90° above the plane, while the oblique view (Figure 3D) was 25° above the plane. The oblique view rendered the landmarks as three-dimensional objects rather than two-dimensional objects. The oblique view of the space is like the static circle view in that there is a fixed perspective albeit not at the same angle above the plane.

The oblique view experienced by Experiment IV subjects allows for some interesting comparison. Some researchers have argued that the key to predicting when a cognitive map will have an orientation bias is knowing if the learning experience provided a single or multiple orientations (Evans and Pezdek, 1980; MacEachren, 1992). A comparison of the results for subjects who learned from the static circle display versus the spinning circle display also supports this argument. Given that the subjects with an oblique view encoded their cognitive maps from a single perspective with north at the top of the display, one might expect their cognitive maps to have a fixed-orientation bias (Figure 2A). The plot of the mean reaction times does not suggest this is true (Figure 6). Like the data for those who learned from the spinning circle, the BOTTOM quadrant for the oblique view subjects is the faster than the left or right quadrants. The Oblique View subjects had significantly longer reaction times than either the static circle subjects (t(5.32), P>t=0.0002) or the spinning circle subjects (t(3.76),P>t=0.003). The shape of the plot for the oblique view subjects' reaction times resembles an "M" with LEFT and RIGHT means being higher than TOP and BOTTOM means (Figure 6 and Figure 2C). The highest reaction time (RIGHT quadrant, 7567 msec) was significantly different from the lowest time (TOP quadrant, 6096 msec), but no other pairs were significantly different. The oblique plot is not as flat at the spinning circle plot, but both have mean reaction times for the BOTTOM that are faster than would be expected if mental rotations were being used by the subjects to determine if a trial was a "correct" or "mirror" category. Although the oblique view subjects took longer to respond than any other group (Figure 6), their mean accuracy was the highest (84.6%) and approximately equal to the subjects considering maps of South Carolina (84.5%) (Figure 7).

The individual subject data suggest a tendency toward higher reaction times and lower percentage correct values for the RIGHT quadrant (Figure 8). The distribution of lowest reaction times exhibits the "M" shape distribution with the TOP quadrant having the highest percentage followed by the BOTTOM quadrant.

The important finding here is that a single perspective learning experience can produce cognitive maps without the usual orientation bias. This result relates to Presson and Hazelrigg (1984; 1989) finding that a single oblique view of a real environment produced cognitive maps that violated the equiavailability principle allowing subjects to respond faster to objects in front of them than to objects behind them. Taken together the results suggest that oblique views of environments and maps are not encoded as a single fixed-orientation image. All of our experiments provide *secondary* learning experience, and yet some produced cognitive maps with an apparent orientation bias, i.e., South Carolina map and static circle experiments, and others, i.e., spinning circle and oblique view, did not. An argument that only *primary* learning experiences produce orientation-free cognitive maps is not supported (Sholl, 1987). It would appear to be true that some *secondary* experiences produce cognitive maps without fixed-orientation biases (MacEachren, 1992). Our findings suggest that observing a single perspective oblique view of a map produces a cognitive map without the usual fixed-orientation bias. The "M" shaped distribution does suggest a possible bias of vertical (Top-Bottom) over horizontal (Left-Right).

Navigation Versus the Oblique View

The mean reaction time for subjects who learned by viewing simulated navigations of the space (6266 msec) was faster than the mean time for subjects the learned an oblique view of the space (6980 msec). The difference, however, was not significant (t(-1.64),P>t= 0.14). Subjects in both groups considered three-dimensional landmarks and produced the two highest mean reaction times for the five experiments. The plot of the mean reaction times for navigation subjects by quadrant is an allusion of the shape expected if images were being rotated to north at the top (Figure 2A). The navigation group, however, had no information regarding the space's cardinal directions and it seems unlikely that they would be able to encode a cognitive map with a north-at-the-top orientation bias. It is possible that the natural major axis of the distribution of landmarks could provide the basis for forming an image with an orientation bias. A number of subjects considered the red landmark to be important because it was closest to the edge of the space. Considering the red landmark as being in the top of the space would produce an image with north as the TOP quadrant. Although some subjects may have imposed some type of structural axes to the space, a simpler explanation is that there is no actual north-at-the-top bias in the data. For the navigation subjects a comparison of the BOTTOM reaction time mean (6870 msec) with the TOP reaction time mean (5900 msec) indicates they are not even marginally significantly different (t(1.11), P>t=0.30). In addition, the individual subject data does not suggest distributions within the quadrants for either reaction times or percentage correct that is consistent with a rotation argument (Figure 8). Although there appears to be a pattern to the means they are not statistically different from a flat distribution. This suggests that the cognitive map formed by the navigation subjects did not have a statistically significant orientation bias.

Conclusions

A number of conclusions can be made from these results. First, subjects that encoded spatial information from three-dimensional displays took longer to respond to the identification task than subjects that encoded spatial information from two-dimensional displays. It should be noted that they also took much longer to encoded the displays. We did not record the learning time, but informal observations suggested that subjects in Experiments IV and V generally took much longer to learn the locations of the landmarks in the spaces. This could partially be explained by the additional information available to encode in the third dimension. In addition encoding relationships among the landmarks as verbal propositions probably takes longer than encoding images. When trying to identify the map for a particular trial the complexity and form of the informtion in the cognitive map resulted in longer reaction times (Figure 6).

Second, all of the experiments involved *secondary* learning experiences. Earlier studies that concluded *primary* learning resulted in orientation-free cognitive maps and *secondary* learning experiences resulted in orientation-biased cognitive maps only considered maps with a perspective 90° above the plane. Our results support MacEachren's (1992) study in showing that some unusual *secondary* learning experiences can eliminate the usual orientation biases. Using motion to provide multiple orientations of a map viewed form 90° above the plane, a one perspective oblique view of the space 25° above the plane, and simulated navigations on the plane all produced reaction times that deviated from those expected if images were being rotated back to a fixed perspective.

Third, learning experiences that present multiple orientations appear to eliminate a fixed orientation bias and supported previous results (Evans and Pezdek, 1980; MacEcheren, 1992). Although the evidence is not clearly significant, learning experiences that provided a single oblique view of the space seemed to produce results biased in favor of the vertical over the horizontal axis. Significant differences were only found for the RIGHT quadrant compared to the TOP quadrant. The results may have been an artifact of the particular spatial distribution used in this study. This would be true if triads were not equally difficult in all quadrants. The "M" shaped reaction time plots, however, were similar to those reported by Hintzman, O'Dell, and Arndt (1981) for a different task.

Future studies might consider whether this apparent dimensional bias is generally true for single perspective oblique views and is present for other types of tasks.

Finally, it is interesting that accuracy was about the same for all the experiments. MacEachren (1992) compared subjects who learned from a map with a single orientation with subjects who learned from a map presented in 10 versions with 10 different orientations. He reported that learning from a map with multiple orientations eliminated an orientation bias, but that subjects took longer to respond and made more errors. Our data show that learning from a spinning map with multiple orientations (Experiment III) or learning from multiple route simulations (Experiment V) also seems to eliminate the fixed-orientation bias. Our data show longer response times but no loss of accuracy. The difference between our results and those of MacEachren (1992) is in the time spent learning the space. Our subjects took as much time as they needed to encode the spatial information. Our subjects learning multiple orientations (Experiments III and V) needed more time to encode the information, but could achieve comparable levels of accuracy with those considering similar maps who took less time to encode the spatial information (Experiment II). MacEachren's study had all subjects learn the maps for a fixed amount of time. Increased accuracy can be achieved if more time is devoted to learning multiple orientation maps. The increased reaction times reported by both studies for subjects learning from maps with multiple orientations is probably related to how the information is encoded. Images can be processed in parallel once rotated back to north at the top. This two-stage process is still faster than decoding verbal proposition of how landmarks are related to each other and the center or the edge of the space.

Future studies should consider Taylor and Tversky's (1992A, 1992B) argument that mental models are orientation free no matter how they were encoded. There results found this to be true for inference questions that could not be directly known from the text read to encode the mental model. They argued that subjects might generated biased views of the space to answer question directly answerable from that view. All of our subjects may have actually encoded a higher-order mental model (cognitive map) that was unbiased. The apparent biases present for some subjects in our experiments may be related to the types of questions we were asking. Our questions were not inference questions since they were presented in a form directly related to the learning experience in all cases. Higher order cognitive maps may exist and may be unbiased by the learning experience.

References

Bryant, D., Tversky, B., and Franklin, N. (1992). Internal and external spatial frameworks for representing described scenes. *Journal of Memory and Language* 31, 74-98.

Carpenter, P. and Just, M. (1986). Spatial ability: An information processing approach to psychometrics. In *Advances in the Psychology of Human Intelligence* (R. Sternberg, ed.), pp. 221-252. Hillsdale, NJ: Erlbaum.

Conerway, V. (1991). *The Effects of Complexity on the Mental Rotation of Map Images.* Unpublished MA Thesis, Department of Geography, University of South Carolina.

Evans, G. and Pezdek, K. (1980). Cognitive mapping: knowledge of realworld distance and location information. *Journal of Experimental Psychology: Human Learning and Memory* 6, 13-24.

Franklin, N., Tversky, B., and Coon, V. (1992). Switching points of view in spatial mental models. *Memory and Cognition* 20, 507-518.

Goldberg, J., MacEachren, A., and Kotval, X. (1992). Mental image transformation in terrain map comparisons. Unpublished manuscript.

Hintzman, D., O'Dell, C., and Arndt, D. (1981). Orientation in cognitive maps. *Cognitive Psychology* 13, 149-206.

Holmes, J. (1984). *Cognitive processes used to recognize perspective three- dimensional map surfaces.* M.A. Thesis, Department of Geography, University of South Carolina.

Kahneman, D, Treisman, A. and Gibbs, B. (1992). The Reviewing of object files: object-specific integration of information. *Cognitive Psychology* 24, 175-219.

Kosslyn, S. (1980) *Image and Mind*, Cambridge: Harvard University Press.

Kosslyn, S. and Koenig, O. (1992). *Wet Mind*, New York: The Free Press.

Levine, M. Jankovic, I., and Palij, M. (1982). Principles of spatial problems solving. *Journal of Experimental Psychology: General* 111, 157-175.

Lloyd, R. (1982). A look at images. *Annals of the Association of American Geographers* 72, 532-548.

Lloyd, R. (1989). Cognitive mapping: encoding and decoding information. *Annals of the Association of American Geographers* 79, 101-124.

Lloyd, R. (1993). Cognitive processes and cartographic maps. In *Behavior and Environment: Psychological and Geographical Approaches* (T. Garling and R. Golledge, eds.), pp. 141-169. Amsterdam: Elsevier Science Publishers.

Lloyd, R. and Hooper, H. (1991). Urban cognitive maps: computation and structure. *The Professional Geographer* 43, 15-27.

Lloyd, R. and Steinke, T. (1984). Recognition of disoriented maps: the cognitive process. *The Cartographic Journal* 21, 55-59.

Lowe, D. 1987. The viewpoint consistency constraint. *International Journal of Computer Vision* 1, 57-72.

MacEachren, A. (1992). Learning spatial information from maps: can orientation- specificity be overcome? *Professional Geographer*, 44, 431-443.

Muehrcke, P. (1986). *Map Use: Reading, Analysis, and Interpretation*, Madison: JP Publications.

Neisser, U. (1976). *Cognition and Reality: Principles and Implications of Cognitive Psychology*, San Francisco: Freeman.

Presson, C. and Hazelrigg, M. (1984). Building spatial representations through primary and secondary learning. *Journal of Experimental Psychology: Learning, Memory, and Cognition* 10, 716-722.

Presson, C., DeLange, N, and Hazelrigg, M. (1989). Orientation-specificity in spatial memory: what makes a path different from a map of a path? *Journal of Experimental Psychology: Learning, Memory, and Cognition* 15, 887-897.

Rice, K. (1990). Distorted prism maps: a recognition experiment (abstract) *Cartographic Perspectives* 4, 32.

Sagi, D. and Julesz, B. 1985. "Where" and "what" in vision. *Science*, 228, 1217-1219.

Shepard, R. (1978). The mental image. *American Psychologist* 33, 125-137.

Shepard, R. and Cooper, L. (1983). *Mental Images and Their Transformations*, Cambridge: M.I.T. Press.

Shepard, R. and Hurwitz, S. 1984. Upward direction, mental rotation, and discrimination of left and right turns in maps. *Cognition*, 18, 161-193.

Sholl, M. (1987). Cognitive maps as orienting schemata. *Journal of Experimental Psychology: Learning, Memory, and Cognition* 13, 615-628.

Steinke, T. and Lloyd, R. (1983). Images of maps: a rotation experiment. *The Professional Geographer* 35, 455-461.

Taylor, H. and Tversky, B. (1992A). Descriptions and depictions of environments. *Memory and Cognition* 20, 483-496.

Taylor, H. and Tversky, B. (1992B). Spatial mental models derived from survey and route descriptions. *Journal of Memory and Language* 31, 261-292.

Throndyke, P. and Hayes-Roth, B. (1982). Differences in spatial knowledge acquired from maps and navigation. *Cognitive Psychology* 14, 560-581.

Robert Lloyd and Rex Cammack
Department of Geography
University of South Carolina
Columbia SC 29208, USA

COGNITIVE MAPPING AND WAYFINDING BY ADULTS WITHOUT VISION

Reginald G. Golledge, Roberta L. Klatzky and Jack M. Loomis

Abstract: In this chapter we focus on processes that underlie successful wayfinding by those traveling without the help of vision. Particular emphasis is placed on how travelers develop and use cognitive maps. We discuss both the cognitive mapping process and wayfinding, emphasizing the skills that have to be developed with and without the aid of assistive technology. Both experimental and real world examples are used to illustrate the skills, abilities, and processes needed for successful travel without vision. Throughout, our emphasis is on wayfinding by adult populations.

Background

This section describes the problems encountered in wayfinding by the blind and reviews general theories regarding their underlying spatial competence.

Purpose

There appears to be abundant evidence that people, whether sighted or without vision, are capable of solving practical spatial problems that are formally more complex than their geometric, trigonometric, or mathematical knowledge would allow. As illustrated in a different chapter of this book (Unger et al., this volume), even young children can solve complex spatial problems that require knowledge of advanced mathematics or geometry to operationalize and explain. In the course of solving wayfinding problems, significant subproblems of orientation, location, spatial updating, heading, direction, developing and using frames of reference, and making spatial inferences arise – as for example in shortcutting procedures. Despite the complexities of these tasks, people are successful at finding their way.

Since vision can be used both explicitly and implicitly to extract information about locations and layout of near and far space (Warren, 1978; Warren and Strelow, 1985), the wayfarer denied vision has to operate at a decided disadvantage. A visually impaired traveler wishing to navigate independently in a local neighborhood or community faces

J. Portugali (ed.), The Construction of Cognitive Maps, 215–246.

considerable problems. Among these are remembering and recalling information about environments (i.e., using cognitive maps), sequential learning of path segments and turn angles, dead reckoning or path integration, and spatial updating. Pretravel route planning, path selection, destination choice, travel mode selection, landmark recognition, choice point identification, and obstacle or barrier avoidance are also among the wayfinding skills needed by such a traveler.

There are various ways to alleviate some burdens of travel. Assisted travel negates the need to know how to perform navigational procedures such as dead-reckoning or path integration, landmark sensing, obstacle or barrier recognition and avoidance, route memorization, and mode choice (Foulke, 1971, 1982; Jacobson, 1993; Strelow, 1985; Welsh and Blasch, 1980). If traveling alone, the blind wayfinder must learn about new routes and remember previous ones. For most blind people spatial information about geographic space is usually conveyed through spoken explanations or descriptions. Verbal explanations are, where possible, enhanced by direct acquaintance and experience. Familiarization may be of an environment, as mediated by auditory or tactual experience when traveling through it. As Bentzen (1980) has indicated, these avenues for acquiring information are often inadequate for developing environmental knowledge. She has also pointed out that travelers may have difficulty in recalling instructions because of complexity or unfamiliarity with concepts or technical terms used. Further, it is difficult to verbally code irregular spatial relationships such as unusually shaped buildings, gradually curving walkways, irregularly spaced turns, and inconsistent entrance placement.

Route-following behavior and general environmental knowledge often improve when enroute aids or non-verbal travel planning tools are provided. For example, in 1970 Leonard and Newman showed that travel over a new route using portable route maps produced fewer orientation errors than did travel over the same route using only memory or verbal instructions. More recently, Bell (1994) showed that sighted subjects made fewer errors in learning from maps or traveling routes when survey or segmented strip maps were used in the pre-planning or enroute stage of wayfinding. Bell argued that sequenced strip maps aid route learning and travel by promoting segmentation or chunking of routes.

In this chapter we will focus on processes that underlie successful independent travel by those without vision. Particular attention will be placed on the process of cognitive mapping and the use of cognitive maps in the context of navigation and route following. We will discuss characteristics of wayfinding without vision, emphasizing the skills and abilities required to successfully complete wayfinding tasks and illustrating via experimental and real world examples the status of current thinking about how blind and vision impaired adults handle wayfinding tasks.

Theories of Difference: Sighted versus Non-Sighted

Attempts to determine the spatial competence of blind and vision impaired individuals and to assist their wayfinding capabilities in both natural environments and experimental laboratory settings have continued over much of this century. Various theories have existed with respect to the relationship between blind and sighted individuals, and there have been conflicting evidences concerning the spatial abilities of both congenitally and adventitiously blinded individuals. A pessimistic view of the spatial competence of blind individuals has been expressed by von Senden (1932, 1960), who argued that people who had not experienced vision since birth were deficient in spatial abilities. More optimistic views have been offered by Worchel (1951), Dodds, Howarth, and Carter (1982), Klatzky, Loomis, Golledge, Cicinelli, Doherty, and Pellegrino (1990), Loomis, Klatzky, Golledge, Cicinelli, Pellegrino, and Fry (1993), Rieser, Lockman, and Pick (1980), and Rieser, Guth, and Hill (1982). Following suggestions by Fletcher (1980), Andrews (1983) summarized theories of the spatial abilities of the blind in terms of difference theory, inefficiency theory, and deficiency theory. Most of today's evidence points to the rejection of deficiency theory and supports either inefficiency or difference theory. For a review of literature supporting each theoretical position, see Unger and Blades (this volume), and Loomis et al. (1993).

Some researchers argue that poorer spatial performance of blind persons, when observed, results from the cognitive and memory demands engendered by the need to experience phenomena nonvisually (Warren and Kocon, 1974). Using vision, a person can process spatial data in a continuous, integrative, and gestalt-like manner. Without vision an individual has to actively search the environment in a piecemeal manner to encounter phenomena and to record the motoric responses needed to get from one place to another. This information may be recorded in terms of time or effort as measures of distance. At times it needs the intervention of a sighted observer to explain the feature or features being encountered (e.g., properties of a significant building that acts as a landmark). The result is that the information encoded by the blind traveler is often second hand.

Most verbal descriptions provided to those without vision rely on natural language. Natural language is particularly imprecise in the spatial domain. For example, prepositions used in spatial phrases (such as to the left, to the right, above, below, in front of, behind, near, along, across, close to, and so on) are often referenced to the body of the describer and require rotation or translation to be encoded correctly by the listener. This introduces an additional potential source of error into the cognitive maps of those without vision. Other spatial descriptors, such as measures of distance, angular direction, or fractional amounts of change, are also often poorly described – for example, at one scale a distance may be described as "far" while at another scale the same distance may be described as "close" in natural language terms. Without a magnitude base for interpreting these descriptors or without haptically experiencing the fractional amount, the person without vision faces severe difficulties in relating one bit of encoded verbal information to another (Hill and Blasch, 1980).

While it is now frequently suggested that the cognitive mapping processes of blind persons are exactly the same as for sighted persons, there is evidence that vision allows a person to comprehend, abstract, store, recall, and use the range of configurational, layout, or spatial relational properties of large scale complex environment more effectively than can one who has not experienced them visually. What is not known is just how much information about spatial layout is necessary before an individual without vision can travel confidently and independently. This question is a significant one with respect to the composition of cognitive maps.

Basic Terms

In this section we review basic terms related to wayfinding.

Environmental and Spatial Cognition

Environmental cognition refers to the awareness, impressions, information, images, and beliefs that people have about the different environments in which they live (Moore and Golledge, 1976, p. xii). Spatial cognition is often seen as a subset of environmental cognition and refers to the internalized reflection and reconstruction of space in thought (Hart and Moore, 1973, p. 248). Spatial cognition generally includes the ability to locate features accurately, to put individual features into an accurate configurational structure with consequent comprehension of direction, orientation, and intervening distances among features, to orient a configuration correctly with respect to a global or local frame of reference, to understand the paths or other connectivities that link features, to correctly recognize the names or labels of features at particular places, and to correctly regionalize features. While some of these components of spatial cognition have not yet been examined extensively in the non-vision domain, considerable emphasis has been placed on those of cognitive mapping, cognitive maps, and wayfinding.

Cognitive Mapping and Cognitive Maps

Cognitive mapping has been defined as: "the process composed of a series of psychological transformations by which an individual acquires, stores, recalls, and decodes information about the relative locations and attributes of the phenomena in his everyday spatial environment" (Downs and Stea, 1973, p. 7). The end product of a cognitive mapping process is called a cognitive map (Tolman, 1948).

Cognitive maps are essentially individualized internal representations or models of the worlds in which we live. Cognitive maps are schematized, symbolized, incomplete, and otherwise distorted representations of the different natural, built and socio-cultural environments surrounding us.

The cognitive map is now a widely used term in a multitude of disciplines covering the biological, physiological, psychological, social, and behavioral sciences. Since most of this book is devoted to examining cognitive mapping processes and different types of cognitive maps, we allocate no more space to the elaboration of these definitions but rather assume that they are well known and accepted terms. A recent elaborate discussion of the evolution, nature and use of cognitive mapping and cognitive maps can be found in Kitchin (1994).

Cognitive maps can be assessed by asking subjects to produce their own maps of a spatial layout. For example, Casey (1978) investigated cognitive maps of blind and sighted high school students by having them reconstruct a tactile map of a high school campus. He found that although some blind students made quite well organized and accurate maps of the campus, in general maps by congenitally blind subjects were poorly organized and were not well integrated when compared with the maps made by the blindfolded or partially sighted subjects. Moreover, he indicated that the ability of blind subjects to represent a location and the spatial relations of features in an environment depended strongly on mobility or the extent to which the subjects travel independently throughout the environment.

Casey's study involved providing subjects with a cloth covered game board and 22 wooden models – one for each building structure on the campus – which were labeled with building names in print and Braille. Velcro attached to the bottom of each feature model allowed it to be firmly located on the game board and flexible velcro strips were provided to indicate roads or pathways. Two independent judges evaluated the resulting configurations by marking a line whose end points were labeled low and high (accuracy). The result showed that maps made by the partially sighted subjects were significantly superior in organization and accuracy to those constructed by the congenitally blind. The partially sighted also included significantly more features on their maps than did their blind counterparts (16 versus 12, out of the 22 features provided). Time to completion did not differ significantly between the two groups. Characteristics of the sample maps published by Casey showed that congenitally blind subjects had a tendency to linearize slightly curved paths, that the maps tended to be segmented and chunked rather than being integrated, and that features along well traveled routes were better represented than those along less well-traveled routes or those infrequently experienced.

Quantitative-scaling methods have also been used to reconstruct cognitive maps. For example, Lockman, Rieser, and Pick (1981) used nonmetric multidimensional scaling to uncover the latent spatial structure in subjective interpoint distances between three landmarks from a well known environment. Using triadic comparisons in which subjects were asked which two out of any three places were closest together or furthest apart, the totality of the responses was input to a nonmetric MDS to extract latent spatial structure of the configuration of places. In a manner similar to that suggested by Golledge et al. (1974), in a study designed to develop latent configurational structures of the distribution of familiar landmarks in Columbus, Ohio, the subjective configurations were then rotated and stretched or shrunk using the program CONGRU (Olivier, 1970) which then measured the coincidence between the transformed MDS output and a Euclidean map of the locations.

Wayfinding and Mobility

Passini, Dupré, and Langlois (1986) argue that "Spatial orientation and wayfinding are the foundations of mobility" (p. 904). It has long been accepted that the ability to move freely and independently is a prerequisite to maintaining an acceptable quality of life for blind or vision impaired individuals. Without such independence they cannot achieve full social integration and often become dependent charges on family, friends, or society in general.

Wayfinding refers to a person's ability, both cognitive and behavioral, to find his or her way from a specified origin to a specified destination. A person undertaking a wayfinding task requires the following information: reference points for identifying current location and a destination; cues to signal the spatio-temporal transition between origin and destination; information used to maintain a heading; information signaling when to turn; and a mechanism for generally determining the direction in which one must proceed or a mechanism for defining one's position with respect to a home base. The latter is called path integration.

The term mobility is often used as a synonym for wayfinding. Mobility, however, often carries the implication that movement may be physically restricted or prevented by an impairment, whereas wayfinding is generally seen more as a cognitive skill than a physical ability. Nevertheless, there is considerable overlap in the use of the terms. For example, Strelow (1985) argues that wayfinding in the absence of vision depends on the selection processes and path-following strategies already familiar to the wayfarer, on the nature of the task confronting the potential traveler, and on the type of information that is initially available or can be collected along the way.

Strelow uses the term mobility to describe wayfinding skills, defining it as: "The skill of traveling through the spatial environment, avoiding obstacles, and traveling directly or indirectly towards goals..." (p. 226). He argues that mobility is a characteristic of both animal behavior and human behavior. He then expands the term and uses it to describe the overall process of guidance by which travelers move through space, a concept often described by the term "navigation." Strelow, among others, points to the paradox of mobility having a complex basis in visual processes while at the same time being performed successfully in the total absence of vision. Strelow argues that this is possible because the process of cognitive mapping, or developing mental representations of environments, is a general one that is not constrained to vision.

Veering

Veering is a tendency to depart from linearity when traveling. Cicinelli (1989), in a historical review, reports on the studies by Schaeffer (1928), who observed blindfolded travelers walk, run, swim, row, and drive automobiles. Building on his earlier research with animals and simpler life forms such as marine larvae, worms, and mollusks, Schaeffer found that humans veered in clock-spring spiral pathways of greater or less regularity when attempting to pursue a straight path. Schaeffer suggested that there was a deep-seated spiraling mechanism in the nervous system that when used in the absence of vision, produced such veering.

Lund (1930) disputed this result, arguing that veering was the result of physical asymmetry between the two sides of the body. He showed that when blindfolded travelers were asked to report whether a path just traveled was straight or curved, they were insensitive to the curvature. It was suggested that this curvature was equivalent to their natural veer and consequently would not be observable. An asymmetry in body proportions cannot explain all veering, however. D'Oliveira (1939) examined the amount of veer displayed by pilots in flight formation with their veering tendency when asked to walk a straight path while blindfolded. The pilots tended to veer in the same direction when walking blindfolded as they veered when flying a plane.

Veering is observed in blind and sighted people. Rouse and Worchel (1955) tested 18 congenitally blind people by asking them to walk 100, 200, and 300 feet on a flat field devoid of vegetation except for grass underfoot. The subjects performed in 1 of 4 conditions: a) blindfolded, eliminating any light cues; b) blindfolded wearing a hood to eliminate wind, sun, or facial tactile cues; c) blindfolded with the ears plugged eliminating sound; and d) blindfolded wearing a hood and earplugs thus eliminating all cues. The ability to walk a straight line was found to be unaffected by the presence of external cues such as sun, wind or sound. At 100 feet subjects veered an average just over 11 degrees; at 200 and 300 feet average deviation was 18.38 degrees and 23.34 degrees respectively.

Cratty (1965), and Cratty and Williams (1966) reported on the attempts of blindfolded sighted subjects and blind adults to walk in a straight line across a 140 yard field of grass.

A gridlike pattern of white chalk was laid down on the field in 10 yard increments to allow paths to be plotted and veer angle estimated. Subjects wore earplugs and a black, opaque fabric hood covering head and shoulders. Unlike Banerjee (1928), who had found that veering errors occurred more often to the left than to the right, Cratty and his coworkers failed to find such a difference. Cratty found that 65.2% of his subjects were homotropic (veering in the same direction on each trial), 25% were heterotropic (veering in opposite directions on such trials), and 9.8% produced exact patterns. As a group, subjects veered an average of 36.9 degrees per 100 feet. Rouse and Worchel also found that veering tendencies were fairly consistent in direction for subjects across trials and conditions, but not all subjects veered in the same direction. In another study, Harris (1967) showed that blind subjects who are highly anxious veered more than a less highly anxious blind group. More recently, Cicinelli (1989) found no distinct tendency to veer either to the left or the right, and reported veering angles similar to those found by Cratty.

Frames of Reference and Spatial Orientation

A frame of reference is a context for defining the position of points in space. It can be local and relational, as with respect to landmarks or street systems, or might be related to a global and widely accepted frame of reference such as traditional geographic latitude and longitude coordinate systems and the cardinal compass directions. Spatial orientation refers to a person's ability to relate personal location to environmental frames of reference. Rieser et al. (1982) suggest that the major components for spatial orientation include (i) knowledge of spatial layout of destinations and landmarks along the way, (ii) the ability to keep track of where one is and in which direction one is heading, and (iii) comprehension of the organizing structural principles embedded in a given environment. Without the first component, travel can take place only by using search and exploration strategies. Without the second, the traveler will quickly become lost. Without the third, needless effort is expended in trying to memorize every component that can be encoded as a simple pattern (e.g., a grid-like street system).

It is sometimes assumed that blind or visually impaired travelers have little need for a frame of reference. This is not so. Denying blind people contextual information may cause them to reduce their activity schedules, limit their behavior paths, alter efficient destination selection procedures, and oversimplify their local knowledge base.

Wayfinding in the Blind

This section reviews basic components of wayfinding and evaluates their performance by blind individuals.

Representing Current Location

To represent one's current location is to place oneself in a cognitive map. In this section we look at the ability of blind and vision impaired persons to locate themselves in an environment by imagining they are at specific locations and then illustrating where other environmental features were located. For example, Dodds, Howarth, and Carter (1982) compared congenitally and adventitiously blind eleven year old children to examine the development and structure of their mental maps and to determine if they were capable of identifying their positions within them. Sighted subjects were also used in a pilot study to provide a comparative base. In one task, the subjects learned two routes consisting of a walk around two separate blocks. After the first familiarization trial, subjects carried a board containing a pointer and a scale marked in five degree intervals. Before and after each turn on the route they set the pointer towards home and a variety of target goals located elsewhere on the route. On each trial each subject made ten pairs of orientation responses and drew one map. Pointing validation was undertaken by placing the pointer scale at selected locations on the hand-drawn map such that home and target goal directions could be estimated. This process, called construct validation, involved using such orientations as predictors for the pointer responses obtained during the next trial. Significant correlations were obtained among the sighted students, thus implying that a valid means of inferring mental representation had been developed. A reliability test showed that drawings produced under blindfolded conditions by the sighted students similarly showed highly significant correlations between subjective and objective orientations and location of target features. Using the measure of absolute pointing error in degrees (i.e., difference between subjective and objective directions), Dodds et al. showed that adventitiously blind children performed better than did congenitally blind ones and that all the children were better oriented at some route points than at others. After the first trial, only one congenitally blind subject could produce a route drawing consistent with sighted conventions, and consequently perform the orientation and location tasks. This implied that blind subjects had difficulty in identifying their locations and their position relative to other places (and vice versa).

Results also showed significant qualitative differences in performance between congenitally and adventitiously blind children. Three of the four congenitally blind children appeared to adopt self reference spatial coding systems. None of the adventitiously blind children adopted this referencing procedure. However, one congenitally blind child did perform as well as members of the adventitious group. This implies that while visual experience is not a necessary precondition for the ability to represent spatial layout and use it for orientation purposes, it does assist in the development of coding strategies designed to represent spatial information.

Dodds et al. further found that there was an overall increase in pointing errors as physical distance increased. They argued that errors increased with the number of cognitive operations required. They also came up with the interesting conclusion that the drawings of the children who were capable of drawing reliably were more accurate representations of the route features than was pointing. This was not the case among the sighted subjects in the pilot study. Dodds et al. further found that congenitally blind subjects had great difficulty in using spatial inference to combine the two routes into a single reliable diagram.

Progressing on a Target Route

While some general strategies associated with following a route can be found among the blind and vision impaired population as a result of orientation and mobility (O&M) training, individual heuristics that consist of modifications of generally accepted O&M practices are quite common. When following a selected path from an origin to a destination, blind travelers will often follow linear features such as the edge of a sidewalk, a grass verge, a fence, a curb, or the walls or hallways of a building. This is called "shorelining." When sidewalks are not readily available, traffic noise and motion can provide some assistance in keeping the blind traveler off a main traffic artery. As with the sighted, some blind travelers rely on cardinal directions to which they can orient on the basis of environmental features such as sun angle or wind direction. Others use local frames of reference including street systems for linear reference modes and landmarks for polar coordinate modes. Others count steps, street corners, obstacles, or regularly occurring environmental features such as telephone poles or fire hydrants. Still others use temporal indicators usually based on a "feeling" that they have walked far enough and should be close to a cue, choice point, or a destination. As technology increases the type of supplemental temporal information available to the blind traveler (e.g., talking wristwatches and clocks), the temporal dimension appears to be playing a more significant role in comprehending environmental layout, in estimating the appropriate amount of time to be spent walking a particular path segment, and in providing some indication as to when completion of a trip should be imminent (e.g., when riding a bus, taxi, or a suburban train).

Perhaps the simplest design for a path following heuristic is to segment a route by focusing on critical choice points and remembering the number of segments and the turn angles that occur between a given origin and destination. This emphasizes route type learning which indeed is the most common form of spatial learning used by blind and vision impaired travelers. Involved in such a representation is an egocentric frame of reference, the definition of segment sequence and segment length, and a feeling for the integration of space and time required to follow a preselected path between a given origin and destination. This form of wayfinding strategy lends itself to improvement with repetition. It does not help develop a capacity to choose alternative routes, to take shortcuts when available, to develop successful search strategies when well known routes are blocked by unexpected obstacles such as construction, or to assist in the substitution of one destination for another in real time as travel is being undertaken.

Spatial Updating

Spatial updating appears to be an important component of all wayfinding capabilities. In other words, for successful travel the traveler must know from moment to moment where he or she stands in relation to places in the environment. Spatial updating can be undertaken in a variety of ways, such as:

1. Recognizing that one is near a known landmark or distinctive environmental pattern whose location is well known;
2. Reconstructing previous movements in order to mentally retrace a path and infer one's current location;
3. Perceptually relating locomotion to environmental knowledge.
4. Establishing object-to-object relations using anchorpoints and polar vectors, grid line routinized searches or perimeter to perimeter searches.

With respect to the first technique, a traveler can spatially update by finding an object or environmental feature that can serve as a reference. To date, no definitive study exists of the strategies that are used by persons with little or no vision with respect to how they find objects. Although some systematic spatial search procedures such as grid line following and perimeter to center exploration are described in orientation and mobility texts, there appears to be no detailed simple list of strategies that can be potentially used.

Tellevik (1992) examined some search strategies by having subjects on consecutive days find objects in a room after learning their locations from the first day and being required to search for rearranged objects on the second day. Tellevic's results showed that in the first day strategies of perimeter and grid line search were dominant, whereas in the second day, anchor point vector strategies or reference point strategies were used.

Hill et al. (1993), found when comparing videotaped strategies with self confessed strategies that only the best blind performers were able to articulate the strategies that they used successfully. Of the best performers, an object-to-object strategy (anchorpoints) was used by twelve of the forty-two performers, while perimeter-to-object and perimeter-to-perimeter search were used by nine and eight performers respectively. Grid search and home-base-to-object strategies were used by five performers, and a mental image and cognitive map strategy was used by eight performers. The majority of the worst performers said they may have used either anchorpoint or perimeter-to-perimeter search.

Hill et al. (1993) suggested that there are multiple stages in the search process, the perimeter search commonly occurring in the initial phase and then either self-to-object, perimeter-to-perimeter, or object-to-object (anchorpoint) strategy being consequently used.

Updating by Path Integration

Path integration is a strategy in which the traveler senses self velocity (heading and speed) and integrates to obtain moment-to-moment position within some coordinate system and bounding frame of reference (Gallistel, 1990; Mittelstaedt and Mittelstaedt, 1982). Although this does not rely explicitly on external positional information (such as optical, acoustical, chemical, or radio signals) some signals may be involved in the sensing of self velocity. There is a considerable literature showing that some non-human species have an uncanny ability in performing path integration, albeit with optical flow as one input (e.g., Mittelstaedt and Mittelstaedt, 1982; von Saint Paul, 1982; Müller and Wehner, 1988).

Rieser et al. (1986) conducted an experiment using a spatial updating task that relies on path integration. In one task subjects were guided from an origin to each of a set of objects in a room, then were taken to a new origin and asked to point at the objects (the locomotion task condition). This was designed to assess the extent to which subjects were able to perceive where they walked in relation to the arrangement of targets in the experimental setting (i.e., the use of a perceptual strategy to update position). In the second task subjects were asked to imagine standing at a specific location and then to point to where each of the other objects would be with respect to that imagined location (the imagination task condition). This assessed the extent to which participants were able to figure out new directions and use a computational strategy to update position. Both blindfolded sighted and blind and vision impaired subjects reported the imagination task as being much more difficult. Here, however, Rieser et al. reported that the early blind subjects found both locomotion and imagination tasks difficult, whereas the blindfolded sighted and late blind adults found the locomotion task comparatively simple. These results suggested that early blind people were deficient at perceptual updating when traveling from place to place. In a follow-up study, however, Loomis et al. (1993) found considerable individual variation in task performance, with some of the congenitally blind performing extremely well in both real and imagined locomotion conditions. They suggested that one possible explanation for the difference in results could be found in the nature of subjects used; Loomis et al. used only subjects that were independent travelers in their local area, while the subjects of Rieser et al. were not similarly constrained.

Another task that relies on path integration is pathway completion, a task in which the subject follows an outbound multi-segment pathway and attempts to return directly to a home base without vision (Klatzky et al. 1990; Loomis et al. 1993; Worchel, 1951). In our own research we have assumed that subjects use some sort of metric representation in performing the given task. This assumption is supported by the systematic dependence of subjects' pathway completion responses upon quantitative manipulations of the configuration of the outbound paths (Loomis et al. 1993).

One possibility for a minimal metric representation would indicate only relative location of the origin in terms of the turn and distance that would be needed for the observer to return home (Fujita, Loomis, Klatzky, and Golledge, 1990; Müller and Wehner, 1988). These parameters constitute a "homing vector", which could be updated with each step of the observer. Such a representation is history free, for there is no record of the path the observer took to arrive at the current location. At another extreme a metric representation might be what has been called a "survey representation" of the task environment. Within some common coordinate system is represented the origin of locomotion and the observer's path, including current location. Other locations could be represented as well, thus allowing the observer to plan direct paths to any of them. Evidence against the history-free hypothesis comes from the time to initiate pathway completion, which we found increased with the complexity of the previously traversed route. If subjects had been continuously computing a homing vector and discarding information about the path, then the latency to start homeward should have been independent of pathway complexity. In light of this result, we concluded that subjects probably develop and use representations of their outbound paths.

We have modeled pathway completion in blind and sighted by assuming the operation of four internal processes (Fujita, Klatzky, Loomis, and Golledge, 1993). These are (i) perception of each of the pathway segments and turns by sensing and integrating the translational and rotational velocities respectively; (ii) formation of an internal representation of the outbound paths; (iii) computation of the direct path back towards the origin; and (iv) execution of the completed path. Path integration is subsumed in the early processes.

Errors in triangle completion responses could result from any of the component processes. We argued, however, that output processes alone (i.e., (iii) and (iv) above) cannot entirely account for errors in pathway completion. Our model (Fujita et al. 1993) proposes that the early stages can account for much of the error in completion of simple triangular pathways. Description of the representation encoded by a subject is assumed to be sufficient to characterize subject performance. The model provides a mechanism for computing the internal representation that corresponds to a simple multi-segment path. Thus, it allows the computation of a hypothetical encoding function which relates actual values of segment lengths and turns to internalized encoded values. Errors in pathway completion are predicted by these derived encoding functions with a high degree of accuracy. Tests of the model found that subjects encoded both turns and extents with considerable regression towards the mean of the values they had experienced. The group encoding functions for distance and turn were linear with non-zero intercepts and slopes substantially less than one. This indicated that subjects overestimated small values and underestimated large ones – a result that appears frequently throughout the spatial cognition literature. It might be hypothesized that much of this tendency reflects imprecision in proprioceptive sensing and would be reduced if not eliminated if subjects had visual information about self motion, even if they did not have vision of the pathway as a whole. However, we have also found (Klatzky et al. 1990) that people are able to replicate walked linear paths and single turns well, suggesting that proprioceptive sensing is not the root of the problem in human path integration. The translation from sensory data to a cognitive map may be more critical.

A version of pathway completion, the triangle completion problem, is one of the most widely accepted tasks for indicating the possible existence of complex spatial understanding in blind individuals (Worchel, 1951). In this problem an individual walks two legs of a triangle, turning the included angle, and then is required to travel back directly to the origin while not retracing the experienced legs. This can only be achieved by making a correct final turn angle and correctly estimating and walking the distance back to the origin. This task has been accomplished by leading people over the first two triangle legs (Worchel, 1951). It has also been undertaken by having subjects haptically experience two legs of a triangle and report about the third (Bambring, 1976; Klatzky, Golledge, Loomis, Cicinelli, and Pellegrino, 1994).

Computing a Novel Path

From their pathway representation, travelers can be assumed to compute a trajectory to some desired location. This computational process might be equivalent to cognitive trigonometry in that it introduces no systematic source of error. Note that measurement of an image (analogous to perceptually accessing angles and leg lengths when a path is viewed on a map) is not distinguishable from an abstract trigonometric computation. Alternatively, as Fujita et al. (1993) noted, subjects might use weak heuristics with systematic biases in order to solve their problem. For example, in the task of completing a triangle after traversing two legs and a turn, an example of a rule that could be used would be: "if the initial turn was small, turn almost 180 degrees and go a distance equal to the sum of the two leg lengths; if the turn was very large, turn only a little and go a distance equal to the first leg minus the second." The Fujita et al. model of triangle completion assumed the first of these alternatives--the encoded values are fed into a computational process which, unlike a heuristic, introduces no further systematic error. However, this is unlikely to hold for completing complex configurations.

Spatial Abilities Without Vision

In this section we review a variety of spatial tasks in which the ability of the blind has been evaluated.

Wayfinding Abilities

Using experimental groups consisting of congenitally totally blind, adventitiously totally blind, and subjects with weak visual residue, together with a control group of sighted and blindfolded sighted individuals, Passini, Proulx, and Rainville (1990), tested their spatio-cognitive competence with respect to wayfinding. Eight basic wayfinding tasks were defined. Each represented a particular spatio-cognitive operation that was identified as an important part of the wayfinding process. The first set of tasks consisted of learning a new route, returning to the origin from the destination, and combining previously learned routes into new combinations. Passini et al. also involved subjects in learning a route on a small scale model and executing it in the real setting, thus examining the degree to which successful transference can be made from a preprocessing device to real world execution. A second set of tasks involved pointing to specific locations visited while undertaking travel, taking shortcuts, and undertaking mental rotations of a route learned from a map. A third set of tasks involved reproduction of layout for a given setting by comprehending the organizational principles (such as symmetries) involved in the setting's structure. Tasks were executed in a labyrinthine layout, which allowed Passini et al. to control the level of difficulty as well as limiting extraneous perceptual factors.

Passini et al. provide strong evidence that wayfinding and other forms of spatio-cognitive competence can be achieved by those without vision or by subjects without previous visual experience. The congenitally totally blind subjects tended to perform better than the adventitiously totally blind and the sighted blindfolded groups but not as well as the sighted group or the group with the visual residue. The visually impaired groups tended, however, to require more time to complete the tasks than either of the sighted groups.

Passini et al. also argue that wayfinding presupposes planning and requires active decision making and execution of travel plans. All this takes place within a continuous information processing environment. The wayfinding task itself is defined in terms of spatial problem solving, together with the processes of spatial decision making and implementation of travel plans via spatial behaviors. The ability and opportunity to preprocess environmental information and to set up a plan of travel before wayfinding per se is considered significant. Gärling and Golledge (1989) have also stressed the significance of developing and implementing travel plans for success in wayfinding. The success of blind people in the tasks devised by Passini et al. indicates that planning can be done by the blind.

Wayfinding abilities of the blind have been studied in natural contexts. In one of the most comprehensive studies of everyday wayfinding practices in blind and vision impaired persons, Passini et al. (1986) catalogued an elaborate set of activities involved in wayfinding. First, they assembled profiles of the members of groups of congenitally and adventitiously blind subjects, including subjective evaluations of their own mobility, their own assessment of what constituted difficult settings, the variety of technical or human aids used in their wayfinding activities, the information they used during actual wayfinding behavior, their history of mobility and orientation instruction, their evaluation of the magnitude of difficulty associated with different physical obstacles and dangers likely to be encountered during wayfinding and their cognitive mapping ability. They also examined subjects' abilities to represent the structure and feature content of large-scale environments, their evaluation of situations that produced spatial disorientation, their attitude toward wayfinding tasks, and the nature of the environmental cues or the physical structure of pathways that contributed most to successful wayfinding and safety. Passini et al. deliberately did not include individuals with known neurophysiological problems such as brain lesions or Alzheimer's Disease. Their initial observation was that all persons regardless of which group to which they belonged were able to develop a wayfinding proficiency. This was defined as the ability to move unaided throughout a specific environment. As partial proof, Passini et al. identified a particular client who was a totally blind piano tuner who reached his clients daily using public transportation for the larger distances and a braille map to lead him to specific final destinations.

In the analysis of their survey responses, Passini et al. found that the most difficult places experienced by the visually handicapped included the following: Shopping complexes; department stores; hotel lobbies; train and bus stations; airports; parking lots; open spaces and park land areas; and other places which were either too crowded, too exotic, or which lacked distinctive auditory reference points to assist with wayfinding. Indoors, the subjects indicated that important information identifying reference points was often masked by sound absorbing materials such as carpets or acoustic tiles or when a high level of mixed background noises existed.

Another common difficulty was the inability of blind and vision impaired persons to understand the layout of the environments through which they had to find their way. This appears particularly true when layouts are ambiguous, such as where specific patterns are repeated multiple times over a given area of space (e.g., repetitively landscaped blocks, segments of stores arranged exactly the same way, and so on). But, on the other hand the repetitive example of a rectangular street system with uniform size blocks was often given as an example of a simple but highly commended organizing principle.

Passini also found that individuals with some residual vision relied extremely heavily on that residual during locomotion. All of the subjects used auditory cues as reference points when they were distinct and memorable enough as both location and directional indicators (e.g., the direction of traffic on streets).

Learning Layouts

Hollyfield and Foulke (1983) examined spatial learning of novel routes using adults over twenty years of age. Two sighted groups were used, one performing with vision and the other blindfolded. Another group of adventitiously or late blind persons with at least twelve years of sighted experience and a group of early blind with less than one year of sighted experience were also used. The route covered five city blocks and was about half a mile in length. Initial exposure was via an experimenter guided tour followed by five subject trials. Subjects' verbal comments along the route were tape recorded during each trial and after each trial a modeling kit was used to map the route. On the final trial a subject was stopped at a midway point, instructed that an obstacle blocked the route, and was requested to find an alternate route to the destination. Results showed that the early and late blind performances were comparable though not as good as the sighted groups. Initially, blindfolded sighted subjects were very poor, but they improved to approximately equal the adventitious blind by later trials. The detour task significantly interfered with performance in all the non-visual groups.

Taking Short Cuts

In previous work we have examined congenitally blind, adventitiously blind, and blindfolded sighted subjects in three different types of wayfinding tasks (Loomis et al., 1993). These included simple reproduction and estimation of turns and distances, table top tasks involving spatial skills (Klatzky et al., 1994), – which included tasks such as mental rotation – and more complex locomotion tasks requiring spatial inference (e.g., undertaking shortcutting activities to complete a trip). Group differences were not found for some of the simple locomotion tasks such as maintenance of heading, turn reproduction, and distance reproduction. One might expect to see differences in performance between those who have had vision and those who have not, when tasks that require an understanding of the abstract properties of space, that require spatial inference, or that appear to require the solution of mental geometric or trigonometric problems. One such task is pathway completion, a form of short cut.

Manual Apprehension of Space

Spatial tasks are often conducted at the scale of the table top, by manipulation of objects (see Klatzky et al., 1994, for review). Cognitive mapping has been explored in what is called manipulatory space (Lederman, Klatzky, Collins and Wardell, 1987) by undertaking tasks that require knowledge of interpoint distances and orientation between features that were not explored directly. For example Cleaves and Royal (1979) had subjects learn a maze which consisted of right angle turns by following a path through it with a finger, then pointing directly from the start to sets of locations within the maze (e.g., the first turn or the end point). Such tasks were pursued either with the original orientation imagined or after imagining a rotation of the task environment. In this study the congenitally blind had higher spatial error and the degree of error was correlated with the duration of blindness. In contrast to this, Brambring (1976) found that the performance of congenitally blind subjects was less subject to distortion than blindfolded sighted or adventitiously blind subjects. In this case the task was feeling the sides of a right triangle and estimating the lengths of the hypotenuse (i.e., triangle completion). Although the variance of the results was highest among the congenitally blind, Bambring's work clearly showed that this group was capable of spatial imagery.

Other studies relevant to cognitive mapping using data from manipulation include that by Marmor and Zabeck (1976), who looked at mental rotation problems and found no significant difference between congenitally and adventitiously blind and blindfolded sighted subjects in terms of the relation between response time and angular disparity between a rotated and standard figure held in the hands. Descriptive statistics, however, showed there was a general tendency for a lower response time for the blindfolded sighted group and a tendency towards highest and largest response time for the congenitally blind group. Similar results were found in rotational tasks by Carpenter and Eisenberg (1978).

Many studies have used haptic recognition of shape as the performance variable in manipulatory space. Dodds (1983) found that recognition errors increased more for the congenitally blind than other groups when the angular distance between the target shape and a duplicate included in a set of distracters was greatest. But even with such findings, there is no doubt that even congenitally blind individuals can perform shape recognition tasks, even when some shapes are rotated. Rotation is commonly accepted as one of the critical dimensions of spatial ability (Eliot and McFarlane-Smith, 1983; McGee, 1979; Lohman, 1979). Thus, the Dodds et al. result provides more evidence of the presence of a comparable set of spatial abilities in people without vision as can be found in those with vision.

Overall this research tends to indicate that blind subjects are capable of performing a wide variety of tasks at the tabletop level in a manipulatory space. In general, however, their response times are slower than blindfolded sighted subjects. Given that the tasks can be successfully performed, and that clear evidence exists of reasonable rotational, orientation, and perspective viewing abilities, very strong evidence exists that the imagery process is present in blind as well as sighted people, but there are usually observable differences in the way that both groups perform.

Use of Auditory Cues in Following a Path

Strelow and Brabyn (1982) investigated the degree to which blind and blindfolded sighted subjects were able to use natural auditory sensing to locate and follow a travel path between obstacles. Their conclusions were two-fold:

1. Blind subjects showed a greater ability than blindfolded sighted subjects to use auditory cues for wayfinding, and
2. Auditory guidance is inferior to visual guidance and deteriorates significantly when small targets are used to define the travel path.

The difference in performance between blind and blindfolded sighted subjects was said to be a consequence of the strategy undertaken by many blind to develop echo location skills for use in estimating proximity, whereas this strategy was a novel idea and unused by the blindfolded sighted people. As target size decreased, the potential for using echo skills to locate and avoid obstacles likewise diminished.

Multiple Task Assessment

Klatzky et al. (1994) reported on the results of congenitally and adventitiously blind, and blindfolded sighted subjects performing a variety of manipulative and ambulatory tasks (using both tabletop manipulatory and locomotion tasks) and then analyzed the entire data set relating to those tasks in an attempt to investigate commonalities in task performance. In this study congenital and adventitious blind and blindfolded sighted subjects recruited from the Los Angeles and Santa Barbara branches of The Braille Institute of America, all of whom were experienced independent travelers in their own environment either by walking or using public transportation, were matched in terms of age and educational level. For each subject testing (including rest periods) took about 4 hours. The session began as an interview for personal data collection, followed by the tabletop tasks which took approximately 30 to 45 minutes and were administered in a random order counterbalanced across subjects. The tasks included same/different judgments of rotated stimuli, assembly of whole shapes from parts and estimation of distances. Feedback was not given to any task. After a break, locomotion tasks were undertaken. These included simple locomotion, including maintenance of a heading and the walking of a constant distance, replicating a walked distance, replicating and estimating a turn, complex locomotion involving completing a triangle after walking two legs, and consequently completing or retracing a two or three sided figure. There were thus both simple and complex locomotion tasks. The simple tasks involved a linear segment or turn and could be performed with direct proprioceptive memory. The complex task combined 2 or more segments with turns between them and required some degree of spatial inference. Generally the blind and blindfolded sighted subjects performed the tasks quite well. In tasks requiring assembly of parts into a particular shape, errors and noncompletions were very similar across groups averaging 28%, 25%, and 27% for the congenital, adventitious, and blindfolded sighted groups respectively. Response times were slightly lower for the adventitious group and somewhat higher for the blindfolded sighted group. Whereas Worchel (1951) showed that performance on similar types of assembly tasks produced significantly more error by the blind population (47% versus 26% respectively), the Klatzky et al. (1990) task showed comparable error rates but slightly higher response times by the congenitally blind subjects. In a rotation task, correct responses were obtained by 62%, 70%, 46% for congenitally, adventitiously, and blindfolded sighted groups respectively. Performance in the rotation task was quite poor compared with other studies of haptically mediated imagined rotation (Carpenter and Eisenberg, 1978; Hollins, 1986). Error rates were higher (about 40% here versus 5% in the other studies) and response latencies greater (about 25 seconds versus 2 to 3 seconds in the other studies). Klatzky et al. found no evidence that response latency increased with the orientation difference between the 2 stimuli being compared. This was markedly different from results from other studies. Results generally, however, slightly favored the blind groups over the blindfolded sighted group.

There were likewise few significant differences in task performance among the groups on locomotion tasks. Using absolute magnitude of veer or simply a measure of the ability to perform simple replications and estimations in turns and linear segments, no group differences were discovered. In the triangle completion task, both the turn toward the origin and the distance to complete the third leg after the turn produced a pattern of systematic regression to the mean. Subjects tended to over respond when the required distance or angle was small and under respond when it was large. These tendencies did not vary significantly by groups. Retrace performance was similar across groups. With a more complex completion problem, all groups found that a 3 segment pathway involving a crossover of 2 legs was particularly difficult.

Combining all these results and subjecting them to a factor analysis isolated 3 factors which accounted for 67.3% of the variance. Factor 1 reflected competency in the locomotion task and involved spatial inference rather than replication. Factor 2 correlated with performance on the tabletop tasks. Factor 3 included the reproduction of simple linear segments and turns. Overall, the results seemed to indicate that past visual experience was not necessary for successful completion of a variety of spatial tasks that drew on different abilities and different spatial processes. Perhaps the most significant outcome overall was the lack of difference among groups across the tasks and the failure to isolate specific spatial deficits among the blind.

However, by discriminant analysis, 7 of the 12 congenitally blind, 8 of the 13 adventitiously blind, and 8 of the 12 blindfolded sighted, were correctly classified into the appropriate group. Thus classification on the basis of performance on these tasks was almost twice as accurate (62%) as chance expectations, and those errors that were made tended to place individuals into adjacent rather than remote groups. This indicates that there were at least some detectable differences among the groups.

How can Wayfinding and Mobility be Improved?

In this section we consider ways to improve wayfinding by the blind.

Information Needed

A blind traveler needs two different types of information to assist with wayfinding and to provide necessary safety. First, there is information about the proximal environment which is represented in one's cognitive map; this provides spatial information about local cues and identifies obstacles. Information may be required about the sequence of driveways passed on a given route, the number of telephone poles or doors bypassed, the location of building entrances, the identification of curbs and street intersections, and local effects such as noises, smells, wind direction, sun angle, and so on (Welsh and Blasch, 1980; Wiedel, 1983; Dodds, 1988; Tatham and Dodds, 1988).

Second, information is needed about larger scale geographic space and this may include knowledge of the location of near and more distant buildings, information about changing terrain, and the definition of and differentiation among path segments. This type of information is generally built up via exploratory search or repetitive travel behavior in a local environment. The information is coded and stored as part of the traveler's cognitive map. And while the person with sight can simply scan an environment, identify critical cues by unique colors or shapes or dominance of visible form or specific function, the blind traveler must either learn about these second hand or not at all. With auditory perception becoming the major substitute for the long distance information processing otherwise provided by vision, interpretation of natural sounds, their auditory localization, the estimation of distance and direction between and among simultaneously or sequentially occurring sounds, and the ability to unpack the information contained in verbal descriptions are the major tools of the blind traveler. There is every evidence that these are less effective than vision in terms of precision, memory load, or cognitive demands.

Orientation Aids

"Orientation aids are tools to be used by visually impaired persons to develop or enhance their understanding of basic spatial relationships, to facilitate a comprehension of specific travel environments, to refresh their memory of routes and areas, to further their skill in independent route planning, to enable them to travel independently in unfamiliar areas, and to add to their knowledge and enjoyment of physical space." (Bentzen, 1980, p. 291). Traditionally orientation aids have been subdivided into three categories: Models, graphic aids, and verbal aids. Models are the tactual devices, usually three dimensional, which represent scaled down versions of real environments or symbolized representations of environments. Graphic aids are primarily tactile or visual and include maps, diagrams, graphics, sketches, and raised drawings. Verbal aids include directly given or recorded auditory descriptions of routes and nearby environments, often emphasizing sequences of segments, choice points, and turns associated with traveling specific routes. If undertaking orientation and mobility training with the help of a specialist, blind or vision impaired travelers may be exposed to all of these aids and be advised of a best match between specific devices and specific tasks. Often the aim of an O&M specialist is to teach a blind or vision impaired person the skill of matching orientation aid with the problem situation and with their personal abilities. The emphasis, however, is usually on using what is provided or what is taught rather than undertaking independent and innovative experimentation with or without such aids. This conservative criterion is invariably adopted because of concern for the welfare and safety of the traveler.

There are, of course, many human and technical aids designed to help a blind traveler. Traveling with a sighted guide or friend requires only the ability to move freely when guided. Somewhat more independent travel is possible by using a long cane or one of the many cane variants (e.g., walking stick, short cane, or laser cane). In the United States about ten thousand blind travelers use guide dogs and more than one hundred thousand use long canes to provide the independence and freedom of movement that is often denied because of lack of vision. In recent years, sympathetic communities the world over have been erecting auditory "You Are Here" maps, placing electronic strips along sidewalks or in hallways in buildings, locating talking signs, providing cassette tapes with directions for travel between local landmarks, strategically locating brailled maps or signs, providing auditory signals to identify pedestrian crossings on street systems, or displaying tactual, auditory, or combined tactual and auditory maps at strategic locations such as in railway stations or airports in attempts to assist the independent blind traveler. An example of such an auditory tactual device is NOMAD (Parkes and Dear, 1990). This auditory tactual information processing system is quite useful for learning layouts of the interior of buildings or larger environments, can be used to help learn specific paths, or simply can provide cue and layout information prior to travel.

Brailling of signs or number systems in elevators or near doors helps that proportion of the population that reads Braille to comprehend and use their environment more effectively. In some cases electronic guidance systems have been built into hallways or along sidewalks to assist the blind or vision impaired traveler. Here, special canes or sensors activate a sonic or laser beam within a defined buffer zone and allow the traveler to home to a destination by following those particular beams (Preiser, 1985). Preiser has experimented with the possibility of giving verbal messages to a blind traveler through a small portable speaker by activating such messages from electronic guidance devices laid in the floor. In Ottawa, Canada, sonar traffic lights have been installed which warn approaching blind pedestrians of potentially hazardous street crossings. Some experiments are also being undertaken in terms of the laying of textured floor tiles that can be followed using a cane or simply the soles of one's feet. Similar tactual tile devices are now commonly being used to warn of proximity to hazardous features such as bike paths. Campuses, such as the University of California, Santa Barbara, where more than twelve-thousand bicyclists regularly circumnavigate and cross the campus, have effectively used these warning tiles to reduce the possibility of the blind or vision impaired person accidentally wandering onto a bike path that is distinguished from the surrounding area only by color coding. In Montreal, the practice has been adopted of ensuring that the corners of sidewalks at intersections are identified by a series of grooves about fifteen millimeters deep and twenty-five millimeters wide. These work as long as the cane guided traveler uses a sweeping technique in which the cane is in constant contact with the ground, but is of less use where the more frequently used "touch and go" procedures are used. And, of course, in areas subject to colder winter temperatures, ice and snow often fill the grooves and make them unusable for a significant part of the year.

Obstacle Avoidance

Of all the problems faced by the blind traveler, obstacle avoidance and avoidance of collisions are usually rated as the most significant. Obstacles to the blind traveler often include environmental features that are thought by those with vision to be highly desirable, alluring, environmentally enhancing, or necessary. Common obstacles include low hanging signs, things that jut out from walls such as fire extinguishers or water fountains, half-open doors, benches or planters located seemingly at random in the midst of sidewalks, stairs without railings, telephone booths that cover only the upper torso, and hazards such as bicycles attached to poles, benches, or other free standing environmental features, or toys or furniture scattered indiscriminately across the floor indoors. In some areas of the country obstacles are seasonal with banks of snow or patches of ice in winter becoming significant obstacles for the unwary blind pedestrian. Curb cuts which allow imperceptible descent to the street, although of critical importance for wheelchair users, unfortunately often allow the blind traveler no warning of an approaching curb and may result in intrusion into a street where fast moving traffic produces a significant hazard. Apart from the many features of the built environment added to facilitate interactive behavior and mobility for sighted people, the blind traveler also faces the obstacles of irregular terrain, steep slopes, and unprotected pools of water, mud, marsh, and so on.

In addition to the personal guidance and navigational aids outlined earlier, there has been a long history of development of tactual aids to assist in obstacle avoidance. These include devices using sonic beams such as the Mowat Sensor, the Sonic Guide, or the Nottingham Obstacle Avoider (Dodds, 1988). Recent modifications of these devices have been developed in several different countries. The essence of such systems, however, is to help the traveler avoid danger. Simple everyday environmental elements such as stairs, curbs, unprotected edges, overhanging obstacles such as tree limbs or signs, and protruding obstacles such as telephone booths, water fountains, benches, and vegetation filled planters, while not appearing to be dangerous to the sighted traveler, can seriously and negatively affect the safety of a person without vision. Obstacle avoiders, such as the sonic devices mentioned above generally have a limited spatial range. Canes and some sensors have a maximum range of three to eight feet from the body; the Sonic Guide is useful up to about eighteen to twenty feet. In other words, the obstacle avoider is useful in the near environment but gives little information about location and layout of features scattered throughout larger environments. These devices are not designed to help the blind individual identify distant landmarks or to assist in developing comprehension of the layout of features in the environment through which they pass. To assist in learning about and traveling through geographic spaces, navigation or guidance aids are needed. To produce an understanding of an environmental setting, structure, or layout, the blind traveler has to integrate information obtained from multiple linear segments which make up the different routes by which they traverse the environment. Integrating this information is an extremely difficult cognitive task unless some tool such as a tactile map or plan is made available to help show spatial relations among features and paths within the environment. In terms of both the limited route knowledge developed by the blind traveler and the limited layout comprehension they are able to develop, reference points or landmarks and choice points, become critically significant.

Navigation Aids

Unlike obstacle avoiders which emphasize the proximate environment, a navigation aid provides geographical information about the general environment through which travel must take place. It also can be of assistance in planning routes and selecting path segments that are as obstacle or barrier free as possible. Overviews of navigational aids that have been developed for blind travelers can be found in Brabyn (1985), Collins (1985), and in Wiedel (1983), or Tatham and Dodds (1988). Most of these devices provide information about the proximal environment using either passive or active sensing such as video cameras, lasers, long canes, or ultrasound. Environmental information is often displayed in acoustic or tactile display form. Recently interest has developed about the possibility of using a combination of satellite based locating systems (Global Positioning Systems or GPS), and spatial databases and analytical functions (Geographic Information Systems or GIS), to answer locational questions, to provide some information about the local environment and to help select travel paths. In these devices, travel for the blind is usually identified by verbal commands (e.g., proceed ahead 100 yards, turn left 90 degrees then proceed for 200 yards in that direction). An alternative idea is to project a beacon in front of a traveler with the beacon guiding the individual along a path defined within a computerized spatial database (Loomis, 1985).

Recent experiments have resulted in the production of a prototype of such a system called the Personal Guidance System (PGS). A unique feature of this prototype is the use of a virtual acoustic display as the interface (Loomis, Hebert, and Cicinelli, 1990; Loomis, Golledge, Klatzky, Speigle, and Tietz, 1994). Such a system provides an answer to the "Where Am I" problem by allowing determination of position in geographic coordinate space with considerable accuracy (e.g., within one meter). The user need not engage in substantial cognitive processing in order to know where he or she is at any particular point in time. The locator unit (i.e., the GPS receiver) gets signals from a set of satellites (the NAVSTAR system) which, although usually distorted by random noise to prevent unauthorized use, can be made more accurate using the differential mode of operation. In this mode a fixed base station provides error correction signals by radio link and thus provides for more accurate localization of the mobile receiver. However, GPS signals, whether corrected or not, can be lost because of the shadowing effects of large buildings (e.g., as in an urban downtown area), high hills, or a forest canopy (in rugged or remote natural environments).

With such a device, unique or commonly available databases can be used for planning regular routes or for experimenting with novel routes. For example, comprehensive Data Bases which include detailed street systems for every urban area in the United States are contained in the Census Departments' TIGER files. Commercial companies have downloaded segments of these files, adding software that allows documenting and path selection to take place (e.g., Street Atlas, USA). Although such readily available databases do not have the detail required for a vision impaired pedestrian, it may soon be possible to add additional information on an "as needed" basis within a specific neighborhood or community of a given traveler. Piggybacking the specific needs of a blind traveler onto an existing database might make the Personal Guidance System adaptable to many different environments and different circumstances.

As opposed to the PGS, other devices involve placement of emitters at targeted sites in the environment which announce their location to a suitably equipped nearby traveler (Kelly, 1981). Collins (1985) proposed a guidance system using global positioning capability somewhat similar to the PGS we are developing.

Accommodation of the Visually Impaired to Wayfinding Problems

The research reviewed here generally shows that prior visual experience is not necessary for the adequate performance of wayfinding tasks. This implies that the cognitive mapping abilities of the blind and vision impaired are adequate for such tasks. Yet they must accommodate to their reduced capabilities in other respects in order to find their way.

The difficulty of solving the problem of learning spatial layouts and remembering specific paths through them forces many blind travelers to rely on a small number of familiar routes (simplification) in order to move around an environment. This tactic is often encouraged in orientation and mobility training, where learning of segment sequences and remembering specific actions such as direction changes or path crossings at specific choice points, provides a safe and simple method for locomoting effectively, efficiently, and safely. Even with such a strategy, however, there are times when additional assistance is required. For example, a well known route may be blocked by a temporary barrier – as in the case of construction. To circumnavigate the barrier, an individual may have to offset a known path by a particular angle and distance. Sometimes this may require exploring new areas, as would be the case if one had to detour around a block that was temporarily obstructed in order to resume travel along a known path. Such problems frequently deter those without vision from traveling extensively in their local environments. This in turn inhibits the development of skills that would help to make them independent travelers. Their overall confidence in dealing with the environment is thus lessened, the number and variety of activities in which they participate drops, and quality of life is seriously impaired.

Obviously things that could be done to improve the mobility and confidence of the blind traveler would add considerably to their well being, reduce the emotional stress and frustration usually felt by individuals constrained in their social, economic, and other interactions, and potentially could make such individuals a better integrated and more productive part of any community. As Kitchin (1994) has noted, cognitive maps are such an important part of everyday activity that their development and use cannot be ignored. This appears to be essentially true for disabled groups who often suffer from poverty-stricken cognitive maps even at the best of times.

When tactual aids can be used as surrogates for vision, their use requires additional time to sense, store, recall, and interpret the information. Thus navigation may be slower, more cautious, less graceful, and perhaps even less effective than movement undertaken with vision. It may be less effective because there is no opportunity to bypass unimportant cues, modify veering tendencies, or to take shortcuts. Without prior knowledge or experience of the route, navigation may be successful but inefficient. The sensed obstacles may be avoided by circuitous loops; multi-segments and multi-turn angle routes may have to be followed to avoid crossing a busy intersection that may produce a more efficient, shorter or quicker route. The selection of a route may be enhanced by previewing, and wayfinding performance may be enhanced by preprocessing that defines cues and choice points so that an opportunity to update general information or alter the order of route segments may occur. Such devices can be summarized in the well-known strip maps prepared for automobile travel but just as easily produced for pedestrian travel (Golledge, 1991; Bell, 1994).

A major problem facing the blind traveler is learning about a new environment in which he/she wishes to navigate. Research that we have summarized gives rise to recommendations for some general strategies blind people might use to learn about a novel location so that they can travel within it. They include:

1. Conduct systematic exploration.
2. Choose a clearly defined origin (e.g., significant landmark or choice point).
3. Try to define the nature of the functional structure embedded in the local environment (e.g., the regularity of street systems and size of blocks, etc.)
4. During initial and consequent exploration, define critical landmark cues (whether they be single objects, intersections, linear segments, small neighborhoods, or other geographic components) to be used as anchorpoints for the cognitive representation of the environment.
5. Develop a cognitive representation which embeds in it layout and functional characteristics.
6. Choose path selection criteria (e.g., minimize effort, minimize time, maximize aesthetics, minimize obstacles, minimize turns, etc.) and select a route.

If these steps are followed, a destination should be clearly imaged and defined, and it should then be possible to select a path that leads to the destination.

Summary

The blind and vision impaired adult faces life with greater challenges because of sensory impairment. This does not, however, mean that fundamental spatial abilities are absent, nor does it mean that activities such as exploration, search, shortcutting, and wayfinding generally are beyond his or her capabilities. The research reviewed in this chapter confirms the presence of such abilities and, in some circumstances, shows no significant difference in spatial task performance between those with and without vision, even when vision has been absent since birth.

We hope that, by presenting this review, we have suggested some directions for fruitful areas of research effort by readers of this book that could materially and positively effect the individual well-being of people without vision and their more complete and natural integration into society as a whole.

Acknowledgment: This paper was produced as part of a larger project that involves investigation of navigation ability and developing a Personal Guidance System for blind travelers. We acknowledge the support of NIH/NEI Grant #EY09740 during the preparation of this paper.

References

Andrews, S.K. (1983). Spatial cognition through tactual maps. In *Proceedings of the 1st International Symposium on Maps and Graphics for the Visually Handicapped* (J. Wiedel, ed.). Washington, DC: Association of American Geographers, pp. 30-40.

Bell, S.M. (1994). Cartographic presentation as an aid to spatial knowledge acquisition in unknown environments. Thesis submitted in partial satisfaction of the requirements for the degree of Master of Arts in Geography. University of California Santa Barbara, Santa Barbara, California.

Bentzen, B.L. (1980). Orientation aids. In *Foundations of orientation and mobility* (R. Welsh and B. Blasch, eds.). New York: American Foundation of the Blind, pp. 291-345.

Brambring, M. (1976). The structure of haptic space in the blind and sighted. *Psychological Research* 38, 283-302.

Banerjee, M. (1928) Blindfold description of distance. *Indian Journal of Psychology* 3, 95-99

Brabyn, J. (1985). A review of mobility aids and means of assessment. In *Electronic spatial sensing for the blind – contributions from perception, rehabilitation, and computer vision* (D.H. Warren and E.R. Strelow, eds.). Boston: Martinus Nijhoff Publishers, pp. 13-27.

Carpenter, P.A. and Eisenberg, P. (1978). Mental rotation and the frame of reference in blind and sighted individuals. *Perception and Psychophysics* 23, 117-124.

Casey, S.M. (1978). Cognitive mapping by the blind. *Journal of Vision Impairment and Blindness* 72, 8: 297-301.

Cicinelli, J. (1989). Veer as a function of preview and walking speed. Thesis submitted in partial satisfaction of the requirements for the degree of Master of Arts in Psychology. University of California Santa Barbara, Santa Barbara, California.

Cleaves, W.T. and Royal, R.W. (1979). Spatial memory four configurations by congenitally blind, late blind, and sighted adults. *Journal of Visual Impairment and Blindness* 73, 13-19.

Collins, C.C. (1985). On mobility aids for the blind. In *Electronic spatial sensing for the blind: Contributions from perception, rehabilitation, and computer vision* (D.H. Warren and E.R. Strelow, eds.). Dordrecht: Martinus Nijhoff Publications, pp. 35-64.

Cratty, B.J. (1965). *Conceptual thresholds of nonvisual locomotion: part 1.* Department of Physical Education Monograph (NIH Grant #NB05577-0251) Los Angeles: University of California.

Cratty, B.J. and Williams, H.G. (1966). *Perceptual thresholds of nonvisual locomotion: part 2.* Department of Physical Education Monograph (NIH Grant #NB05577-0251) Los Angeles: University of California,

D'Oliveira, E. J. (1939). Place of the pilot in formation flight. *Reviste Medicina Latina Americana* 24, 1232-1235.

Dodds, A.G. (1983). Mental rotation and visual imagery. *Journal of Visual Impairment and Blindness* 77, 16-18.

Dodds, A.G. (1988). *Mobility training for visually handicapped people*, London: Croom Helm.

Dodds, A.G., Howarth, C.I. and Carter, D. (1982). The Mental Maps of the Blind; The role of previous visual experience. *Journal of Vision Impairment and Blindness* 76, 1: 5-12.

Downs, R.M., and Stea, D. (eds.) (1973). *Image and environment: Cognitive mapping and spatial behavior.* Chicago: Aldine.

Eliot, J., and McFarlane-Smith, I.M. (1983). *An international directory of spatial tests*, Oxford: NFER-Nelson Publishing Company.

Fletcher, J.F. (1980). Spatial representations in blind children, 1: Development compared to sighted children. *Journal of Visual Impairment and Blindness* 74, 10: 381-385.

Foulke, E. (1971). The perceptual basis for mobility. *Research Bulletin of the American Foundation for the Blind* 23, 1-8.

Foulke, E. (1982). Perception, cognition and the mobility of blind pedestrians. In *Spatial abilities: Development and physiological foundations* (M. Potegal ed.). New York: Academic Press, pp. 55-76.

Fujita, N., Loomis, J.M., Klatzky, R.L., and Golledge, R.G. (1990). A minimal representation for dead-reckoning navigation: Updating the homing vector. *Geographical Analysis* 22, 4: 326-335.

Fujita, N., Klatzky, R.L., Loomis, J.M., and Golledge, R.G. (1993). The encoding-error model of pathway completion without vision. *Geographic Analysis* 25, 4: 295-314.

Gallistel, C.R. (1990). *The Organization of Learning*, Cambridge, MA: MIT Press.

Gärling, T., and Golledge, R.G. (1989). Environmental perception and cognition. In *Advances in environment, behavior, and design, Volume 2* (E.H. Zube and G.T. Moore, eds.). New York: Plenum Press, pp. 203-236.

Golledge, R.G. (1991). Tactual strip maps as navigational aids. *Journal of Visual Impairment and Blindness* 85, 7: 296-301.

Golledge, R.G., Rivizzigno, V.L., and Spector, A. (1974). Analytical methods for recovering cognitive information about a city. Paper presented to IGU Regional Conference, Palmerston, New Zealand.

Harris, J. C. (1967). Veering tendency as a function of anxiety in the blind. *American Foundation of the Blind Research Bulletin* 14, 53-63

Hart, R.A., and Moore, G.T. (1973). The development of spatial cognition: A review. In *Image and environment: Cognitive mapping and spatial behavior* (R.M. Downs and D. Stea, eds.). Chicago: Aldine, pp. 246-288.

Hill, E., and Blasch, B.B. (1980). Concept Development. In *Foundations of Orientation and Mobility* (R.L. Welch and B.B. Blasch, eds.). New York: American Foundation for the Blind.

Hill, E.W., Rieser, J.J., Hill, M.M., Hill, M., Halpin, J., and Halpin, R. (1993). How persons with visual impairments explore novel spaces: Strategies of good and poor performers. *Journal of Visual Impairment and Blindness* October, 295-301.

Hollins, M. (1986). Haptic mental rotation: more consistent in blind subjects? *Journal of Visual Impairment and Blindness* 80, 950-952.

Hollyfield, R.L., and Foulke, E. (1983). The spatial cognition of blind pedestrians. *Journal of Visual Impairment and Blindness* 5, 204-209.

Jacobson, W.H. (1993). *The Art and Science of Orientation and Mobility*, New York: AFB Press..

Kelly, G.W. (1981). Sonic orientation and navigational aid (SONA). *Bulletin of Prosthetics Research* 1, 189.

Kitchin, R.M. (1994). Cognitive maps: What are they and why study them? *Journal of Environmental Psychology* 14, 1: 1-19.

Klatzky, R.L., Golledge, R.G., Loomis, J.M., Cicinelli, J.G., and Pellegrino, J.W. (1994). Performance of blind and sighted in spatial tasks. *Journal of Vision Impairment and Blindness* (in press).

Klatzky, R.L., Loomis, J.M., Golledge, R.G., Cicinelli, J.G., Doherty, S., and Pellegrino, J.W. (1990). Acquisition of Route and Survey knowledge in the absence of vision. *Journal of Motor Behavior* 22, 1: 19-43.

Lederman, S.J., Klatzky, R.L., Collins, R., and Wardell, J. (1987). Exploring environments by hand and foot: time based heuristics for encoding distance in movement space. *Journal of Environmental Psychology: Human Learning, Memory, and Cognition* 13, 606-614.

Leonard, J.A., and Newman, R.C. (1970). Three types of ""maps" for blind travel. *Ergonomics* 13, 165-179.

Lockman, J.J., Rieser, J.J., and Pick, H.L. (1981). Assessing blind travelers' knowledge of spatial layout. *Journal of Visual Impairment and Blindness* 7, 321-326.

Lohman, D.F. (1979). *Spatial ability: Review and re-analysis of the correlational literature*. Aptitude Research Project, Report #8, Stanford University.

Loomis, J.M. (1985). *Digital map and navigation system for the visually impaired*. Unpublished manuscript. Department of Psychology, University of California, Santa Barbara.

Loomis, J.M., Hebert, C., and Cicinelli J.G. (1990). Active localization of virtual sounds. *The Journal of the Acoustical Society of America* 88, 4: 1757-1764.

Loomis, J.M., Golledge, R.G., Klatzky, R.L., Speigle, J.M., and Tietz, J. (1994). Personal guidance system for the visually impaired. In *Proceedings of the First Annual International ACM/SIGCAPH Conference on Assistive Technologies*, Marina Del Rey, California, October 31-November 1.

Loomis, J.M., Klatzky, R.L., Golledge, R.G., Cicinelli, J.G., Pellegrino, J.W., and Fry, P.A. (1993). Non-visual navigation by blind and sighted: Assessment of path integration ability. *Journal of Experimental Psychology, General* 122, 1: 73-91.

Lund, F.H.(1930). Physical asymmetries and disorientation. *American Journal of Psychology* 42, 51-62.

Marmor, G.S., and Zabeck, L.A. (1976). Mental rotation by the blind: does mental rotation depend on visual imagery? *Journal of Environmental Psychology: Human Perception and Performance* 2, 515-521.

McGee, M.G. (1979). Human spatial abilities: Psychometric studies and environmental, genetic, hormonal, and neurological influences. *Psychological Bulletin* 86 (5), 889-918.

Mittelstaedt, H., and Mittelstaedt, M-L. (1982). Homing by path integration. In *Avian navigation* (Papi and Wallraff, eds.). Berlin: Springer-Verlag, pp. 290-297.

Moore, G.T., and Golledge, R.G. (1976). Environmental knowing: Concepts and theories. In *Environmental knowing* (G.T. Moore and R.G. Golledge, eds.). Stroudsburg, PA: Dowden, Hutchinson and Ross, pp. 3-24.

Müller, M., and Wehner, R. (1988). Path integration in desert ants, Cataglyphis Fortis. *Proceedings of the National Academy of Sciences of the United States of America* 85, 5287-5290.

Olivier, D. (1970). Metric for comparison of multidimensional scaling. Unpublished manuscript.

Parkes, D., and Dear, R. (1990). *NOMAD: An interacting audio-tactile graphics interpreter.* Reference manual, Version 2.0. Institute of Behavioral Science, University of Newcastle, NSW, Australia.

Passini, R., Dupré, A., and Langlois, C. (1986). Spatial mobility of the visually handicapped active person: A descriptive study. *Journal of Visual Impairment and Blindness* 80 (8), 904-907.

Passini, R., Proulx, G., and Rainville, C. (1990). The spatio-cognitive abilities of the visually impaired population. *Environment and Behavior* 22 (1), 91-118.

Preiser, W.F.E. (1985). A combined tactile/electronic guidance system for visually impaired persons in indoor and outdoor spaces. *Preconference proceedings of the International Conference on Building Use and Safety Technology*, pp. 49-53.

Rieser, J.J., Guth, D.A., and Hill, E.W. (1982). Mental processes mediating independent travel: Implications for orientation and mobility. *Journal of Visual Impairment and Blindness* 76 (6), 213-218.

Rieser, J.J., Guth, D.A., and Hill, E.W. (1986). Sensitivity to perspective structure while walking without vision. *Perception* 15, 173-188.

Rieser, J.J., Lockman, J.J., and Pick, H.L. Jr. (1980). The role of visual experience in the mental representation of spatial layout. *Perception and Psychophysics* 28 (3), 185-190.

Rouse, D.L. and Worchel, P. (1955). Veering tendency in the blind. *The New Outlook for the Blind* 4, 115-119.

Schaeffer, A.A. (1928). Spiral movement in men. *Journal of Morphology* 45, 293-298.

Strelow, E.R. (1985). What is needed for a theory of mobility: Direct perception and cognitive maps'lessons from the blind. *Psychological Review* 92 (2), 226-248.

Strelow, E.R., and Brabyn, J.A. (1982). Locomotion of the blind controlled by natural sound cues. *Perception* 11, 635-640.

Tatham, A.F., and Dodds, A.G. (1988). *Proceedings of the Second International Symposium on Maps and Graphics for Visually Handicapped People.* Nottingham, England: University of Nottingham.

Tellevik, J.M. (1992). Influence of spatial exploration patterns of cognitive mapping by blindfolded sighted persons. *Journal of Visual Impairment and Blindness* 92, 221-224.

Tolman, E.C. (1948). Cognitive maps in rats and men. *Psychological Review* 55, 189-208.

von Saint Paul, U. (1982). Do geese use path integration for walking home? In *Avian navigation* (F. Papi and H.G. Wallraff, eds.), Berlin: Springer-Verlag, pp. 298-306.

von Senden, S.M. (1932). *Space and Sight: The Perception of Space and Shape by the Congenitally Blind Before and After Operation*, Glencoe, Il: The Free Press.

von Senden, M. (1960). Space and sight. (Translated from the 1932 edition by P. Heath). Glencoe, IL: Free Press.

Warren, D.H. (1978). Perception by the blind. In *Handbook of Perception, Volume 10* (E. Carterett and N. Friedmann, eds.). New York: Academic Press, pp. 65-86.

Warren, D.H., and Kocon, J.A. (1974). Factors in the successful mobility of the blind. *American Foundation for the Blind Research Bulletin* 28, 191-218.

Warren, D.H., and Strelow, E.R. (eds.) (1985). *Electronic spatial sensing for the blind – contributions from perception, rehabilitation, and computer vision.* Boston: Martinus Nijhoff Publishers.

Welsh, R.L., and Blasch, B.B. (eds.) (1980). *Foundation of orientation and mobility*, New York: American Foundation for the Blind.

Wiedel, J. (ed.) (1983). *Proceedings of the First International Conference on Maps and Graphics for the Visually Impaired.* Washington, D.C.: The Association of American Geographers.

Worchel, P. (1951). Space perception and orientation in the blind. *Psychological Monographs: General and Applied* 65, 1-27 (Whole No. 332).

Reginald G. Golledge
Department of Geography
and Research Unit in Spatial Cognition and Choice
University of California Santa Barbara
Santa Barbara, California 93106

Roberta L. Klatzky
Department of Psychology
Carnegie-Mellon University
Pittsburgh, Pennsylvania 15213

Jack M. Loomis
Department of Psychology
University of California Santa Barbara
Santa Barbara, California 93106

THE CONSTRUCTION OF COGNITIVE MAPS BY CHILDREN WITH VISUAL IMPAIRMENTS

Simon Ungar, Mark Blades and Christopher Spencer

Abstract: The way in which children who have visual impairments construct cognitive maps of their environment is of considerable theoretical and practical importance. It sheds light on the role of sensory experience in the development of spatial cognition which can in turn suggest how spatial skills might be nurtured in visually impaired children. In most of the studies reviewed here, groups of children who lost their sight early in life perform less well on a variety of spatial tasks than sighted children or children who lost their sight later in life. We will argue that it is not the lack of visual experience in itself which produces this pattern, but rather the effect of lack of vision on the spatial coding strategies adopted by the children. Finally we will discuss a number of methods for encouraging visually impaired children to use coding systems which are appropriate for the construction of flexible and integrated cognitive maps, with particular reference to the use of tactile maps.

Introduction

The spatial abilities of visually impaired people have been a focus of study within psychology for both theoretical and practical reasons. We can discover a great deal about the nature of spatial representation in general by studying cases of sensory deprivation – for instance, whether spatial representations are necessarily based on visually derived codes. In practical terms, a greater understanding of the way in which visually impaired people represent space is important for the development of methods for improving their spatial skills.

It is now generally believed that visually impaired people can acquire spatial concepts and representations through their intact senses, but there is still some debate about the level of representation which can be constructed on the basis of non-visual information. A number of authors suggest that visually impaired people are limited to a relatively fragmentary and inflexible representation of the environment. Others, however, believe that the representations of visually impaired people are limited only by their experience of the environment and that even totally congenitally blind people have the potential to form integrated representations of an environment if provided with sufficient appropriate experience.

J. Portugali (ed.), The Construction of Cognitive Maps, 247–273.

This chapter focuses on the construction of spatial representations by children with visual impairments. We will consider the ways in which visually impaired children come to understand the spatial structure of their environments and their potential for acquiring more powerful representations. For convenience, the literature in this area will be considered under two general headings – small-scale space and large-scale space – although these are intimately related. We shall also report some of the work on spatial representation in visually impaired adults where this clarifies or strengthens findings from the literature on children.

Theoretical Background on Spatial Development

Foulke (1982; Foulke and Hatlen, 1992) argued that vision is the 'spatial system par excellence', on the basis that no other sense has such scope or clarity. Fletcher (1980) has identified three broad theoretical positions which have existed in the literature on the spatial concepts of visually impaired people. The first of these, which Fletcher calls the 'Deficiency Theory', is exemplified in the work of von Senden (1932) who concluded that spatial concepts are impossible in people who have been blind from birth, and that visual experience for some part of life is essential for even a minimal understanding of space. This theory is now primarily of historical interest as more recent research has largely discredited Senden's position.

The second theoretical strain identified by Fletcher is the Inefficiency Theory which suggests that people who are blind from birth develop concepts and representations of space but that they are functionally inferior to those of the sighted and the late blind. This theory continues to receive support, especially from researchers in the Gibsonian tradition (e.g. Rieser, 1990; Rieser et al., 1986) who have found that the congenitally blind have difficulty mentally updating their position in an environment experienced through locomotor exploration. Researchers have also shown that the congenitally blind tend to construct representations of environments experienced through locomotion in terms of linear routes consisting of sequences of paths linked by decision points. Such a representation would not support the inferences about the relative locations of places that are possible with an overall, survey representation of space.

The third theoretical position is the Difference Theory which proposes that the visually impaired may build up a set of spatial relations, which are functionally equivalent to those of the sighted, but that they do so more slowly and by different means. For example, Juurmaa (1973) suggested that the visually impaired do attain a spatial competence equivalent to that of the sighted by mid-adolescence. Juurmaa proposed that this delayed development of spatial cognition in the visually impaired relative to the sighted may, in part, account for the differences found between these groups when peers are compared on spatial tasks. Juurmaa also argued that the poor performance of the visually impaired relative to blindfolded sighted participants on many tests of spatial competence is attributable to the use of experimental stimuli which are highly familiar to the sighted but less so for the visually impaired participants. For example, in a study by Worchel (1951), participants were led along two sides of a triangle and then asked to return to the start. Congenitally blind participants performed less well than blindfolded sighted participants. The sighted participants may have had a much stronger mental image of a triangle prior to the experiment than the visually impaired participants.

There are, of course, alternatives to vision. The haptic-proprioceptive system can provide precise spatial data, but only within the scope of the body itself, and therefore, as a blind person moves through the environment encountering objects, sounds, underfoot textures etc., this sequence of sensory impressions must be actively cognitively constructed to form an ordered representation. The auditory system also provides access to stimuli, and over a more extensive range, but it is much less useful for precise localisation.

Juurmaa (1973) claims that although the visually impaired lack the quantity and quality of spatial experience which is constantly available to the sighted individual, the developmental delay could nevertheless be minimised if visually impaired children are provided with enough experience of the right kind from an early age. Millar (1988) agrees that senses other than vision are less adequate for coding spatial relational information, but argues that the potential of the visually impaired to acquire a fully integrated representation of space is no less than that of the sighted.

Representation of Small-Scale Spaces: Crossmodal interaction

The study of the interrelationship between the sensory modalities has important implications for theories of spatial cognition of visually impaired people. Central to this issue is whether sensory information is coded in the brain in a way that is specific for each modality or whether information is re-coded in more uniform way, and independently of any specific modality (Millar, 1981; Warren, 1984). According to the former belief, environmental information would retain the features of the sensory modality through which it was apprehended and given Foulke's (1982; Foulke and Hatlen, 1992) analysis this would predict that visually impaired people, relative to people with sight, are impaired in the ability to encode spatial information.

A number of studies on cross-modal functions in visually impaired and sighted children (Hermelin and O'Connor, 1971; Hermelin and O'Connor, 1975; Hermelin and O' Connor, 1982; Millar, 1981; O'Connor and Hermelin, 1972) have shown that the modality used in tasks influences the way in which information is encoded. For instance,

O'Connor and Hermelin (1972) presented congenitally totally blind and blindfolded sighted children with sequences of auditory stimuli which were separated in space. They randomly varied the relationship between the temporal and spatial sequencing of presentation of the stimuli. Both groups tended to report the 'middle' stimulus as being the temporally middle rather than the spatially middle stimulus. In contrast, sighted children performing an identical task with visual rather than auditory stimuli tended to choose the spatially middle stimulus. This result suggests a general tendency to structure information differently in the two modalities; auditory information being structured according to temporal occurrence and visual information being structured spatially.

It has also been shown that visual experience influences the coding of tactile and kinaesthetic information in non-visual tasks. Hermelin and O'Connor (1971; 1975) presented congenitally totally blind and sighted children with two tasks which examined the tendency to code spatial information according to self-referent or external cues. In both tasks it was found that most of the visually impaired children used a coding strategy with reference to their own body. Sighted children, in contrast, tended to code spatial position and movement within an external frame of reference. However, the fact that a small number of visually impaired children did use external cues suggests that visual experience is not a necessary requirement for the development of an exterocentric system of reference. A number of other studies which have found group differences between early blind and sighted participants on spatial tasks also report a number of early blind people who perform similarly to the sighted (Casey, 1978; Dodds et al., 1982; Fletcher, 1980).

This phenomenon is discussed by Millar (1975; 1976; 1979; 1981; 1988) in her extensive analysis of the representation of tactile and kinaesthetic spatial information by visually impaired children. Millar (1988) argued that spatial information can be derived from hearing, touch and movement and that the visually impaired therefore have the potential to acquire concepts and representations of the spatial domain equivalent to those of the sighted. Millar (1988) emphasized the importance of recognizing the basic potential of children to acquire certain skills before interpreting the performance of children in tests of those skills.

Millar (1988) proposed a model of sensory deprivation in which the 'type and reliability of spatial information' (p. 72) available to visually impaired children differs from that available through vision and these differences in the quality of experience can prompt the child to organize spatial information by different coding strategies from those which arise from visual experience. In this respect, it is important to reassess the concept of 'ability' as being a level of competence in a certain, spatial or other, skill. The lack of vision and the resulting difference in the quality of experience of space lead the child to approach a task in a different way, using different strategies. The tendency for congenitally blind children is to use self-referent and movement coding strategies because, in the absence of vision, these are generally highly efficient for spatial tasks. Consider, for instance, the task of repeatedly locating a cup of tea placed in a constant position on a desk as you remain seated at the desk in a constant position. With vision it may be more natural to encode the cup's position relative to other objects on the desk. In the absence of vision, this strategy would involve locating the reference objects by touch each time you wanted to take a sip. It would be far more efficient in this case simply to encode the cup's position relative to your own body coordinates or according to a particular, reliably reproducible arm movement. Strategies are seen by Millar as "optional forms of coding" which differ in the types of information selected (e.g. relationships between locations in space or relation of locations relative to the body mid-line) and the coding heuristics appropriate for a particular type of information (e.g. external frame of reference, self-referent, movement). The strategies are optional in the sense of being interchangeable, although they are not identical. Visual experience prompts children to attend to external cues (e.g. the interrelationships between locations) and this is the case both for sighted children performing the tasks blindfold and for late blinded children. Congenitally blind children tend to neglect such cues and thus adopt different strategies.

Studies on mental imagery in sighted people have suggested that representations of perceptible objects and events are picture-like or are at least in some way analogous to visual perception. Producing mental images of words facilitates their recall, and images can be mentally scanned and rotated just as such events would occur in perception (e.g. Kosslyn et al., 1978; Neisser and Kerr, 1973; Paivio, 1986; Shepard and Metzler, 1971). However, as Paivio (1986) points out, imagery can be derived from all the sensory modalities and therefore even congenitally blind people could, in principle, form mental images of objects based on their intact sensory modalities (especially audition and touch). In fact all of the studies cited above have been adapted for congenitally blind people to determine whether vision is necessary for mental imagery (Carpenter and Eisenberg, 1978; Kerr, 1983; Marmor and Zaback, 1976; Zimler and Keenan, 1983). In general, visually impaired participants in these studies perform very similarly to sighted participants, with the exception that reaction times tend to be rather slower for congenitally blind participants. This suggests that visual experience is not necessary for mental imagery (i.e. the representation of the spatial structure of objects and events) but that visual experience might facilitate the manipulation of images.

The findings from mental imagery tasks provide further support for the argument that visually impaired children can acquire spatial representations which are functionally equivalent to those of sighted people. Millar (1982) has extended her argument to suggest that "the various types of 'imagery' are optional coding strategies" (p. 119) in the sense that they are interchangeable. For instance imagining a rotation in terms of an arm or hand movement can provide just as adequate a basis for performance on a mental rotation task as imagining the rotation as if it was visually presented.

Representation of large-scale spaces

So far our discussion has focussed on studies performed in small scale spaces. A number of studies have compared visually impaired and sighted people's understanding of large scale environments. Golledge (1993) notes a number of commonalities in the learning of small-scale and large-scale spaces, such as hierarchical organisation and clustering. In contrast to the learning of small scale environments, which can be perceived at a single glance with vision but must be sequentially explored by touch, when learning a large-scale environment, visually impaired and sighted people alike are faced with the task of integrating information over time. In this important sense, the task of acquiring a representation of a large scale environment ('...one whose structure is revealed by integrating local observations over time, rather than being perceived from one vantage point.' Kuipers, 1982, p.203) is formally similar for the sighted and for the visually impaired. There are thus no a priori grounds for assuming that representations of large scale spaces acquired from vision or haptics-proprioception should differ in structure. For instance, within his model of spatial representation, Kuipers (1982) has stressed that views (the basic elements of the spatial representation) 'need not be visual images: A blind person's views could be auditory, tactile or even olfactory.' (p. 213) The integration of such elements into an environmental representation is in principle similar for visually impaired and sighted people.

However, the literature concerned with small-scale space indicates that visual experience has some influence on the way in which spatial information is encoded or organized, even when sighted controls perform the task blindfold. It is possible that such processing differences also apply to large-scale spatial tasks, even though the immediate task constraints are more similar for visually impaired and sighted people. Within the literature on large-scale environments, a few studies have focussed on the understanding of familiar environments but most have considered participants' ability to construct representations of novel environments, for example real buildings or experimental layouts.

Familiar Environments

Bigelow (1991) explored young visually impaired children's representations of the layout of their homes and neighbourhoods. Totally blind, partially sighted and sighted children in two age groups (mean ages: 4.7 and 6.0 years) were asked to point to locations in their homes and neighbourhoods, and their pointing responses were scored according to three criteria. The response was either in the Euclidean direction of the named location

(Euclidean), along the first segment of the functional route to the location (route) or in neither of these directions. The partially sighted and sighted children mastered the tasks within the fifteen month testing period, mostly on the first session. In contrast, the totally blind children failed to master most of the tasks within the study period. Analysis of the children's errors showed that totally blind children more often pointed along the route to each location (route response) rather than towards the location itself (Euclidean response). This finding is supported by a number of studies (Byrne and Salter, 1983; Casey, 1978; Lockman et al., 1981; Rieser et al., 1980) using a variety of methods for externalizing the spatial representations of visually impaired and sighted adults of familiar environments, which found that the early blind tended to code the relative location of places according to their functional separation, suggesting a representation based on routes rather than an integrated configuration.

Ungar (1994, Experiment 4) examined the spatial knowledge of a familiar space in eighteen visually impaired children (aged 6 to 12.5 years). The children's judgements of the relative distance between nine locations in their school (see Figure 1) were tested using the method of triadic comparisons (Rieser et al., 1980). The locations were named in sets of three and each child was asked which two locations were the furthest apart, which two were the closest together (the intermediate pair can be inferred from these two questions). A balanced incomplete design was used (Burton and Nerlove, 1976) in which a set of forty eight triads were presented, each pair of locations appearing four times.

The pair judged to be furthest apart was assigned a value of 2, the pair judged to be closest together was assigned a 0 and the remaining pair was assigned a 1. Each child judged each pair of locations in the context of four different triads. The scores of these four judgements were summed to give an overall value ranging from 0 to 8 for each pair of locations (i.e. for each inter-location distance). Two sets of values were also generated for the 'ideal Euclidean participant' and for the 'ideal functional participant'. These were generated by completing the triads test on the basis of Euclidean and functional measurements taken from a scale map of the school and its grounds (the Euclidean distance is the straight-line or 'crow's flight' path between two locations whereas the functional distance is the normal path of travel between two locations).

Two error scores were calculated for each of the children, one relative to Euclidean distances and the other relative to functional distances. For each pair of locations the child's mean value was subtracted from that of both the ideal Euclidean participant and the ideal functional participant. This yielded two measures of error - one against a Euclidean baseline and one against a functional baseline.

Relative to the Euclidean baseline, the scores of the totally blind children were significantly higher (i.e. worse) than those of the children with residual vision. The analysis of scores relative to the functional baseline yielded no significant effects, but scores were generally lower for this analysis than for the Euclidean analysis.

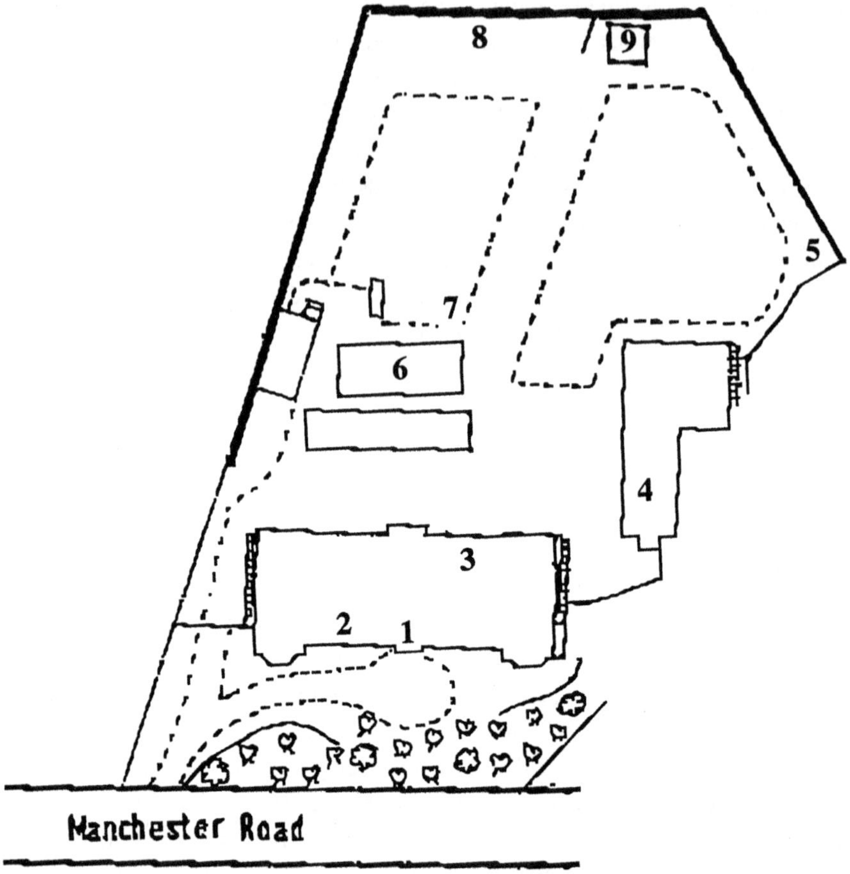

Figure 1: Layout of the school and grounds used to test children's knowledge of a familiar environment by Ungar (1994, Experiment 3) 1– Entrance steps, 2 – Dining room. 3 – Staff room, 4 – Assembly hall. 5 – Skittle alley, 6 – Play area, 7 – Sand pit, 8 – Boat, 9 – Swings.

Rank order correlations were performed to compare the distance judgements of each age and visual status group with those of the ideal Euclidean and functional participants. As Euclidean and functional distances are themselves intrinsically correlated, this correlation was partialled out. On the whole, children's relative distance judgements correlated more highly with the functional baseline than with the Euclidean baseline. However, in the case of the younger residual vision group, a correlation with the Euclidean baseline was obtained. The result from the younger residual vision group did not concur with the findings of Rieser et al. (1980) with adults. In their study, Rieser et al. found that both visually impaired and sighted adults' estimates correlated highly with functional distances unless participants were specifically instructed to give straight line distances (in the latter case, sighted and late blind adults shifted to Euclidean estimated while congenitally blind adults continued to give functional estimates).

Therefore, correlations were performed for each child's judgements relative to the Euclidean and the functional baseline. Apart from one child whose judgements were correlated with neither baseline, the judgements of all the children correlated with either one or other of the baselines but not both. Only four of the children made judgements which were correlated with the Euclidean baseline; all four had residual vision and three were in the younger group.

In order to gain some impression of the mental representations underlying children's relative distance judgements, the data were analysed using the multidimensional scaling procedure ALSCAL (SPSS release for the Macintosh). The input for this program is a matrix of dissimilarities - in this case judged relative distances - and the output is the two dimensional configuration of locations which best fits the input dissimilarities (i.e. the program comes as close as possible to the monotonic relationship between metric distance and ordinal dissimilarity between locations which would exist in a two dimensional space). While this representation does not necessarily provide a full externalization of a participant's mental representation of space it effectively displays a person's impression of the relative distances between locations.

For all participants taken together a picture emerged in which functional distances were exaggerated. We would also suggest (speculatively) that the functional distances were based on habitual paths of movement by the children which were observed over three years of the author's work at the school. The Entrance Steps and the Dining Room were represented as relatively close in space at the entrance to the school. Staff Room was represented as lying on the path from these to the outside. From the side entrance to the school, two paths appear to diverge which cross the expanse of tarmac playground behind the school. One of these leads to the Assembly Room block and the adjacent Skittle Alley, while the other leads to the Play Area and Sand Pit. A relatively shorter distance leads from the Assembly Hall / Skittle Alley group to the Swings and from the Play Area / Sand Pit group to the Boat. The Swings and the Boat were generally located relatively close together.

Overall the results were consistent with those obtained by Rieser et al. (1980) for adult visually impaired participants. In particular, the error scores for all groups were lower relative to the functional than to the Euclidean baseline and the children's judgements were highly correlated with functional distances in the space for all groups.

The MDS analysis provided a visualisation of the children's functional bias in estimating relative distances. The indoor locations were generally represented very close together while distances across the open expanse of the playground were exaggerated.

This may be because children need to concentrate harder on maintaining a bearing across an open space than when walking along a corridor and thus distances across open spaces may be experienced as longer. Apart from these functionally exaggerated distances, the plots for the groups bore a close resemblance to the arrangement of the locations in the school suggesting that all the children had a good impression of relative distances from their extensive experience in the school environment.

When asked to estimate distances and directions in familiar environments, visually impaired children tend to respond on the basis of their functional experience of the environment rather than from an integrated representation. This result cannot be taken as evidence that visually impaired children are incapable of forming integrated representations of an environment as participants may have chosen to respond in this way although they were capable of providing straight line estimates. The fact that some of the children (some with very little sight) in the study described above spontaneously gave straight line estimates serves to underline this cautious interpretation of the results.

Constructed Environments

One problem in testing people's knowledge of familiar environments is that it is impossible to control for individual differences in experience. Rather than use familiar environments, a number of studies have tested children in novel environments – either an experimental environment constructed in the laboratory (Fletcher, 1980; Landau et al., 1984; Rieser, 1990) or an unfamiliar part of the real world (Dodds et al., 1982; Leonard and Newman, 1967; Ochaíta and Huertas, 1993).

Landau and her colleagues (Landau, 1986; Landau et al., 1981; Landau et al., 1984) carried out a series of studies with a single congenitally blind child, Kelli, between the ages of 3 and 5 years. In one study, Landau et al. (1984) familiarized Kelli with an experimental space (see Figure 2) by guiding her from a homebase position to one object and then from the homebase to the other object. Kelli was then led to the first object again and asked to walk to the second object. Kelli's trajectories were recorded on video and a number of measures were obtained from an analysis of the tapes, such as her initial heading as she left the first object and her final position. Landau et al. (1984) reported that Kelli's performance on the task was above chance and that she performed comparably to a group of blindfolded sighted children. Landau et al. suggested that Kelli was able to integrate information from the familiarisation phase into an overall impression of the relative positions of objects in the layout. In other words, that she formed a representation of the experimental space which was similar to that of the sighted children.

There have been several criticisms of Landau's methodology (Liben and Downs, 1989; Millar, 1988). Liben and Downs (1989) argue that Landau et al. (1986) overstated Kelli's accuracy in the inference tasks – for example Kelli's routes were not sufficiently straight to assume that she possessed accurate angular knowledge about the position of the target, and her distance errors were relatively large considering the size of the experimental room and the placement of targets. Millar (1988) pointed out that if Kelli was aware of or calculated the bearing of the target object at the outset of each route, one would not expect her to adjust her heading during movement. The fact that she did adjust her movement on many trials suggests that she had some means of updating her position and may have used sound cues or even some residual vision. In the light of these criticisms and the fact that only a single visually impaired child was tested, the results of Landau et al. (1984) should be interpreted with caution.

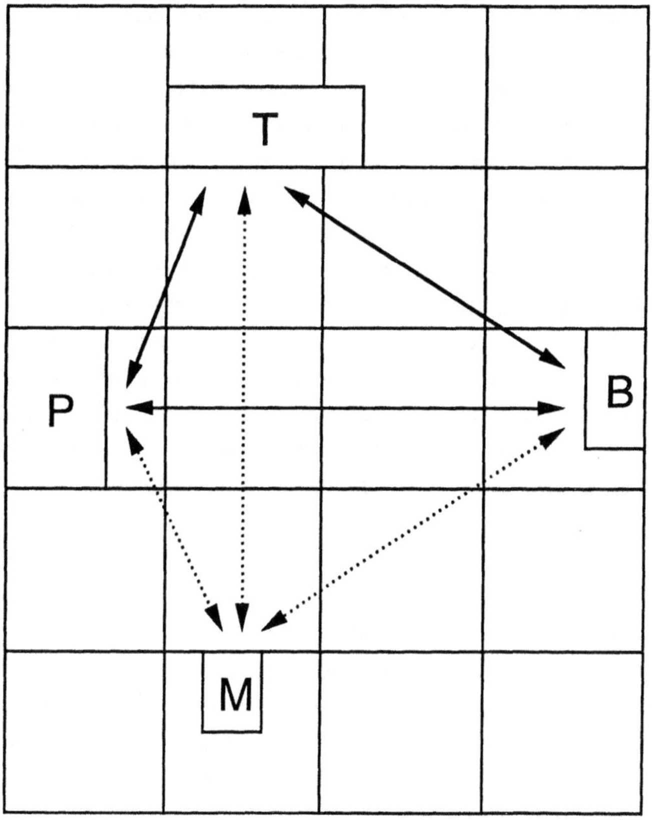

Figure 2: Layout used by Landau et al. (1984). M – Mother, P – Pillows, T – Table, B – Basket.

◀——▶ – Test Routes ◀⋯⋯▶ – Trained Route

Another study with young children is reported by Rieser (1990). An experimental layout was constructed consisting of eight buckets evenly spaced around a four foot diameter circle. One of these (the target) contained an interesting toy. Totally congenitally blind children (22 to 44 months old) were led from the centre of the circle to the target bucket and allowed to play with the toy. Then the children were asked to step back from the target bucket, returning to the centre of the circle. After this the children were turned by the experimenter on the spot through either 90° or 270°. The children were then asked to turn to face the target once again. All the children consistently reversed the rotation in order to face the target bucket despite the fact that the simplest solution in the 270° rotation condition would have been to continue turning in the same direction. In contrast, a group of sighted children tested in a darkened room consistently turned in the shorter direction to face the target. Rieser suggested that children coded their change in position in terms of their own body movement rather than in relation to the layout of external space.

Working with older children, Fletcher (1980) introduced visually impaired children and adolescents (7 to 18 years) to the layout of a room using either a scale model or direct experience. The participants' exploration of the space was either guided by the experimenter or unguided. Following exposure to the space, participants' knowledge of the room was tested with two types of questions, 'route' questions which tested participants' knowledge of spatial relations directly experienced during exploration and 'map' questions which required participants to go beyond their direct experience and infer spatial relationships between places not linked during exploration. For the congenitally totally blind children, accuracy was higher for route questions than for map questions– in contrast to a sighted control group who performed similarly on the two measures in a separate analysis – indicating that the congenitally blind children were unable to integrate information from their exploration of the room into an overall impression of the layout of objects.

Ungar et al. (in press, Experiment 2 - Exploration Condition) introduced congenitally totally blind and partially sighted children (aged 4 to 12 years) to a layout of six objects (see Figure 3) by walking them from a central point to each of the objects in turn, returning to the central point after each object had been visited. The children were then asked to aim a pointer from the central point to each of the other objects and from two of the object locations to all the other objects. Overall, the congenitally totally blind children were considerably less accurate than the partially sighted children.

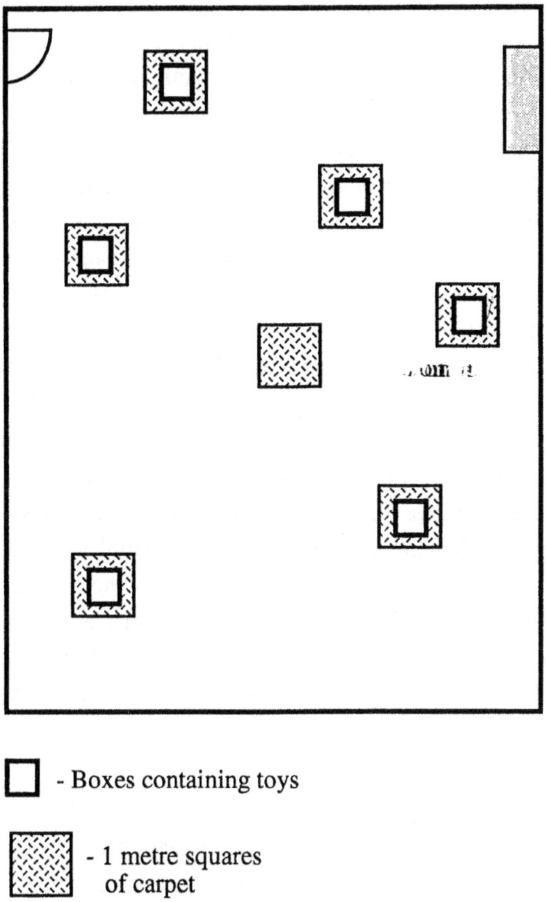

Figure 3: Example of the layout used by Ungar et al. (in press).

Novel Environments

Dodds et al. (1982) introduced congenitally and late totally blind children (mean age: 11.5 years) to a short urban route by leading them along it four times. As they walked the route children were repeatedly asked to make pointer estimates to a number of locations along the route. Overall, errors in direction estimation increased with distance from the target, but this effect was considerably greater for the congenitally blind children, who were less accurate overall than the late blind children. This finding suggests that visual experience facilitated the construction of coordinated spatial representations of locomotor displacements. As all the children were able to walk the route, Dodds et al. argued that the congenitally blind children must have formed an egocentric or self-referent representation of the route, but were not able to form an overall survey-level representation of the layout.

In another study, Ochaíta and Huertas (1993) familiarized visually impaired children and adolescents (from 9 years to 17 years) with a route linking seven landmarks in a real environment (school grounds or a public square) by leading them along it once. On three subsequent days, each participant led the experimenter along the route. If a participant required assistance from the experimenter this was noted. At the end of each session, participants were asked to construct a scale model of the space and to perform a distance ratio estimation task based on the between - landmark distances. No differences were found between congenitally blind and adventitiously blind groups. However there were differences between the four age groups with only the older participant groups showing evidence of a coordinated representation of the travelled environment. Ochaíta and Huertas (1993) concluded that although congenitally blind participants achieved a coordinated understanding of the environment, their ability to do so was delayed relative to the established chronologies for sighted children (e.g. Hart and Moore, 1973). Ochaíta and Huertas suggested that the sudden increase in spatial understanding achieved by visually impaired children at adolescence may be due to 'the competence in abstract and propositional reasoning (which is assumed to be reached at this age)' and that '[it] is possible that in adolescence, verbal reasoning may 'remediate' some of the problems caused by the lack of vision in understanding figurative and spatial problems' (1993, p.40).

Strategies for Coding Large-Scale Layouts

The studies reported above might be taken to imply that visual experience is necessary for forming coordinated, survey representations of the environment. However, such an interpretation would ignore the possibility, emphasised by Millar (1988), that alternative (non-visual) spatial coding strategies might be available to visually impaired children who are provided with appropriate experiences. The use of alternative coding strategies was highlighted in a study by Rieser et al. (1982; 1986). They tested the ability of congenitally totally blind, later blinded and blindfolded sighted adults to keep track of their position relative to a number of landmarks as they moved or imagined moving through an experimental layout of objects. The participants learned the layout by walking with an experimenter from the start point to each of the landmarks in turn, returning to the start each time. The participants were then tested in two experimental conditions. In the locomotion condition, participants were led by a circuitous route to one of the experimental landmarks and asked to aim a pointer at each of the other landmarks in turn. In the imagination condition participants made pointer estimates from the start point but were asked to imagine that they were standing at one of the experimental landmarks.

The sighted and the adventitiously blind groups performed very accurately in the locomotion condition but less accurately in the imagination condition. In contrast, the early blind performed both conditions at the same level which the other groups achieved in the imagination condition

The response latencies of the sighted and adventitiously blind were longer for the imagination condition than for the locomotion condition, whereas the latencies for the early blind group in both conditions were similar to those of the other groups in the imagination condition. Rieser et al. (1982) suggested that this pattern of results reflected differences in the way the task was performed by the groups with and without visual experience. In the locomotion condition, the previous visual experience of the sighted and adventitiously blind groups afforded them a sensitivity to the changing perspective structure of the environment and thus allowed them to update their position automatically as they moved. In the imagination condition, without the locomotor information to support automatic updating, these groups had to resort to a strategy of calculating the relative positions of the landmarks. The early blind group, with similarly long latencies and high errors for the locomotion and the imagination conditions, appear to have used a calculation strategy in both conditions.

Although the study of Rieser et al. (1982; 1986) is interpreted within a Gibsonian framework, their results and interpretation provide some support for Millar's (1988) model of sensory deprivation, as the congenitally blind participants apparently brought to the task modes of encoding spatial information which differed from those used by the sighted participants. However this model does not distinguish between Fletcher's (1980) Inefficiency and Difference theories; it is possible that the preferred strategies of people who have had no visual experience of space are necessarily inferior or less efficient for performing spatial tasks. Some light is thrown on this distinction by the fact that some of the early blinded participants in the Rieser et al. (1982; 1986) task performed identically to the other groups in terms of both errors and latencies. If we assume, with Rieser et al. (1982), that these individuals were necessarily using different strategies for performing the task from the later blind and the sighted participants, this finding would provide support for the Difference theory; these participants may have been using different strategies which were functionally equivalent to (i.e. just as efficient as) the strategies of the participants whose performance was based on previous visual experience.

However Loomis et al. (1993) in a replication of the Rieser et al. (1982; 1986) study failed to reproduce the pattern of errors of the earlier study; there were no significant group differences in error scores and all but one of the early blind participants performed at the level of the sighted participants. However response latencies of the early blind were higher than those of the sighted which would suggest a speed/accuracy trade-off consistent with the use of a calculation strategy by the visually impaired participants.

Overall, the evidence for a general spatial impairment in early blinded children is inconclusive. It appears that task and procedural differences can result in poorer performance by early blind groups or similar performance between early blind and sighted groups. Furthermore in most of these studies, individual visually impaired participants performed well within the range of sighted and late blind groups. Thus a strong Deficiency theory can be rejected. It seems more plausible that there are differences in the ways of coding space between groups or even between individuals which are more or less appropriate or adequate for the task demands.

Coding Systems Re-Examined

The distinction between self-referent and external coding strategies has been subjected to much analysis in previous research. It has been argued that young children are bound to egocentric spatial coding systems which are supplanted by external reference systems later in development (Piaget and Inhelder, 1956; Piaget et al., 1960; Siegel and White, 1975). Piaget's (Piaget and Inhelder, 1956; Piaget et al., 1960) theory of the development of spatial cognition also assumes that visually impaired children's reduced opportunities for interaction with objects in the environment would result in a delay in the acquisition of higher level, non-egocentric spatial representation (see Fraiberg, 1977). A number of studies indicate that early blinded people, even in adulthood, are generally restricted to a representation of the environment at the route level (i.e. in terms of sequences of landmarks or decision points). Carreiras and Codina (1992) have termed this latter position the 'visual representation hypothesis' because it implies that vision is necessary for the formation of configurational representations of the environment. A sequential representation, while permitting efficient travel along well known routes, does not easily support inferences about the relative locations of places not linked in a learned route; this information would have to be actively calculated at the time of retrieval and this could impose severe limitations on a person's mobility.

Three types of evidence suggest that a rigid Piagetian model is inadequate. Firstly it has been shown that even very young children can use external cues to code object positions when the experimental setting makes those cues highly salient for the children and therefore there is no reason to propose two distinct modes of encoding space, one (external or geocentric) which supplants the other (egocentric) at some point in development (Acredolo, 1982; Bremner, 1978; Presson and Ihrig, 1982; Presson and Somerville, 1985; Rieser, 1979). Secondly, it is clear that self-referent coding strategies are used by adults when it is appropriate to do so, for instance when external cues are unreliable (Millar, 1988; Presson, 1987). In this respect, Millar (1988) cites the example of a person sitting in a train looking out of the carriage window as the next-door train begins to move off. Here it is egocentric (vestibular) information which tells us that we are not moving. Thirdly, Millar (1988) argues that congenitally blind children, rather than exhibiting a delay in an invariant developmental sequence of levels of spatial representation, are simply using strategies which are most appropriate to the performance of the majority of spatial tasks in the absence of vision.

Carreiras and Codina (1992) have termed the alternative position the 'amodal representation hypothesis' which they characterize thus:

It is assumed that internal spatial representation is not linked to any specific sensory modality. According to this hypothesis, blind persons are able to preserve and process spatial images in a similar way to that used by sighted persons, although such processing may require less time when vision is involved. If given sufficient training, blind persons are assumed to be able to acquire a configurational spatial representation, and solve spatial problems with strategies similar to those employed by sighted persons (p. 55).

This theory leaves open to speculation the possibility that the early blind can acquire representations or coding strategies equivalent to those of people who have had visual experience, if they are given sufficient training, preferably from an early age (Juurmaa, 1973; Millar, 1988). The amodal representation hypothesis suggests that there would be differences between visual experience groups at the stage of constructing representations of the environment, but all the studies carried out in novel environments have controlled exposure to the experimental layouts such that all groups received the same quality and quantity of experience in them. In the studies of familiar environments it cannot be known how the groups differ in experience of independent mobility. It is recognized that lack of vision imposes limitations on a person's independent mobility and thus, although the early blind might require more experience of an environment to acquire representations equivalent to those of the sighted it is possible that the visually impaired participants in the studies described above had less independent experience than the sighted.

It seems likely, from the foregoing discussion, that the construction of spatial representations may be mediated by a number of optional (i.e. interchangeable) strategies each of which may be more or less adequate for a given spatial task. This suggests that visually impaired children can form coordinated global representations of the environment if they can be encouraged to use the appropriate (i.e. externally based) coding strategies. A number of possibilities for interventions to facilitate the spatial cognition of the visually impaired have been discussed and some of these will be presented in the next section.

Early Intervention

That visually impaired people clearly have the potential to achieve the same level of efficiency in spatial tasks has been shown in the study of Rieser et al. (1982; 1986) and in a number of other studies in which visually impaired individuals performed at or even above the level of sighted participants in spatial tasks (Casey, 1978; Dodds et al., 1982; Fletcher, 1980). However group comparisons in these and other studies often show that the early blind are impaired relative to the sighted on many spatial tasks, especially those conducted in large scale spaces which involve locomotion (Rieser et al., 1982; 1986; Rieser et al., 1980; Ungar et al., in press). Thus it would seem that in general the coding strategies or spatial representations of the early blind, while adequate for many small scale tasks, are generally inadequate, or at least inappropriate, for large scale tasks. Warren (1984) draws the following implications from the literature:

In order to make available to the child the widest range of effective strategies of spatial information processing, he or she must be brought to dissociate strategies from modes of experience and their natural constraints. Tactual experience, for example, must be structured in such a way as to lend itself easily to external reference systems, so that those systems become flexibly available to the child, along with the internal ones (p.88).

Warren suggests two lines of research which should be pursued to find effective methods of remediation. Firstly, an attempt should be made to identify the aspects of early experience which give rise to differences in the ways in which children code spatial relations. Secondly, procedures should be devised to train visually impaired children to use an external spatial framework or to form configurational representations. Similar recommendations have been made by Millar (1988), Juurmaa (1973) and Carreiras and Codina (1992) among others. Juurmaa in particular stresses the importance of intervention from the earliest years. So far no research has been carried out in the first of these areas but a number of methods have been proposed to facilitate the development of children's understanding of the relationships between objects in the environment.Training children to be more aware of their 'body image' (Cratty and Sams, 1968) is one of the main traditional methods thought to facilitate the development of spatial cognition in young blind children. The conceptual basis for this method is the belief that once a child has mastered concepts such as laterality, verticality and the arrangement of objects in space with respect to her own body, these will form the basis of an understanding of space external to the child. Although such techniques are widely used by Orientation and Mobility teachers they have not been adequately validated and one recent study (Morsley et al., 1991) suggested that there is in fact no relationship between a child's body image and her general spatial skills.

Another approach has been to substitute vision with an electronic device which converts optical information about objects in the environment into auditory or tactile information. One such device, the Sonicguide™ has been used in several studies with visually impaired infants and young children. It was hypothesized that providing children with auditory information about objects and surfaces in external space from an early age would facilitate their general understanding of the environment. In one study by Aitken and Bower (1982; 1982) three congenitally blind infants were given frequent sessions wearing the Sonicguide™ by their parents. The youngest of the three infants showed a number of spatially oriented behaviours (such as reaching and grasping) at approximately the appropriate age for sighted children whereas the other two infants apparently did not benefit from the Sonicguide™ at all. Warren (1984) cited a number of studies with similarly inconclusive results.

Another potential means of bringing the layout of external space to the child is through a tactile map. A tactile map can provide a vicarious source of spatial information which preserves all the interrelationships between objects in space but presents those relationships within one or two hand-spans. The relevant information is presented clearly (irrelevant 'noise' which may be experienced in the actual environment, is excluded); with relative simultaneity (a map can be explored rapidly with two hands and with less demand on memory); and without other difficulties associated with travel in the real environment (e.g. veering or anxiety). Furthermore, if maps can compensate to some extent for the limitations of the visually impaired, they may form a crucial component of mobility training (Gilson et al., 1965; Yngström, 1988).

Tactile maps may have important benefits in both the short term and in the long term. Liben (1981) draws a distinction between 'abstract' and 'particular' levels of spatial thought, the former refers to a person's ability to construct and use mental spatial representations, while the latter refers to the quality of a person's representations of specific environments. Tactile maps can usefully be used to introduce children to particular spaces, such as a classroom or a play ground, but the use of maps in general, and especially the exercise of relating maps to the environment which they represent, can improve the child's abstract level spatial thought in the long term, for instance by encouraging the child to adopt externally based coding frameworks for structuring spatial representations of the environment.

Research on Tactile Maps

Most of the existing research related to tactile maps has been concerned with the design and production of such maps (e.g. Aldrich and Parkin, 1987; Bentzen, 1977; Bentzen and Peck, 1979; Berlá, 1981; Dacen-Nagel and Coulson, 1990; Horsfall and Vanston, 1981; Parkin et al., 1988). In recent years a few studies have considered how visually impaired adults and adolescents use a map (Andrews, 1983; Berlá, 1973; Berlá and Butterfield, 1977; Berlá et al., 1976; Berlá and Murr, 1975; Brambring and Weber, 1981; Dodds, 1988; Gladstone, 1991; Golledge, 1991; Spencer and Travis, 1985) or a model (Carreiras and Codina, 1992; Fletcher, 1980; Herman et al., 1983).

Research with Adults

Carreiras and Codina (1992) asked congenitally blind, blindfolded sighted and sighted adults to explore a model of layout of streets with a number of places (landmarks) represented by discriminably textured pins. The adults were asked to make distance estimates between places on the model; both functional (route) and straight line (map) distance estimates were required. In order to be accurate on straight line distance estimates, the adults had to infer configurational level information from their sequential exploration of the model (as for Rieser et al's [1982; 1986] 'experimental' questions and for Fletcher's [1980] 'map' questions). There was no difference between the groups on the estimation of functional distances but the congenitally blind group was less accurate than the sighted groups on straight line distance estimates.

The studies by Carreiras and Codina (1992) and Fletcher (1980) (reported above) examined the mental representation of a tactile model but did not address the problem of transferring such information to the space represented by the map. Only a few studies have examined the spatial behaviour of visually impaired people who have learnt about a space from a tactile representation (i.e. a map or a model). For example, Herman, Herman and Chatman (1983) asked visually impaired teenagers and adults to learn a large-scale layout of four objects from a table-top model. The participants' knowledge of the layout was tested by asking them to walk between locations in the large-scale space. The results suggested that the congenitally blind participants could form a configurational representation of the layout from the model which they could then use to navigate in the real space. However Warren (1984) has pointed out that the errors of these participants were very high and may not have been different from chance levels. Therefore the experiment of Herman et al. (1983) does not provide unambiguous evidence that transfer from the model to the real space occurred.

Bentzen (1972) introduced six congenitally and adventitiously blind adults to a novel environment (the campus of the Perkins School for the Blind in Boston) using a tactile map of the area. The participants were asked to travel a planned route carrying the map. Though there were too few participants to perform statistical analyses, Bentzen reported that all the participants completed the route with a high degree of accuracy and two of the six participants completed the route without help from the experimenter. The other participants became disoriented at least once, but were able to pick up the route when led back to the correct course.

In a similar study, Brambring and Weber (1981) asked 27 'blind' participants to learn a novel built environment (a network of streets in Marburg) using either direct exploration, a verbal description or a tactile map of the area. The participants were then required to walk certain routes in the area. It was found that participants learned the area more quickly and their wayfinding performance was better with the map than with the other methods of familiarisation.

Overall, studies with visually impaired adults suggest that tactile maps can facilitate the construction of cognitive maps. Furthermore, tactile maps may be a more effective means of familiarizing visually impaired people with an environment than direct locomotor experience. Tactile maps can apparently provide the visually impaired reader with a more integrated and global impression of the environment.

Research with Children

Gladstone (1991) presented three congenitally blind children (9, 12 and 13 years of age), five congenitally blind adults and 12 blindfolded sighted adults with tactile maps of routes varying in overall length (8 metres or 16 metres) and route complexity (two turning points or six turning points) and allowed them unlimited time to learn each map. Participants were then taken to a large empty space and led from the origin point to the first turning point of a given map to demonstrate the scaling factor between that map and the large scale space. They then returned to the origin point and were asked to walk the route that they had just learned. This procedure was repeated for sixteen different routes.

Visually impaired and sighted participants took the same amount of time to complete each type of route. Overall, the visually impaired adults made more accurate turns (absolute angular error) than the visually impaired children and sighted adults on all types of route. Visually impaired children were as accurate in turning as the sighted adults on the short and long routes and were only slightly less accurate on the complex maps. Therefore, Gladstone's (1991) participants clearly could transfer information from the map to behaviour in the environment. Although Gladstone's task involved carrying the map and thus did not require forming a survey representation of the experimental space, all the visually impaired participants reported forming 'a representation of the map as a whole' (p. 31). Four of the visually impaired adults and all the children reported 'using a strategy of counting steps and relating this to the distance on the map' (p. 31).

Given the fact that several authors have stressed the importance of introducing maps and map concepts to children from the earliest possible age (Gilson et al., 1965; Kidwell and Greer, 1972), it is surprising that there has been little consideration given to whether young visually impaired children might benefit from using a tactile map.

One of the very few studies with young children was by Landau (1986) who asked Kelli (at four years of age) to use a simple map of an experimental space which depicted her own position and the location of a target object. Kelli's performance in walking to the target was above chance level indicating that she could use the map to determine the location of the object in space and Landau (1986) concluded that blind children from as young as four years may be able to understand and use a simple map to guide movement through the environment. She suggested that this ability is directly supported by the 'spatial knowledge system' which is used for other spatial behaviour. This claim is akin to the proposition that maps are 'transparent'– windows on the large-scale environment – which simply specify the spatial relations between places with minimal requirement for interpretation.

The evidence from the few studies with visually impaired children suggests that they have the potential to learn about, understand and use simple maps to perform orientation tasks in the environment. However, until recently the evidence about visually impaired children's map use was based on small groups of older children (e.g. Gladstone, 1991) or single case studies (e.g. Landau, 1986). Therefore we carried out a number of experiments considering the potential of visually impaired children from five to twelve years to understand and use maps (Spencer et al., 1989; Spencer et al., 1992; Ungar, 1994; Ungar et al., 1993; Ungar et al., in press; Ungar et al., 1992).

In one study, we compared the performance of visually impaired children (aged from 5 to 11 years) who were asked to learn about an environment either by directly exploring that environment or by being shown a tactile map of it (Ungar et al., in press). The environment consisted of a number of familiar toys arranged randomly around the floor of a large hall (see Figure 3). Tactile maps were constructed showing the location of all the toys.

Both the totally blind and the partially sighted children were able to understand and use the map. Most importantly, we found that the totally blind children learnt the environment more accurately from the map than from direct exploration. The results of this study demonstrated the importance of tactile maps for helping young totally blind children to form an impression of the space around them.

The general finding from our work is that young visually impaired children do have the potential to understand and use tactile maps. In some studies (e.g. Ungar, 1994; Ungar et al., 1992) it was found that the strategies which children used to perform the map tasks affected their performance; visually impaired children who adopted effective tactile strategies often performed as well as or better than sighted and partially sighted children.

Studies on the use of tactile maps by visually impaired children have produced similar results to those with adults. Tactile maps can provide visually impaired children with an impression of the layout of the environment and thereby facilitate the construction of accurate and integrated cognitive maps. This finding is particularly significant considering the relatively poor performance of visually impaired children in forming coherent cognitive maps from direct experience on the environment. If visually impaired children are trained to use tactile maps effectively, they might form the basis for improving the general spatial skills of these children and in particular the construction of cognitive maps.

Conclusion

This chapter summarized research related to the construction of spatial representations of the physical environment in visually impaired children. The evidence from groups of early blinded, late blinded and sighted children suggests that visual experience facilitates the construction of spatial representations. However, we have argued that visual experience may not be a necessary requirement for the ability to form integrated, global impressions of the environment. Rather, children's modes of experience of the world may prompt the use of different strategies for coding information. These strategies are interchangeable (i.e. a tactile strategy could replace any given visual strategy) but strategies are more or less appropriate or effective for certain tasks.

If visually impaired children can be encouraged from an early age to adopt appropriate strategies this will improve the quality of their spatial representations. For instance, for learning the layout of large-scale environments, visually impaired children should be encouraged to adopt an external reference system rather than one based purely on their bodies or their own movements. We have discussed two means (electronic aids and tactile maps) by which visually impaired children can be encouraged to code the positions of objects in the world relative to each other.

Acknowledgments: The authors gratefully acknowledge the continuing financial support from the Economic and Social Research Council. We also thank the staff and pupils of Tapton Mount School in Sheffield and The Royal Blind School in Edinburgh for their kind cooperation.

References

Acredolo, L.P. (1982). The familiarity factor in spatial research. In *New Directions for Child Development. Volume 15* (R.Cohen, Jossey-Bass: San Francisco, California.

Aitken, S. and Bower, T.G.R. (1982). Intersensory substitution in the blind. *Journal of Experimental Child Psychology* 33, 309-323.

Aitken, S. and Bower, T.G.R. (1982). The use of the Sonicguide in infancy. *Journal of Visual Impairment and Blindness* 76, 91-100.

Aldrich, F.K. and Parkin, A.J. (1987). Tangible line graphs: an experimental investigation of three formats using capsule paper. *Human Factors* 29, 301-309.

Andrews, S.K. (1983). Spatial cognition through tactual maps. *Proceedings of the 1st Int. Symposium on Maps and Graphics for the Visually Handicapped*, Washington: Association of American Geographers.

Bentzen, B.L. (1972). Production and testing of an orientation and travel map for visually handicapped persons. *New Outlook* 66, 249-255.

Bentzen, B.L. (1977). Orientation maps for visually impaired persons. *Journal of Visual Impairment and Blindness* 71, 193-196.

Bentzen, B.L. and Peck, A.F. (1979). Factors affecting traceability of lines for tactile graphics. *Journal of Visual Impairment and Blindness* 73, 264-269.

Berlá, E.P. (1973). Strategies in scanning a tactual pseudomap. *Education of the Visually Impaired* 5, 8-19.

Berlá, E.P. (1981). Tactile scanning and memory for a spatial display by blind students. *Journal of Special Education* 15, 341-350.

Berlá, E.P. and Butterfield, L.H. (1977). Tactual-distinctive features analysis: training blind students in shape recognition and in locating shapes on a map. *Journal of Special Education* 11, 335-345.

Berlá, E.P., Butterfield, L.H. and Murr, M.J. (1976). Tactual reading of political maps by blind students: a videomatic behavioural analysis. *The Journal of Special Education* 10, 265-276.

Berlá, E.P. and Murr, M.J. (1975). Psychophysical functions for active tactual discrimination of line width by blind children. *Perception and Psychophysics* 17, 607-612.

Bigelow, A. (1991). Spatial mapping of familiar locations in blind children. *Journal of Visual Impairment and Blindness* 85, 113-117.

Bremner, J.G. (1978). Egocentric versus allocentric spatial coding in nine-month-old infants: factors influencing the choice of code. *Developmental Psychology* 14, 346-355.

Brambring, M. and Weber, C. (1981). Taktile, verbale und motorische Informationen zur geographischen Orientierung Blinder. *Zeitschrift für experimentelle und angewandte Psychologie* 28, 23-37.

Byrne, R.W. and Salter, E. (1983). Directions and distances in the cognitive maps of the blind. *Canadian Journal of Psychology* 37, 293-299.

Carpenter, P.A. and Eisenberg, P. (1978). Mental rotation and frame of reference in blind and sighted individuals. *Perception and Psychophysics* 23, 117-124.

Carreiras, M. and Codina, B. (1992). Spatial cognition of the blind and sighted - visual and amodal hypotheses. *Cahiers de Psychologie Cognitive – European Bulletin of Cognitive Psychology* 12, 51-78.

Casey, S. (1978). Cognitive mapping by the blind. *Journal of Visual Impairment and Blindness* 72, 297-301.

Cratty, B.J. and Sams, T.A. (1968). *The Body-Image of Blind Children*, New York: American Foundation for the Blind.

Dacen Nagel, D.L. and Coulson, M.R.C. (1990). Tactual mobility maps: a comparative study. *Cartographica* 27, 47-63.

Dodds, A. (1988). Tactile maps and the blind user: perceptual, cognitive and behavioural factors. In *Proceedings of the 2nd Int. Symposium on Tactile Maps and Graphics for Visually Impaired People* (A.F. Tatham and A.G. Dodds eds.). Nottingham University: Nottingham University Press.

Dodds, A.G., Armstrong, J.D. and Shingledecker, C.A. (1981). The Nottingham obstacle detector: development and evaluation. *Journal of Visual Impairment and Blindness* 75, 203-209.

Dodds, A.G., Howarth, C.I. and Carter, D.C. (1982). The mental maps of the blind: the role of previous experience. *Journal of Visual Impairment and Blindness* 76, 5-12.

Fletcher, J.F. (1980). Spatial representation in blind children 1: development compared to sighted children. *Journal of Visual Impairment and Blindness* 74, 318-385.

Foulke, E. (1982). Perception, cognition and the mobility of blind pedestrians. In *Spatial Abilities: Developmental and Psychological Foundations* (M. Portegal ed.), pp. 55-76. New York: Academic Press.

Foulke, E. and Hatlen, P.H. (1992). A collaboration of two technologies. Part 1: Perceptual and cognitive processes: their implications for visually impaired persons. *British Journal of Visual Impairment* 10, 43-46.

Fraiberg, S. (1977). *Insights from the Blind*, London: Souvenir Press.

Gilson, C., Wurzburger, B. and Johnson, D.E. (1965). The use of the raised map in teaching mobility to blind children. *New Outlook* 59, 59-62.

Gladstone, M. (1991). Spatial Cognition and Mapping Abilities: A Comparison of Sighted and Congenitally Bind Individuals. Unpublished Honours Dissertation, Edinburgh University.

Golledge, R.G. (1991). Tactual strip maps as navigational aids. *Journal of Visual Impairment and Blindness* 85, 296-301.

Golledge, R.G. (1993). Geography and the disabled - a survey with special reference to vision impaired and blind populations. *Transactions of the Institute of British Geographers* 18, 63-85.

Hart, R.A. and Moore, G.T. (1973). The development of spatial cognition: a review. In *Image and Environment: Cognitive Mapping and Spatial Behaviour* (R.M. Downs and D. Stea eds.), pp. 246-288. Chicago: Aldine.

Herman, J.F., Herman, T.G. and Chatman, S.P. (1983). Constructing cognitive maps from partial information: a demonstration study with congenitally blind subjects. *Journal of Visual Impairment and Blindness* 77, 195-198.

Hermelin, B. and O'Connor, N. (1971). Spatial coding in normal, autistic and blind children. *Perception and Motor Skills* 33, 127-132.

Hermelin, B. and O'Connor, N. (1975). Location and distance estimates by blind and sighted children. *International Journal of Experimental Psychology* 27, 295-301.

Hermelin, B. and O'Connor, N. (1982). Spatial coding in children with and without impairments. In *Spatial Abilities: developmental and physiological foundations* (M. Portegal ed.). New York: Academic Press.

Horsfall, R.B. and Vanston, D.C. (1981). Tactual maps: discriminability of shapes and textures. *Journal of Visual Impairment and Blindness* 75, 363-367.

Juurmaa, J. (1973). Transposition in mental spatial manipulation: a theoretical analysis. *American Foundation for the Blind Research Bulletin* 26, 87-134.

Kerr, N.H. (1983). The role of vision in "visual imagery" experiments: evidence from the congenitally blind. *Journal of Experimental Psychology: General* 112, 265-277.

Kidwell, A.M. and Greer, P.S. (1972). The environmental perceptions of blind persons and their haptic representations. *New Outlook* 66, 256-276.

Kosslyn, S.M., Ball, T.M. and Rieser, B.J. (1978). Visual images preserve metric spatial information: evidence from studies of image scanning. *Journal of Experimental Psychology: Human Perception and Performance* 4, 47-60.

Kuipers, B. (1982). The "map in the head" metaphor. *Environment and Behaviour* 14, 202-220.

Landau, B. (1986). Early map use as an unlearned ability. *Cognition* 22, 201-223.

Landau, B., Gleitman, H. and Spelke, E. (1981). Spatial knowledge and geometric representation in a child blind from birth. *Science* 213, 1275-1278.

Landau, B., Spelke, E. and Gleitman, H. (1984). Spatial knowledge in a young blind child. *Cognition* 16, 225-260.

Leonard, J.A. and Newman, R.C. (1967). Spatial orientation in the blind. *Nature* 215, 1413-1414.

Liben, L.S. (1981). Spatial representation and behaviour: multiple perspectives. In *Spatial Representation and Behaviour Across the Lifespan: Theory and Application* (Liben, L.S., Patterson, A.H. and Newcombe, N. eds.). New York: Academic Press.

Liben, L.S. and Downs, R.M. (1989). Understanding maps as symbols: the development of map concepts in children. In *Advances in Childhood Development and Behaviour Volume 22* (H. Reese, ed.), pp. 145-201. New York: Academic Press.

Lockman, J.J., Reiser, J.J. and Pick, H.L. (1981). Assessing the blind traveller's knowledge of spatial layout. *Journal of Visual Impairment and Blindness* 75, 321-326.

Loomis, J.M., Klatzky, R.L., Golledge, R.G., Cicinelli, J.G., Pellegrino, J.W. and Fry, P.A. (1993). Nonvisual navigation by blind and sighted - assessment of path integration ability. *Journal of Experimental Psychology - General* 122, 73-91.

Marmor, G.S. and Zaback, L.A. (1976). Mental rotation by the blind: does mental rotation depend on visual imagery? *Journal of Experimental Psychology: human perception and performance* 2, 515-521.

Millar, S. (1975). Spatial memory by blind and sighted children. *British Journal of Psychology* 66, 449-459.

Millar, S. (1976). Spatial representation by blind and sighted children. *Journal of Experimental Child Psychology* 21, 460-479.

Millar, S. (1979). The utilization of external and movement cues in simple spatial tasks by blind and sighted children. *Perception* 8, 11-20.

Millar, S. (1981). Crossmodal and intersensory perception in the blind. In *Intersensory Perception and Sensory Integration* (R.D. Walk, and H.L. Pick, eds.). New York: Academic Press.

Millar, S. (1981). Self referent and movement cues in coding location by blind and sighted children. *Perception* 10, 255-264.

Millar, S. (1982). The problem of imagery and spatial development in the blind. In *Knowledge and Representation* (B. de Gelder, ed.). London: Routledge and Kegan Paul.

Millar, S. (1988). Models of sensory deprivation: the nature/nurture dichotomy and spatial representation in the blind. *International Journal of Behavioural Development* 11, 69-87.

Morsley, K., Spencer, C. and Baybutt, K. (1991). Is there any relationship between a child's body image and spatial skills ? *British Journal of Visual Impairment* 9, 41-43.

Neisser, U. and Kerr, N. (1973). Spatial and mnemonic properties of visual images. *Cognitive Psychology* 5, 138-150.

O'Connor, N. and Hermelin, B.M. (1972). Seeing and hearing in space and time. *Perception and Psychophysics* 11, 46-48.

Ochaíta, E. and Huertas, J.A. (1993). Spatial representation by persons who are blind: a study of the effects of learning and development. *Journal of Visual Impairment and Blindness* 87, 37-41.

Paivio, A. (1986). *Mental Representations: A Dual Coding Approach.* Oxford: Oxford University Press.

Parkin, A.J.and Aldrich, F.K. (1988). Remembering tangible line graphs: How important is prior visual experience? In *Proceedings of the 2nd Int. Symposium on Tactile Maps and Graphics for Visually Impaired People* (A.F. Tatham and A.G. Dodds, eds.). Nottingham University: Nottingham University Press.

Piaget, J. and Inhelder, B. (1956). *The Child's Conception of Space.* London: Routledge and Kegan Paul.

Piaget, J., Inhelder, B. and Szeminska, A. (1960). *The Child's Conception of Geometry.* London: Routledge and Kegan Paul.

Presson, C.C. (1987). The development of spatial cognition: secondary uses of spatial information. In *Contemporary Topics in Developmental Psychology* (N. Eisenberg, ed.). New York: Wiley.

Presson, C.C. and Ihrig, L.H. (1982). Using mother as a spatial landmark: evidence against egocentric coding in infancy. *Developmental Psychology* 18, 699-703.

Presson, C.C. and Somerville, S.C. (1985). Beyond egocentrism: a new look at the beginnings of spatial representation. In *Children's Searching: the development of search skills and spatial representation* (H. Wellman, ed.). Hilsdale, NJ: Erlbaum.

Rieser, J.J. (1979). Spatial orientation of six-month-old infants. *Child Development* 50, 1078-1087.

Rieser, J.J. (1990). Development of perceptual-motor control while walking without vision: the calibration of perception and action. In *Sensory-Motor Organizations and Development in Infancy and Early Childhood* (H. Bloch and B.I. Berenthal, eds.). Kluwer Academic Publishers.

Rieser, J.J., Guth, D.A. and Hill, E.W. (1982). Mental processes mediating independent travel: implications for orientation and mobility. *Journal of Visual Impairment and Blindness* 76, 213-218.

Rieser, J.J., Guth, D.A. and Hill, E.W. (1986). Sensitivity to perspective structure while walking without vision. *Perception* 15, 173-188.

Rieser, J.J., Lockman, J.J. and Pick, H.L. (1980). The role of visual experience in knowledge of spatial layout. *Perception and Psychophysics* 28, 185-190.

Senden, S.M.von (1932). *Space and Sight: the Perception of Space and Shape in the Congenitally Blind Before and After Operation.* Glencoe, IL: Free Press.

Shepard, R.N. and Metzler, J. (1971). Mental rotation of three-dimensional objects. *Science* 171, 701-703.

Siegel, A.W. and White, S. (1975). The development of spatial representations of large-scale environments. In *Advances in Child Development and Behaviour. Volume 10* (H.W. Reese, ed.). New York: Academic Press.

Spencer, C., Blades, M. and Morsley, K. (1989). *The Child in the Physical Environment: the Development of Spatial Knowledge and Cognition.* Chichester: Wiley.

Spencer, C., Morsley, K., Ungar, S., Pike, E. and Blades, M. (1992). Developing the blind child's cognition of the environment - the role of direct and map-given experience. *Geoforum* 23, 191-197.

Spencer, C. and Travis, J. (1985). Learning a new area with and without the use of tactile maps: a comparative study. British *Journal of Visual Impairment* 3, 5-7.

Ungar, S.J. (1994). The Ability of Young Visually Impaired Children to Use Tactile Maps. Unpublished PhD, University of Sheffield.

Ungar, S.J., Blades, M. and Spencer, C. (1993). The role of tactile maps in mobility training. *British Journal of Visual Impairment* 11, 59-62.

Ungar, S.J., Blades, M., Spencer, C. and Morsley, K. (in press). The use of maps by visually impaired children to estimate directions in the environment. *Journal of Visual Impairment and Blindness.*

Ungar, S.J., Foy, K., Blades, M. and Spencer, C. (1992).Visually impaired children's memory for a map-like tactile layout. Paper presented at British Psychological Society - Developmental Section Annual Conference, Edinburgh, 11-14th September 1992.

Warren, D.H. (1984). *Blindness and Early Childhood Development.* New York: American Foundation for the Blind.

Worchel, P. (1951). Space perception and orientation in the blind. *Psychological Monographs* 65, 1-28.

Yngström, A. (1988). The tactile map: the surrounding world in miniature. In *Proceedings of the 2nd Int. Symposium on Tactile Maps and Graphics for Visually Impaired People* (A.F. Tatham and A.G. Dodds, eds.). Nottingham University: Nottingham University Press.

Zimler, J. and Keenan, J.M. (1983). Imagery in the congenitally blind: how visual are visual images? *Journal of Experimental Psychology: learning, memory and cognition* 9, 269-282.

Simon Ungar, Mark Blades and Christopher Spencer
Department of Psychology,
University of Sheffield,
P.O. Box 603,
Sheffield S10 2UR.

LANGUAGE AS A MEANS OF CONSTRUCTING AND CONVEYING COGNITIVE MAPS

Nancy Franklin

Abstract: We frequently rely on language, in conjunction with or in the absence of perceptual experience, to convey spatial information. Production of such descriptions involves a host of processes, including selection of important elements to communicate, temporal structuring of the elements, selection of frames of reference and perspectives, and verbal regularization of spatial relations. The addressee is then faced with the challenge of forming a spatial model from linearly ordered simple expressions. The literature shows that memory representations for these described configurations bear some functional resemblance to perceptually derived representations. However, their construction is not guaranteed, and even when mental models are created, they do not always appear to have perception-like properties. How we organize memory for described spatial information appears to depend not only on the structure of the described space itself, but also on cues from the text, characteristics of our typical interactions with space, and the nature of the expected task.

Introduction

Various characteristics of cognitive maps have now been established that show them to be both basic and versatile. They are formed in nonhumans as well as humans (Tolman, 1932), in the blind as well as the sighted (Kennedy, Gabias and Heller, 1992), in children as well as adults (Pick and Riser, 1982), under circumstances of high as well as low working memory load (Sholl, 1993), and after a single exposure. They can be transformed, for example, through imagined rotation (Maki, Maki and Marsh, 1977). One's perspective on them can be modified (Holyoak and Mah, 1982). And they can be acquired through a variety of media.

Language is a particularly interesting acquisition medium, and representations constructed through language are referred to as *spatial mental models*. In the literature, these descriptions typically refer to relatively small configurations of objects, such as those arranged on a tabletop or computer screen (e.g., Ehrlich and Johnson-Laird, 1982; Mani and Johnson-Laird, 1982; Hayward and Tarr, in press; Payne, 1993), or configurations of objects in relatively small spaces, such as rooms (e.g., Franklin and Tversky, 1990), buildings (Taylor and Tversky, 1992b), or small outdoor areas (de

J. Portugali (ed.), The Construction of Cognitive Maps, 275–295.

Vega, 1994a, 1994b). Because so much of the literature draws from configurations on these scales, our discussion will rely heavily on work pertaining to smaller configurations. Where possible, we will pay particular attention to larger-scale environments for which it would be most appropriate to use the term "cognitive map".

Before describing how mental models are constructed and searched, and before describing their similarities to cognitive maps derived from perceptual experience, we need to justify the claim made above that construction of cognitive maps from description is a particularly interesting phenomenon. This argument is based on several observations about language and space: (1) Language is by its very nature transient, while the space that it describes perpetuates in time. (2) Language is linear, while the maps it describes are 2-dimensional, and the situations they describe are 3- or 4-dimensional. (3) An utterance can specify few spatial relations easily, while perceptual experience can be of an unlimited number of relations simultaneously. (4) Perceptual experience allows easily for complex and noncanonical spatial relations, while language consists of only a small set of spatial terms that must be used to capture these relations. (5) Language itself is expressed through propositions, while space is analog, and processing of these different kinds of information is associated with distinct brain regions.

Taken together, these present what may seem to be a hopeless situation for the ability of language to convey spatial information. But in fact, it is used all the time for this purpose, and with great success. As will be shown, cognitive maps derived from description often appear to be organized on spatial principles, similar to those derived from visual or navigational experience. That is, space is translated into description, and the description is translated back into a spatial model. The translation of information in both directions is psychologically interesting, requiring sophisticated cognitive processes for identifying and expressing important relations and for making appropriate spatial inferences.

Why Have Mental Models?

If nontrivial cognitive processing – and possibly even a transformation of representational code – are involved in constructing a mental model, why would a memory system go to all that effort? One strong possibility is that it is a good investment. We can illustrate this by comparing the usefulness of text-based vs. spatial representation for operations that someone might need to perform on spatial knowledge. Almost certainly, the task would involve spatial processing (e.g., search, updating, transformation), and so information directly expressing spatial relationship would have the greatest psychological value. That is, it is useful to know the relative positioning of two objects but usually not useful to retain exactly how that relation was expressed. Whether we heard that the library is west of the restaurant or the restaurant is east of the library, the two are equivalent spatially. It is of great value to use a representational medium by which the two truly are equivalent.

If after hearing that the library is west of the restaurant, we then heard that the restaurant is south of coffeehouse, a mental model encoding the three items would explicitly represent information not given in the text: the relation between the library and coffeehouse (Bryant, Tversky and Franklin, 1992; Denis and Denhiere, 1990; Foos, 1980; de Vega, 1994a). A representation of the situation as an integrated configuration, where relations implicit in the text are encoded into the model, provides a lot of power for subsequent retrieval and spatial problem solving.

A third advantage of a spatial representation concerns the information lost, and thus the efficiency gained. Long-term verbatim memory has little value, unless an error is detected and we need to return to reinterpret the text. These errors, however, are frequently detected soon after the statement in question is read, when textual propositions are still in working memory. Similarly, precise memory for the temporal order in which relations are learned is typically not important to subsequent tasks. So although text-based representations are superior to spatial ones for expressing the serial order in which items were described, spatial relations among objects do not depend on their relative temporal positions in the description. (One noteworthy exception to this is the landmark effect. In spatial situations learned both from primary experience (Sadalla, Burroughs and Staplin, 1980) and from description (Payne, 1993), the first spatial relations learned appear to have privileged, or landmark, status.)

Processing and Spatial Characteristics

The spatial content of the description and the task, not the medium of acquisition, predict the kind of information processing that will be needed after initial learning. In addition, both content and organization of the mental model are important to the outcome of operations on it. One class of operations, for example, involves mental simulation of events. The fidelity of a thought experiment like "What would happen if I take a step beyond the edge of this cliff?" determines the accuracy of predicting the corresponding action's outcome.

Relative accessibility for different components of a mental model seems to reflect their relative salience. That is, spatial organization of the memory representation allows for retrieval whereby information that needs to be accessed more frequently or more urgently, is. Spatial characteristics of mental models are generally demonstrated with various kinds of timed tasks, usually verification tasks. Patterns of response times typically resemble those for comparable perceptual situations or uphold theoretical predictions regarding spatial representation.

Distance

Perhaps the spatial characteristic for which the most evidence has been amassed is distance, because so many paradigms lend themselves easily to studying distance effects. One such paradigm involves mental scanning, the time-course of which appears to mirror perceptual inspection. Just as longer distances take longer to travel or to scan physically, mentally scanning tends to be faster for shorter than for longer paths in cognitive maps derived from map viewing (Kosslyn, Ball and Reiser, 1978) and from description (Denis and Cocude, 1989; Denis and Zimmer, 1992).

In Denis and Cocude (1989), subjects learned the positions of several features on a fictitious island from text. Once the configuration was known, subjects were to mentally focus on a feature identified by the experimenter and, when a second feature was named, to mentally scan to it. Subjects pressed a button to indicate when the scan was complete, and the time they took to press the button was a linear function of distance in the referent environment. Distance was not confounded with separation of items in the text, so the most parsimonious interpretation is that this was a spatial effect.

A second paradigm produces inverse distance effects, sometimes known as symbolic distance effects (Moyer and Bayer, 1976). This paradigm is based on the argument that in comparison tasks, large differences are easier to detect than smaller ones. Such an argument has been supported with configurations learned from description, using distance comparison (Denis and Zimmer, 1992) and direction comparison (Federico and Franklin, 1994). Differences in response time for these tasks indicate that the retrieval process used by subjects is something akin to visuo-spatial inspection. Although this effect is the inverse of a distance effect produced from scanning, both rely on spatial representation and perceptual-like processing. Not only is this spatial quality interesting and important, but it is also impressive. Representations derived from description require more constructive inference than those derived from viewing in order to produce "distance" so that when examined, they behave in some ways like real, extended space.

A third paradigm used for demonstrating distance as an organizational principle of mental models is priming. If subjects are using a textual representation, then what is proximal in memory, and what should constitute an optimal prime, would be an item near the target textually. But if subjects are using a spatial representation, then proximity is defined spatially, and the best prime should be the object nearest the target in the referent situation. Spatial priming has been found for perceptually-derived representations (e.g., McNamara, Ratcliff and McKoon, 1984), and similar effects have been shown for description-based ones (Denis and Zimmer, 1992; Federico and Franklin, 1994; Wagener, Wender and Wagner, 1990). This effect cannot be attributed to order of description, since it occurs even when spatial and textual proximity are uncorrelated.

In one of the best demonstrations of spatial priming in mental models (Denis and Zimmer, 1992), subjects viewed the outline of an island with crosses designating where objects will be located. They then studied a description that identified the object in each position. In an old/new recognition test that followed, verification time depended on the spatial proximity of the item named in the previous trial, as if attention was shifted smoothly through continuous space. There was perceptual input at learning, though, and it is unclear whether and to what degree that contributed to the effect. The jury is still out for spatial priming effects in mental models, but so it is for spatial priming in general (Sherman and Lim, 1991).

As suggested by the above examples, one way to produce distance effects is to cause attention to be transformed from one position to another. Once a position is foregrounded, or under the focus of attention, it is more readily available for inspection (Morrow, 1985). Since readers generally keep their attention on the main protagonist of a story, this can be accomplished by describing a target object as being near or far from the protagonist and then causing the reader's attention to shift to the object by asking about it. In Glenberg, Meyer and Lindem (1987), verification times to say that these objects were in the story were faster in the "nearby" than in the "distant" condition, as if subjects were searching something analogous to a physical scene. This here/there effect has been extended to multiple levels of distance (e.g., Morrow, Greenspan and Bower, 1987; but see Sharp and McNamara, 1991). It has also been extended to other ways in which one's attention might be focused, for example, by thinking about a place even when not there (e.g., Morrow, Bower and Greenspan, 1989).

Textual signals for foregrounding. There are aspects of text that influence allocation of attention. For example, in Glenberg et al. (1987), the organization of the described events is around a protagonist, and any described motion by this character is a signal to shift attention. If multiple protagonists exist in a description, one is usually signalled as the main character, for example, through the use of main clauses. Addressees foreground that character in their representation, enhancing the character's availability until attention shifts away (Morrow, 1985). So, there are devices for foregrounding characters, which then become highly available. There are also common ways to foreground locations themselves with language, through selection of prepositions and grammatical devices (Stenning, 1978). Subjects tend to treat sentences involving a protagonist "walking into" one room from another as locating the current position of the protagonist at the destination room (Morrow, 1990). On the other hand, if the protagonist is "walking toward" one room from another, subjects infer the current location to be near the source room or somewhere along the route.

Although the evidence is overwhelmingly for spatial representation in mental models, it is not known whether precise distances are stored explicitly. When actually driving, you may have a sense of how far you've driven but may not notice that your odometer has advanced 2.5 miles. Similarly, descriptions often give precise distances between locations, but they don't have to, and even if the information is presented explicitly, it may not be represented that way. As is true for cognitive maps derived from other sources, subjects may have to rely on inferential processes to make distance estimates. With his analog timing model, Thorndyke (1981) has suggested a means by which this can be accomplished. According to Thorndyke, distance is often estimated indirectly by mentally scanning the to-be-judged route, obtaining an estimate of mental scan time, and converting this scan time into a distance estimate. That is, people can make use of the monotonic relation that exists between distance and mental scan time.

A cognitive map structured according to something like physical distance allows for this mechanism of spatial inference. Any characteristic, such as clutter along the route, that slows scan time should increase distance estimates. Thorndyke's arguments and evidence pertain to environments learned from maps, but they can be extended to situations where learning is through description. Such effects have not yet been demonstrated for mental models, but because scanning effects appear for these representations, effects of spatial complexity on distance estimates are predicted.

Direction

For retrieval of various directions around oneself, accessibility is predicted by the relative salience of these directions, as characterized by the spatial framework analysis (Franklin and Tversky, 1990). According to this approach, representation of egocentric direction is organized with respect to three axes: front/back, head/feet, and right/left. These axes can be compared according to their relative functional and perceptual salience. For an upright observer, the head/feet axis is correlated with gravity, which induces physical and behavioral asymmetries for both oneself and other objects in the world. Because of these asymmetries, the head/feet axis is characterized as the most salient, and is predicted to be most accessible to memory. Front/back is next, with asymmetries that stem from the functional and perceptual importance of one's front. It is predicted by the model to be retrieved more slowly than head/feet but faster than left/right, which is associated with the fewest asymmetries and is therefore characterized as the least salient of the three axes.

The spatial framework analysis has been upheld for memory of perceptually acquired situations (Bryant and Tversky, 1991; Hintzman, O'Dell and Arndt, 1981; Bryant, Tversky and Lanca, 1994; Sholl, 1987) and situations learned from description (Franklin and Tversky, 1990; Bryant, et al., 1992; de Vega, 1994a). It also extends beyond predictions of response times to other indices of differential salience among directions. It has been shown, for example, to predict differences in the conceptual size of physical front, back, right, and left regions around oneself (front being the largest), and to predict differences in the psychological resolution of these regions (Franklin, Henkel and Zangas, in press). Interestingly, however, search of the perceptual environment itself does not produce search times consistent with the spatial framework analysis (Bryant and Tversky, 1991). This supports the characterization of the spatial framework as a model of how people conceptualize space, not how they perceive it.

Though direction is highly salient in immediate environments, it is also important to the structure of large-scale, even very large-scale, environments. Most notably, the north-south-east-west frame of reference bears some resemblance to the egocentric spatial framework. It is generally applied to larger configurations, and it is generally associated with a survey rather than an embedded perspective on the environment. The resemblance to the spatial framework seems to stem from the convention of aligning north with an egocentric "forward" direction when reading maps. For environments learned through maps, the relative accessibility of north, south, east, and west are similar to those for front, back, right, and left, respectively (Loftus, 1978; Maki et al., 1977).

Perspective

As observers, our physical relationship with a real environment can vary. We can stand in various locations or adopt various postures, changing our perspective within the environment. We can pop out of the environment to take a view from an airplane or hilltop, or we can adopt such a survey perspective when drawing a map. Similar, even greater, flexibility in perspective occurs for described environments. Whereas one's initial physical placement seems to be primary in determining one's perspective on a physical space, several other factors come into play that affect a reader's or listener's perspective on a described one. One clear factor is the implied perspective from which the environment is described. This is analogous to one's physical position in a real environment, in that the description is consistent with how the environment would be viewed from that position. The description presents the default perspective according to which the mental model can be constructed. This initial perspective on the mental model appears to have privileged status, possibly because it is typically the most familiar (Franklin and Tversky, 1990; de Vega, 1994a). The text may imply a transformation in perspective, and readers can incorporate this change, but comprehension time in such cases is high (Black, Turner and Bower, 1979; de Vega, 1994a).

Perspective may also change in the absence of explicit cues from the text, if one is available that is more compatible with the subject's task than the current perspective is (Franklin, Tversky and Coon, 1992). When retrieving egocentrically defined directions with respect to an observer, subjects appear to adopt the observer's perspective (Franklin and Tversky, 1990; Bryant et al., 1992). In fact, the predicted spatial framework pattern rests on the assumption that subjects adopt this central position in the environment. If there are two described observers, subjects are capable of switching between them to perform the retrieval task (Franklin et al., 1992; de Vega, 1994a), producing a spatial framework pattern of response times with respect to each protagonist. But when the probed observer switches frequently over trials, with dozens of trials, this could be costly. More efficient would be to adopt a single, stable perspective from which all probed relations can be searched. Patterns of response times indicate that subjects do just that (Franklin et al., 1992), but such a perspective is possible only when the two observers are in the same or adjacent environments. When they're in different and nonadjacent environments, the spatial framework pattern of response times emerges, indicating that subjects switch between the embedded perspectives of the observers (Franklin et al., 1992).

The perspective adopted within a mental model at any moment also affects what is in one's focus of attention (e.g., Bly, 1988; see also Denis and Le Ny, 1986). Distance effects are consistent with the possibility that subjects implicitly scan from a survey perspective, though this interpretation is by no means necessary to explain the results. Better evidence exists for the case of embedded perspectives, which imply an imagined view through the protagonist's eyes. In Bly (1988), subjects were induced through the text to adopt the perspective of one of two characters who are looking at each other. Subjects retrieved items associated with the person whose perspective they were adopting more slowly than items associated with the person in their imagined field of view. This is sensible, since what is foregrounded consists of what is in one's focus of attention, and what is in one's focus of attention typically consists of what is in one's field of view.

Survey and route perspectives. Fixed embedded positions are sufficient for relatively small environments. But they supply only limited views of large environments, and so are nonoptimal for exploration or description. For more complete representations of extended environments, there are usually two options. The survey perspective corresponds to a fixed view from outside – usually from above – that captures all important spatial relations among objects. The route perspective is embedded but moving, so that limited sets of spatial relations emerge as a tour through the environment proceeds. Both of these means of acquiring spatial information in the physical world yield complete and relatively accurate cognitive maps (e.g., Presson and Hazelrigg, 1984; Thorndyke and Hayes-Roth, 1982), and both classes of perspective can be used effectively in spatial descriptions (Perrig and Kintsch, 1985; Taylor and Tversky, 1992a, 1992b).

Is a particular perspective, or a particular type of perspective (survey or embedded) encoded as an enduring feature of the representation itself? If so, then adoption of any other perspective would require a transformation on it. This was suggested by the results of Perrig and Kintsch (1985), which showed differences in speed and accuracy of verification by female subjects to be a function of perspective. But subsequent work by Taylor and Tversky (1992b), using less complex and more easily learned environments, showed it not to be the case. They had subjects read about environments, some of which were learned according to a survey perspective and some according to a route perspective. Subjects then judged true-false sentences that were written from a perspective (route or survey) either consistent or inconsistent with the original. If perspective was inherent in mental models, inference statements written from the original perspective would be verified faster than those from the other perspective. No differences were found, suggesting that representation in long-term memory was perspective- free.

But subjects can and do adopt particular perspectives on environments – in fact, predictable ones. This in conjunction with Taylor and Tversky's findings indicates that mental models can be represented at different levels of abstraction. The most abstract form in memory may be perspective-free, but when it is in use, it may take a form associated with a particular perspective. We might refer to this instantiated spatial model as an image of the environment.

We can make a further distinction with respect to the question of perspective, between deictic, intrinsic, and extrinsic frames of reference. This distinction has implications for both memory representation and for comprehension of spatial terms. Many spatial terms, like "above," depend on an adopted frame of reference for interpretation. If selection of reference frame depended strongly on one's orientation in the environment, communicators would risk misunderstanding whenever their perspectives did not match. Although producing and comprehending sentences like "There is a fly above the chair" involve a coordination between language and perception, comprehension is facilitated by additional biases possessed by speaker and listener (e.g., toward extrinsic, or environment-centered and away from deictic, or viewer-based frames of reference) (Carlson-Radvansky and Irwin). Thus, interpretation of language, and the construction or updating of one's mental model, need not be contingent on one's physical or imagined placement in the environment.

Integrating Different Kinds of Information in Memory

Imagine a large elephant next to a small rabbit. Because of the explicit instructions to image and because of the sparseness of the visuo-spatial description, we call this an imagery task (Kosslyn, 1975). But Taylor and Tversky's (1992a) description specifying the relative positions of several animals at a zoo is considered to encourage a mental model. Indeed, the boundary between the two is fuzzy and is complicated by the fact that people draw on a database that includes perceptual memories when constructing representations of even novel scenes from text. Other cases of integration are not hard to find. Construction of cognitive maps often relies on a combination of text and diagrams (Dean and Kulhavy, 1981; Kulhavy, Stock, Woodard and Haygood, 1992; Morrow et al., 1987, 1989). Descriptions often relate to previously navigated territory, or unseen areas connected to known environments, or changes that have occurred in viewed configurations. All of these situations require the integration of spatial information acquired from sources that themselves differ qualitatively.

To complicate matters further, some situations call for us to integrate spatial with nonspatial information (Fincher-Kiefer, 1993; McNamara, Halpin and Hardy, 1992). For example, subjects in McNamara et al. (1992) already had learned a campus layout through direct experience, but in the lab they learned facts (e.g., historical information) about various buildings on campus. Of interest was whether this nonspatial information would be integrated into a spatial model so that a fact about one building would prime the name of a nearby one. This priming effect was observed and was replicated for a fictitious environment that subjects learned by studying a road map. The results suggest that the nonspatial, declarative information can become associated with a representation organized according to spatial characteristics.

Finally, when both textual and perceptual input are presented, they need to be coordinated to produce a single coherent representation, and each is used in interpreting the other. This coordination can lead to interesting results. For example, Tversky and Schiano (1989) showed that verbal interpretive labels can produce systematic distortions in cognitive maps. If the ambiguous diagram in Figure 1 is presented to some subjects as a "map" and to other subjects as a "graph," subsequent attempts at reproduction in a drawing task will lead to greater symmetry than the original for the map subjects, but not the graph subjects. Clearly, verbal information can interact with the spatial information in a deep way such that it affects the contents of subjects' memory representation. Interestingly, the effect occurs even when the original display is present for the drawing task, so perception as well as memory is affected by the verbal label.

The effects are not unidirectional. Assumptions based on one's understanding of spatial situations also affect interpretation of language. For example, the average distances inferred from the word "approaching" differ, depending on whether a sentence describes a luxury liner or a sailboat approaching a dock (Morrow and Clark, 1988). Thus, linguistic and spatial information communicate smoothly and at several levels to produce rich and internally consistent representations.

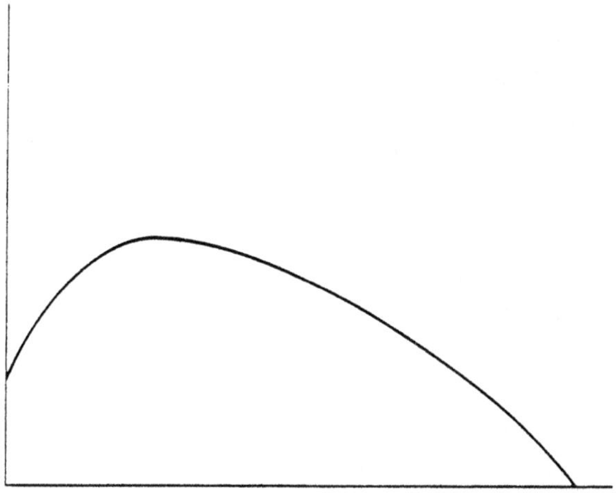

Figure 1. Stimulus used by Tversky and Schiano (1989).

How Spatial Representations From Description Might Differ From Those Acquired Through Perception

There are a lot of similarities in the representation of cognitive maps learned through primary experience and those learned through description. The evidence points to remarkable parallels in the principles governing their organization and governing the way they are learned, searched, updated, and used. And, as we have seen, the evidence points to integration of perception-based and description-based representations.

But they are not the same. Subjects generally learn spatial relations better (Cave, 1993; Denis and Zimmer, 1992) and faster (Denis and Zimmer, 1992) from maps than from description, reminiscent of the picture superiority effect (e.g., Shepard, 1967). Second, cognitive maps learned from text represent spatial relations, and possibly even metric information, as indicated by mental scanning (Denis and Cocude, 1989) and priming effects (Denis and Zimmer, 1992; Federico and Franklin, 1994), but they do not seem to be as rich in metric information as those acquired from navigation, viewing, or map-reading. So they are not comparable in content or fidelity. Third, the spatial framework pattern of response times disappears when subjects search a physically surrounding space (Bryant et al., 1994) or when they learn and are cued at recall with pointing rather than direction labels (de Vega, 1994b). Finally, subjects can readily adopt a neutral perspective over one described as "yours" (Franklin et al., 1992), in contrast to the inescapability of their own perspective in the physical world.

Indeed, representations for spatial descriptions are not limited by what is familiar or even what is possible. Readers can be encouraged to adopt any perspective, even of someone or something else, or to construct a representation of any imaginable situation.

Joe Kim and Andrew Cohen (described by Tversky, Franklin, Taylor and Bryant, 1994) took advantage of this in a study investigating the effect of frame of reference on mental model representation. They had subjects learn about situations (e.g., a gravity-free device at a NASA exhibit) in which either the observer can move relative to the environment or the environment can move relative to the observer. Although these two situations are formally identical, subjects frequently found the latter to be more difficult to comprehend and remember than the former. Although the perceptual counterpart has not been run, it does not seem likely that subjects viewing a physically moving environment would find it overwhelming to perceive the spatial relations or to integrate them into a perceptually-based representation.

Other comparisons between cognitive maps derived from perception vs. description have yet to be thoroughly explored. For example, can spatial relations among items be represented analogically while the objects themselves are not represented fully or analogically? In fact, the evidence for spatial representation that we have discussed is consistent with the possibility that objects themselves are encoded simply as nonimagistic tags (but see Denis and Le Ny, 1986). How objects are represented probably depends in part on the task, working memory load, vividness of the description, and subjects' ability to draw on stored perceptual experiences. But for present purposes, this question is secondary to the question of whether spatial relations are encoded, since what makes for a cognitive map is spatial relations among elements.

The Time-Course of Spatial Representation

When during the course of reading does one build a spatial representation? We can make arguments about when one should. Because construction of the spatial representation should be guided closely by what is said in the description, it should begin before propositional information about the text is lost. The classic experiment by Bransford, Barclay and Franks (1972) showed detailed memory of the text to deteriorate within 15 minutes or so, but a more accurate estimate is that it begins to fade when no longer needed for lexical interpretation, within several propositions (Kintsch and van Dijk, 1978; Sachs, 1967).

Second, the timing should be such that new information from the text can be integrated into a spatial model easily without overwhelming working memory. If a single integrated model can be formed from the description, that representation is preferred to a descriptive one (Radvansky and Zacks, 1991). Construction and updating of a spatial representation should occur ongoingly, as new information arises in the text. And indeed, both textual and situational information seem to be represented in working memory during comprehension (van Dijk and Kintsch, 1983; Johnson-Laird, 1983; Perrig and Kintsch, 1985). Further, there is some evidence for on-line construction of a detailed spatial representation, with reading time higher for sentences describing longer distances than shorter distances (Sharp and McNamara, 1991). Individual reading times for the same sentences outside the descriptive context do not differ, suggesting that the effect is due to processing of a scene, not of text sentences.

Textual information is retained and can be recalled for a short period of time (Speelman and Kirsner, 1990; Taylor and Tversky, 1992b). But even after several minutes it may not be completely lost, even though it fades. Taylor and Tversky (1992b) found verification times for original sentences to be faster than for paraphrases. Because they found both memory for text and evidence of a mental model, we can conclude that subjects are capable of retaining more than one representation.

We now turn to the question of what form spatial models take in long-term memory, which has received almost no attention in the literature. Though the paradigms for studying mental models have varied, most have been used relatively brief periods between learning and test. At these short latencies, subjects read about a scene, typically anticipating tests on spatial relations, and then they receive the test shortly afterward. Because they include spatial inferences, and because they likely draw on limited visuospatial working memory (Baddeley, 1986), spatial models are probably relatively expensive to maintain.

What if the latency period were longer? A spatial model, if it were reconstructed and held until use, might be particularly costly, and thus prove to be a poor representational investment. But memory for text fades relatively rapidly, so a reconstruction of that isn't generally an option. Payne (1993) has argued for a third representation that is created alongside the spatial mental model and that is well-suited for long-term storage. This *episodic construction trace* is a nonspatial, propositional record of important details describing the original construction of the mental model. Because of its relatively sparse propositional (though not textual) form, storage is relatively inexpensive. Because of the kinds of spatial relations it describes, a mental model much like the original can be reinstantiated in working memory when needed. Again, it appears that the translation of information might be a good investment of memory resources. And we have seen that multiple representations of the same information can coexist. Payne (1993) provides empirical support for an episodic construction trace, based on subjects' ranking of how closely test descriptions match the originals. However, the generality of his findings has yet to be determined. For example, they are based on a set of descriptions that are underlearned and that all refer to situations of the same shape. His test induced high working memory load and probably produced spatial interference. All of these factors encourage a propositional representation and discourage use of a spatial one. So, although the question is an important one, the long-term fate of mental models is as yet not clear.

Spatial Descriptions Don't Always Lead to Spatial Representations

Readers are not always called upon to retain a detailed memory of either the situation or the text. By including memory tests in our studies, we encourage subjects to form these representations. Typically, laboratory tasks involve spatial relations, and typically subjects are found to construct mental models.

But this does not tell us what circumstances in less formal settings lead to spatial representations. Will a reader of *Lord of the Flies* construct a cognitive map when the author describes Ralph running through the jungle? Maybe, but the spatial relations among elements may not be highly detailed, since the primary purpose of reading the description is for entertainment (Johnson-Laird, 1983). That is, construction of mental models does not appear to be an automatic consequence of reading spatial descriptions (Glenberg and Langston, 1992; but see also McKoon and Ratcliff, 1992). Even in a lab setting with the standard test probes, representation of scenes learned solely through description is not necessarily spatial and need not, for example, produce distance effects (Sharp and McNamara, 1991). A few factors have been identified that bear on the likelihood that detailed spatial models will be used.

Task Factors. If mental models are constructed in anticipation of spatial tasks, then the nature of the task should affect their construction, and it does (Denis and Denhiere, 1990). When readers expect a verbal recall task, reading times increase (Denis and Zimmer, 1990; Garnham, 1981). That is, when construction of a model would be disadvantageous, subjects appear to use a model of the propositions of the text instead. I've obtained essentially this same finding in my lab using a different paradigm. Work in progress by McNamara, Bower and their colleagues suggests that construction of detailed mental models may depend on the anticipated presence or absence of particular kinds of probes (namely, those that include protagonist names) in the question set.

Textual Factors. Besides the nature of the task, construction and use of spatial representations appear to depend on characteristics of the text as well. In particular, coherent, continuous, determinate descriptions are optimal for producing spatial mental models. Poorly constructed texts can yield spatial representations, but their construction requires longer study time (Denis and Cocude, 1992; Denis and Denhiere, 1990; see also Ehrlich and Johnson-Laird, 1982). In continuous descriptions, each new object is described in relation to one previously described. As a class, continuous descriptions produce faster reading speed, faster listening speed, and better memory than do discontinuous descriptions (Denis and Cocude, 1992; Ehrlich and Johnson-Laird, 1982). Determinacy refers to the quality that all spatial relations necessary for uniquely specifying a configuration are given. Spatial models appear not to be used when a description could refer to multiple specific configurations (Foos, 1980; Mani and Johnson-Laird, 1982). In such a case, the trade-off is between retaining a single, ambiguous description vs. simultaneously constructing several spatial models. Although propositional representations can be indeterminate with respect to spatial relations, mental models cannot be, and so in this case the most efficient format for the representation is propositional.

But indeterminacy does not always deter construction of a mental model. In a paradigm closely paralleling that of Mani and Johnson-Laird's but with more strictly controlled test stimuli, Payne (1993) found no effect of determinacy. In a different paradigm (Radvansky and Zacks, 1991), single sentences indicated simply that a particular protagonist was in a particular scene. In this case, vague representation or default inference-making (Stenning, 1981) is allowable, and the lack of specificity about the location, orientation, etc., does not appear to deter subjects from forming spatial mental models.

Subject Factors. Spatial skill is a good predictor of performance in mental model tasks (de Vega, 1994a; Sharp and McNamara, 1991). Inference latency for spatial information is correlated with independent assessments of both spatial imagery ability and reading comprehension ability (Haenggi, Kintsch and Gernsbacher, 1993). Not surprisingly, construction and use of mental models are demanding on both comprehension and spatial abilities.

The Referent Situation. Finally, mental models are limited in the amount of spatial information that can easily be encoded into them. Complex scenes conveyed through language are difficult to learn, and subjects seem to abandon efforts to construct mental models when reading about them (Perrig and Kintsch, 1985).

Production of Descriptions

Descriptions produced by subjects constitute another source of information for understanding how people conceptualize space. At a gross level, we can distinguish between more stable and less stable aspects of the environment by noting their position within the description. For instance, within a statement presenting the position of one object with respect to another, the latter is signalled as being the more stable, more salient reference object against which the position of the other is established (Miller and Johnson-Laird, 1976; Talmy, 1978)). In sentences of the form *A is next to B* where *A* and *B* do not have equal status, the position of *B* serves as the more salient referent with respect to which *A* is placed. This is why sentences like *The mountain was next to the tree* sound awkward.

In a series of statements, the privileged status of a position is typically signalled by putting it first. In a route description, this is typically an entranceway into the environment (e.g., Ullmer-Ehrich, 1982; Linde and Labov, 1975; Taylor and Tversky, 1994). In a survey description, it is generally a stable, central reference point with respect to which other objects could be placed (Ferguson and Hegarty, 1994; Taylor and Tversky, 1994).

Organization of the Description

Much of the work on production concerns the type of perspective – route or survey – informing subjects' descriptions. (For a broader discussion of issues regarding production, see chapters in this book by Couclelis and by Daniel, Carite, and Denis.) Descriptions typically consist of a mixture of survey and embedded perspectives, but preference varies with features of the environment. For example, if objects fall along a single path, subjects are likely to give a route description.

This has been shown for networks (Levelt, 1982; Robin and Denis, 1991), environments learned through primary experience (Linde and Labov, 1975), environments learned from maps (Cave, 1993; Taylor and Tversky, 1994), and environments learned from text (Cave, 1993). It's particularly interesting that this occurs when subjects learn from maps, since construction of a route-based description requires a transformation in perspective. Apparently, such a transformation is worthwhile, since it simplifies the task (Cave, 1994); the linear nature of language is analogous to one's experience of a journey past a sequence of objects. And a route description can apparently facilitate comprehension, particularly if no visual map is available to the addressee (Ferguson and Hegarty, 1994). For example, subjects given a task of navigating unfamiliar roads perform better if given a route than if given a survey description (Streeter, Vitello and Wonsiewicz, 1985).

Subjects also frequently use survey perspectives in their descriptions (Taylor and Tversky, 1994). One environmental feature biasing subjects toward this type of organization is the presence of one or a few large or salient landmarks with respect to which other objects can be placed. In such a case, subjects prefer to use canonical directions for describing the relative positions of the minor objects with respect to the major ones. On the other hand, if all locations are equally salient, the preference is generally to describe locations with respect to adjacent objects, which leads to a route organization.

Scale of the environment doesn't appear to affect the perspective according to which subjects describe it (Taylor and Tversky, 1994). This is sensible, since all information can be included from either a single survey position or from a moving perspective, and since the feature of size, independent of the environment's content and organization, does not impact the ease with which various types of description can be constructed. Regardless of whether they imply survey or route perspectives, subjects' descriptions tend to have clear, hierarchical structure, reflecting a well-organized spatial model (Taylor and Tversky, 1994) and reflecting a communication strategy that aids addressees in comprehension.

Selection of specific terms. A similar principle seems to hold for selection of deictic phrases: speak so that the addressee can easily understand. Although defaulting to extrinsic frames of reference can reduce ambiguity in some terms, like "above," adopting the appropriate perspective is critical for many deictic terms and in many circumstances. If the perspectives of speaker and addressee do not match, neither will their representations of the environment. And probably a lot of effort would have been expended before this error is noticed. Fortunately, speakers and addressees are sensitive to this potential pitfall, and they are extraordinarily good at negotiating a consistent frame of reference for on-line dialogue (Schober, 1993). This is consistent with Carlson-Radvansky and Irwin's finding discussed earlier on interpretation of spatial terms. Although linguistic terms for describing space are limited (see also Landau and Jackendoff, 1993), we have conventionalized powerful tools for making distinctions between various spatial situations and events. Participants in a conversation can refer to these conventions to limit the possible interpretations of the ongoing description.

Finally, descriptions regarding direction in surrounding space reflect the same biases as do patterns of retrieval times. In Franklin, Henkel and Zangas (in press), subjects described the direction of an object from themselves so that another subject could replicate its position. The frequency with which subjects used the terms "front," "back," "right," and "left" indicated that the front was more highly resolved than the back, right, or left. Higher resolution, or discriminability among positions in a region, reflects higher perceptual and functional status. In addition, emphasis on a given direction term is higher when the object is at or near the pole defined by that term and drops off with angular distance (see also Hayward and Tarr, in press). This corroborates findings from retrieval paradigms that space is organized with respect to canonical axes (Franklin et al., in press; Hayward and Tarr, in press; Huttenlocher et al., 1992), suggesting that the same principles that organize spatial representations and guide retrieval are also expressed in language production.

Conclusions

The linguistic tools for characterizing space were described earlier as being sparse and imprecise. Spatial information often gets simplified, lost, or regularized in the process of being described (Talmy, 1978; Hayward and Tarr, in press). So it isn't surprising that subjects who learn routes from maps produce more accurate reports of the routes than those who learn from descriptions (Cave, 1993). This, however, often doesn't present a problem when environments are learned through description. It usually isn't the goal of either speaker or listener to convey all information about a spatial configuration, particularly at the scale of a large cognitive map. Rather, the critical information is usually limited to the identities and approximate locations of relevant objects.

Given the limitations of language, it's remarkable how successful we are at conveying spatial information with it, and it's remarkable how well equipped the cognitive system is for constructing representations that truly qualify as cognitive maps. Their organization and the cognitive processes that operate on them are well adapted to the kinds of tasks typically applied to spatial knowledge.

References

Baddeley, A. (1986). *Working Memory*, Oxford: Clarendon Press.

Black, J.B., Turner, T.J. and Bower, G.H. (1979). Point of view in narrative comprehension, memory, and production, *Journal of Verbal Learning and Verbal Behavior*, 18, 187- 198.

Bly, B. (1988). Perspective in mental models of text, Unpublished manuscript, Stanford University.

Bryant, D.J. and Tversky, B. (1991). Locating objects from memory or from sight, Paper presented at 32nd Annual Meeting of the Psychonomic Society, San Francisco.

Bryant, D.J., Tversky, B. and Franklin, N. (1992). Internal and external spatial frameworks for representing described scenes, *Journal of Memory and Language*, 31, 74- 98.

Bryant, D.J., Tversky, B. and Lanca, M. (1994). Retrieving spatial relations from observed and remembered scenes, Manuscript submitted for publication.

Carlson-Radvansky, LA. and Irwin, D. E. Frames of reference in vision and language: Where is above? *Cognition* 46, 223, 244.

Cave, C.B. (1993). Effects of information format on learning and reporting navigational routes, Paper presented at the 34th annual meeting of the Psychonomic Society, Washington, D.C.

Dean, R.S. and Kulhavy, R. W. (1981). Influence of spatial organization in prose learning, *Journal of Educational Psychology*, 73, 57-64.

Denis, M. and Cocude, M. (1989). Scanning visual images generated from verbal descriptions, *European Journal of Cognitive Psychology*, 1, 293-307.

Denis, M. and Cocude, M. (1992). Structural properties of visual images constructed from poorly or well-structured verbal descriptions, *Memory and Cognition*, 20, 497-506.

Denis, M. and Denhiere, G. (1990). Comprehension and recall of spatial descriptions, *European Bulletin of Cognitive Psychology*, 10, 115-143.

Denis, M. and Le Ny, J-F. (1986). Centering on figurative features during the comprehension of sentences describing scenes, *Psychological Research*, 48, 145-152.

Denis, M. and Zimmer H. D. (1992). Analog properties of cognitive maps constructed from verbal descriptions, *Psychological Research*, 54, 286-298.

Ehrlich, K. and Johnson-Laird, P. N. (1982). Spatial descriptions and referential continuity, *Journal of Verbal Learning and Verbal Behavior*, 21, 296-306.

Federico, T. and Franklin, N. (1994). Temporal and spatial mental models, Manuscript in preparation.

Ferguson, E.L. and Hegarty, M. (1994). Properties of cognitive maps constructed from texts, *Memory and Cognition*, 22, 455-473.

Fincher-Kiefer, R. (1993). The role of predictive inferrences in situation model construction, *Discourse Processes*, 16, 99-124.

Foos, P.W. (1980). Constructing cognitive maps from sentences, *Journal of Experimental Psychology: Human Learning and Memory*, 6, 25-38.

Franklin, N., Henkel, L. and Zangas, T. (In press). Parsing surrounding space, *Memory and Cognition.*

Franklin, N. and Tversky, B. (1990). Searching imagined environments, *Journal of Experimental Psychology: General*, 119, 63-76.

Franklin, N., Tversky, B. and Coon, V. (1992). Switching points of view in spatial mental models, *Memory and Cognition*, 20, 507-518.

Garnham, A. (1981). Mental models as representations of text, *Memory and Cognition*, 9, 560-565.

Glenberg, A.M. and Langston, W.E. (1992). Comprehension of illustrated text: Pictures help to build mental models, *Journal of Memory and Language*, 31, 129-151.

Glenberg, A., Meyer, M. and Lindem, K. (1987). Mental models contribute to foregrounding during text comprehension, *Journal of Memory and Language*, 26, 69-83.

Haenggi, D., Kintsch, W. and Gernsbacher, M.A. (1993). Situation-based inferences and text comprehension, Poster presented at the 34th annual meeting of the Psychonomic Society, Washington, D.C.

Hayward, W.G. and Tarr, M.J. (in press). Spatial language and spatial representation, *Cognition.*

Hintzman, D., O'Dell, C. and Arndt, D. (1981). Orientation in cognitive maps, *Cognitive Psychology*, 13, 149-206.

Holyoak, K. J. and Mah, W.A. (1982). Cognitive reference points in judgments of symbolic magnitude, *Cognitive Psychology*, 14, 328-352.

Huttenlocher, J., Hedges, L.V. and Duncan S. (1992). Categories and particulars: Prototype effects in estimating spatial location, *Psychological Review*, 98, 352-376.

Johnson-Laird P. N. (1983). *Mental models*, Cambridge, MA: Harvard University Press.

Kennedy J.M., Gabias, P. and Heller, M.A. (1992). Space, haptics, and the blind, *Geoforum*, 23, 175-190.

Kintsch, W. and van Dijk, T.A. (1978). Toward a model of text comprehension and production, *Psychological Review*, 85, 363-394.

Kosslyn, S.M. (1975). Information representation in visual images, *Cognitive Psychology*, 7, 341-370.

Kosslyn, S.M., Ball, T. and Reiser B. (1978). Visual images preserve metric spatial information: Evidence from studies of image scanning *Journal of Experimental Psychology: Human Perception and Performance* 4, 47-60.

Landau, B. and Jackendoff, R. (1993). "What" and "where" in spatial language and spatial cognition, *Behavioral and Brain Sciences* 16, 217-238.

Levelt, W. (1982). Linearization in describing spatial networks, In *Processes, beliefs and questions* (S. Peters and E. Saarinen, eds.) D. Reidel Publishing.

Linde, C. and Labov, W. (1975). Spatial structures as a site for the study of language and thought, *Language* 51, 924-939.

Loftus, G. (1978). Comprehending compass directions, *Memory and Cognition* 6, 416-422.

Maki, R., Maki, W. and Marsh L. (1977). Processing locational and orientational information, *Memory and Cognition* 5, 602-612.

Mani, K. and Johnson-Laird, P. N. (1982). The mental representation of spatial descriptions, *Memory and Cognition* 10, 181-187.

McKoon, G. and Ratcliff, R. (1992). Inference during reading, *Psychological Review* 99, 440-466.

McNamara, T.P., Halpin, J.A. and Hardy, J.K. (1992). The representation and integration in memory of spatial and nonspatial information, *Memory and Cognition* 20, 519-532.

McNamara, T.P., Ratcliff, R. and McKoon, G. (1984). The mental representation of knowledge acquired from maps, *Journal of Experimental Psychology: Learning, Memory and Cognition* 10, 723.

Miller, G.A. and Johnson-Laird, P. N. (1976). *Language and Perception*, Cambridge: Harvard University Press.

Morrow, D.G. (1985). Prominent characters and events organize narrative understanding, *Journal of Memory and Language* 24, 304-319.

Morrow, D.G. (1990). Spatial models, prepositions, and verb-aspect markers. *Discourse Processes* 13, 441-469.

Morrow, D.G., Bower, G.H. and Greenspan, S. (1989). Updating situation models during narrative comprehension *Journal of Memory and Language* 28 292-312.

Morrow, D.G. and Clark, H.H. (1988). Interpreting words in spatial descriptions *Language and Cognitive Processes* 3 275-291.

Morrow, D.G. Greenspan, S. and Bower, G.H. (1987). Accessibility and situation models in narrative comprehension, *Journal of Memory and Language* 26, 165- 187.

Moyer, R.S. and Bayer, R.H. (1976). Mental comparison and the symbolic distance effect, *Cognitive Psychology* 8, 228-246.

Perrig, W. and Kintsch, W. (1985). Propositional and situational representations of text, *Journal of Memory and Language* 24, 503-518.

Pick, H. and Riser, J. (1982). Children's cognitive mapping, In: M. Potegal (ed.), *Spatial abilities*, New York: Academic Press.

Presson, C.C. and Hazelrigg, M.D. (1984). Building spatial representations through primary and secondary learning, *Journal of Experimental Psychology: Learning, Memory, and Cognition* 10, 716-722.

Radvansky, G.A. and Zacks, R.T. (1991). Mental models and the fan effect, *Journal of Experimental Psychology: Learning, Memory, and Cognition* 17, 940-953.

Robin, F. and Denis, M. (1991). Description of perceived or imagined spatial networks, In *Mental images in human cognition* (R.H. Logie and M. Denis, eds.), pp. 141-152. Amsterdam: North-Holland.

Sachs, J. (1967). Recognition memory for syntactic and semantic aspects of a connected discourse, *Perception and Psychophysics* 2, 437-442.

Sadalla, E.K., Burroughs, W. and Staplin, L. (1980). Reference points in spatial cognition, *Journal of Experimental Psychology: Human Learning and Memory* 6, 516-528.

Schober, M.F. (1993). Spatial perspective-taking in conversation, *Cognition* 47, 1-24.

Sharp, D.L.M. and McNamara, T.P. (1991). Spatial mental models in narrative comprehension: Now you see them, now you don't, Unpublished manuscript, Vanderbilt University.

Shepard, R.N. (1967). Recognition memory for words, sentences, and pictures, *Journal of Verbal Learning and Verbal Behavior* 6, 156-163.

Sherman, R.C. and Lim, K.M. (1991). Determinants of spatial priming in environmental memory, *Memory and Cognition* 19, 283-292.

Sholl, M.J. (1987). Cognitive maps as orienting schemata, *Journal of Experimental Psychology: Learning, Memory, and Cognition* 13, 615-628.

Sholl, M.J. (1993). The effect of visual field restriction on spatial knowledge acquisition, *Paper presented at the 34th annual meeting of the Psychonomic Society*, Washington, D.C.

Speelman, C.P. and Kirsner, K. (1990). The representation of text-based and situation-based information in discourse comprehension, *Journal of Memory and Language* 29, 119- 132.

Stenning, K. (1978). Anaphora as an approach to pragmatics, In *Linguistic theory and psychological reality*, Cambridge (M. Halle, J. Bresnan, and G.A. Miller, eds.), MA: MIT Press.

Stenning, K. (1981). On remembering how to get there: How we might want something like a map, In *Cognitive psychology and instruction* (A.M. Lesgold, J. W. Pellegrino, S.D. Fokkema, and J. Glaser, eds.), New York: Plenum Press.

Talmy, L. (1978). How language structures space, In *Spatial orientation: Theory, research, and application* (H. Pick, and L. Acredolo, eds.), New York: Plenum Press.

Taylor, H.A. and Tversky, B. (1992a). Descriptions and depictions of environments, *Memory and Cognition* 20, 483- 496.

Taylor, H.A. and Tversky, B. (1992b). Spatial mental models derived from survey and route descriptions, *Journal of Memory and Language* 31, 261-282.

Taylor, H.A. and Tversky, B. (1994). Perspective in spatial descriptions, Under revision.

Thorndyke, P.W. (1981). Distance estimates from cognitive maps, *Cognitive Psychology* 13, 526-530.

Thorndyke, P.W. and Hayes-Roth, B. (1982). Differences in spatial knowledge acquired from maps and navigation, *Cognitive Psychology* 14, 560-589.

Tolman, E. C. (1932). *Purposive behavior in animals and men*, New York: Irvington.

Tversky, B. Franklin, N., Taylor, H.A. and Bryant, D.J. (1994). Spatial mental models from descriptions, *Journal of the American Society for Information Science*. 45, 656-668.

Tversky, B. and Schiano, D.J. (1989). Perceptual and conceptual factors in distortions in memory for graphs and maps, *Journal of Experimental Psychology: General* 118, 387-398.

Ullmer-Ehrich, V. (1982). The structure of living space descriptions. In *Speech, place, and action* (R.J. Jarvella and W. Klein, eds.) New York: Wiley.

van Dijk, T. and Kintsch, W. (1983). *Strategies of Discourse Comprehension*, NY: Academic Press.

de Vega, M. (1994a). Characters and their perspectives in narratives describing spatial environments, *Psychological Research* 56, 116-126.

de Vega, M. (1994b). Pointing and labelling directions in egocentric frameworks, Manuscript submitted for publication.

Wagener, M., Wender, K. F. and Wagner, V. (1990). Role of routes in spatial memory, Presented at Meeting of the Psychonomic Society, New Orleans.

Nancy Franklin
Department of Psychology,
State University of New York,
Stony Brook, NY 11794-2500.

MODES OF LINEARIZATION IN THE DESCRIPTION OF SPATIAL CONFIGURATIONS

Marie-Paule Daniel, Luc Carité and Michel Denis

Abstract: Speakers or writers who have to describe a spatial configuration to other people are faced with the following problem: The object they have to describe is two- or three-dimensional, but their verbal output is highly constrained by the one-dimensional, linear structure of language. Translating a multidimensional entity into a linear output thus requires the construction of a linear structure, which itself requires a series of cognitive decisions. Some important questions are: Which planning procedures do describers use to produce a description ? Can descriptive strategies be identified in speakers' or writers' productions ? The study reported in this chapter deals with unconstrained descriptions of a spatial configuration. Subjects were presented with the map of a fictitious island bearing nine landmarks and were asked to produce a written description of the map. The descriptions were analyzed using an ATN-based system designed to construct a representation of descriptions and to classify them according to a typology. The texts contained a variety of descriptive sequences, most of which reflected highly systematic structures. The great majority of subjects produced descriptions from a survey perspective. Introductory statements providing addressees with a spatial framework were produced more frequently by subjects who adopted highly systematic descriptive strategies. These subjects also tended to use absolute modes of landmark location, as well as canonical spatial terms referring to cardinal points. Descriptive sequences were considered to reflect the structural organization of the subjects' mental representations of the spatial configurations they had to describe.

Introduction

For the majority of human beings, exploring visually three-dimensional environments and navigating through them are essential for building new spatial knowledge. People with perceptual or motor deficiencies also acquire some form of spatial knowledge and make appropriate use of it, but most do less well than other subjects. Another useful source of knowledge for humans is symbolic information, which is conveyed by analog representations such as geographical maps. The special value of such representations is that they provide people with information whose internal structure "maps" the structure of the represented world. Lastly, humans may also gain new spatial knowledge from the verbal (written or oral) descriptions of environments given by other people. Language is a useful vehicle for conveying a variety of knowledge, of which spatial knowledge is

J. Portugali (ed.), The Construction of Cognitive Maps, 297–318.

only one part. However, linguists and cognitive psychologists have repeatedly pointed to the specific set of problems resulting from the use of language to express spatial knowledge (e.g., Klein, 1983; Tversky, 1991). Verbal descriptions of spatial configurations result from the implementation of a highly arbitrary coding system, which does not preserve the spatial relationships among described elements. Building spatial knowledge from descriptions thus requires subjects to have the capacity to convert linear, proposition-like information into a form which is structurally comparable to visuo-spatial representations. The strategies used for the verbal description of spatial configurations are the major concern of this chapter.

The problems associated with the expression of spatial knowledge in the form of discourse or written text can be assigned to one of two categories. One is mainly concerned with content, and the other with structure. There is no a priori reason for rejecting the assumption that discourse is able to describe the entire content of a visual scene. In principle, some form of exhaustivity can be achieved in discourse, although this is true in general for rather restricted worlds involving a limited set of elements and moderately complex relationships. An exhaustive description of visual scenes or complex spatial entities such as cities is a challenging task, and there are interesting examples in contemporary literature (see, in particular, H. P. Lovecraft's 200-page description of Quebec City, or Georges Pérec's detailed descriptions of an old street in Paris). Such an exercise would, however, be fastidious and practically impossible in natural communicative contexts. Most spontaneous descriptions of spatial environments are intended to convey only a restricted subset of landmarks and relations. An important question for cognitive psychology, then, is that of the mechanisms (deliberate and/or uncontrolled) which govern the selection of landmarks that are eventually incorporated in descriptions of complex environments (cf. Conklin and McDonald, 1982). Selection is an especially relevant aspect of verbal tasks like route descriptions (cf. Brambring, 1982; Denis, 1994; Maass, 1993).

The other issue concerning spatial descriptions pertains to their structure or, more specifically, to the relationships between the structure of the discourse and the structure of described space. The extensive theoretical and empirical work by Levelt (e.g., 1982, 1989) has resulted in this question typically being referred to as the "linearization problem". Language is produced sequentially and generates outputs which are by nature linear. No particular problem should arise when the object being described is itself linear. This is the case of narratives, where discourse linearization results from direct mapping of one structure (the sequence of events) onto another structure (the sequence of verbal outputs). Discourse linearity, then, adheres completely to the linearity of the described entities. But the situation is different when the object to be described is multidimensional, as are most visual scenes, composed of many elements distributed over a spatial array.

Translating a multidimensional entity in the form of a linear linguistic output requires constructing a sequential structure, which is afar from trivial task. This is reflected in the amount of attention that most speakers devote to ordering their information in discourse. For instance, a speaker describing a spatially extended set of landmarks in a complex environment must decide which landmark should be described first, which one should be mentioned next, and so on. A great many different sequences may be produced by different speakers (or even by the same speaker at different times) in describing a given object. There are many ways of describing accurately a given object or configuration by using different modes of linearization. In general, there is not a single structure that will always be perceived as a "good" description. On the other hand, it can easily be shown that there are some structures which are especially difficult to process and recall even by skilled people. This is particularly true of descriptive sequences which violate linguistic principles like referential continuity or preservation of determinacy (e.g., Denis and Denhière, 1990; Ehrlich and Johnson-Laird, 1982; Foos, 1980; Mani and Johnson-Laird, 1982).

The rules which govern discourse linearization have been investigated under experimental conditions which impose explicit constraints on speakers, such as constraints regarding the starting point of the description. In the description of patterns of colored circles connected by horizontal and vertical lines, Levelt (1982) showed that speakers' directional choices do not occur randomly, but that describers prefers strategies which minimizes the number of units they have to store simultaneously in working memory (or the duration of such storage) (see also Bisseret and Montarnal, 1993; Denis, Robin, Zock and Laroui, 1994). When describers of a complex configuration are left entirely free of any constraint on the starting point, their strategies nevertheless show a tendency to adhere to some implicit linear structure. The pioneering work of Linde and Labov (1975) showed that in the description of apartments, the spatial representation is transformed by most speakers into a temporal sequence which maps a linearization process onto the two-dimensional layout of the apartment. A pseudo-narrative thus conducts the addressee on an imaginary tour. Similar "gaze tours" may also be found in room descriptions, which further attest to the prominent role of pragmatic and semantic factors (cf. Ehrich and Koster, 1983; Shanon, 1984).

More recently, descriptions of large-scale spatial environments have been investigated by Taylor and Tversky (1992). Their analysis of descriptions revealed hierarchical structures based on spatial and functional features of the environments and on conventions for sequencing the landmarks. For instance, when describing maps from memory, subjects consistently mentioned landmarks in decreasing order of size, that is, in the description of a town, they first mentioned large natural features such as mountains and rivers, then major highways, then individual buildings. In addition, Taylor and

Tversky collected evidence indicating that the order of description of landmarks is quite similar to the organization revealed when subjects are required to draw maps of the same environments. In particular, subjects tend to recall landmarks in the same order for both tasks.

To summarize, coping with the issue of linearization in the description of spatial configurations is an ideal way for approaching the cognitive factors involved in the choice of sequential structures. Some of these structures are especially difficult to process because they place excessive load on the addressees' working memory, while others are much easier to process. The studies also reveal the wide variety of discursive structures, even for moderately complex materials. The research reported here was designed to account for the processes involved in the elaboration of a verbal message intended to allow an addressee to construct a mental representation of a spatial configuration which was not (and at no time had been) available to his or her perception. We employed an unconstrained situation, in which subjects were presented with the map of a fictitious island containing several landmarks. The subjects were invited to produce written descriptions for an addressee. Analysis of subjects' productions first aimed at documenting the variety of descriptive strategies among a large corpus of descriptions. Further analyses were conducted on the kind of perspective used by describers (route versus survey), the tendency of subjects to introduce descriptions with some preliminary information, the ways landmarks were located, and the use of spatial terminology.

The Experimental Situation

A map of a fictitious island was drawn using an Apple Macintosh computer and SuperPaint software. Figure 1 shows the map which was displayed on the screen of the computer during the experiment. It was square-shaped, and nine landmarks were distributed on it according to a 3 x 3 grid. The resulting configuration was intended to represent the schematic map of an island. In order to avoid problems associated with landmark denomination and its variability across subjects, the corresponding label was written under each landmark.

Subjects were tested individually. They were invited to produce a written description of the map without time limitations. They typed their description on the keyboard and the text was displayed on the lower quarter of the screen, while the rest of the screen displayed the map of the island throughout the experiment. The instructions stressed that the description was intended to provide information to a person who did not know anything about the island and only had a contour of the map available. This person was supposed to acquire exact geographical knowledge of the island.

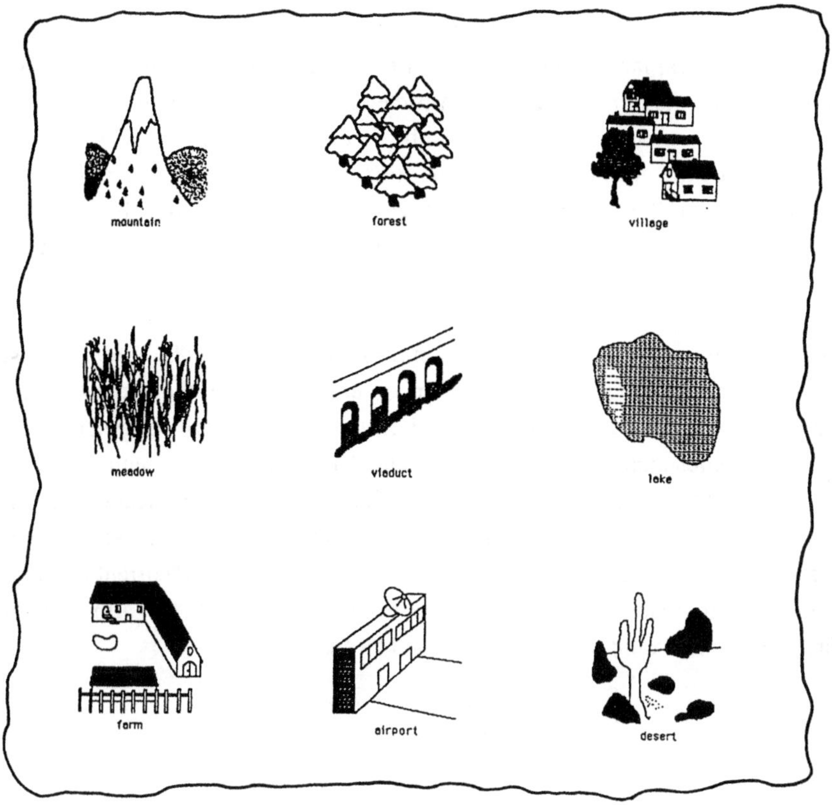

Figure 1: Map used in the experiment

One hundred and eight subjects participated in the experiment. Most of them were undergraduate students in science. The responses of eight subjects were not included in the analysis because these subjects had failed to follow the instructions. In particular, they described the internal features of the landmarks, without mentioning their location in the configuration. Since their descriptions could not be used to fulfill our objectives, the responses of these subjects were discarded. The number of subjects whose protocols were analyzed was therefore 100 (17 females and 83 males).

Since our primary interest was to determine the sequence followed by each subject when describing the location of the nine landmarks on the configuration, we devised a program to analyze and classify subjects' responses automatically. The analysis of protocols relied on an ATN-based (Augmented Transition Network) system designed to construct a representation of each description and to classify it according to a typology. Protocols were analyzed on-line by an analyzer controlled by the ATN, while subjects typed their descriptions on the keyboard.

Like all similar systems, this ATN was a non-deterministic programming language designed for natural language processing. It proved to be an efficient tool for analyzing relatively short texts (10-20 lines) in a restricted context. Its constituent parts were: (a) a morpho-syntactic and a syntactic-semantic knowledge base (or dictionary), which together formed the declarative part of the software; (b) an analyzer, which represented the procedural part that dynamically accessed both knowledge bases; and (c) a preprocessing function, which allowed for correcting misspellings and making distinctions among homonyms to prevent subsequent problems of combinatory explosion.

The first step of processing was a morphological analysis, which resulted in determining the "dictionary entry" form of each word of the processed text, its grammatical type, gender, and number, and, for verbs, their tense and person. The ATN then built a representation of each description. The resulting representations had to be available for automatic processing. They included the mode of location (absolute or relative) of each landmark, as well as the reference landmark (or landmarks), if any. The representation specified which landmark was currently handled ("current landmark") at any point in the description and was used for relative location of further landmarks. Figure 2 shows an example of a short text and its analysis.

"In the northwest of the island, we find a mountain. On going east, we pass through a forest, then we reach a village."

landmark_loc_abs:	
loc_mixt_2:	*west*
loc_mixt:	*north*
landmark:	*mountain*
current_landmark:	*mountain*
landmark_loc_rel:	
referent :	*mountain*
loc_rel_2 :	*null*
loc_rel :	*east*
landmark :	*forest*
current_landmark:	*forest*
landmark_loc_rel:	
referent :	*forest*
loc_rel_2 :	*null*
loc_rel :	*east*
landmark :	*village*
current_landmark:	*village*

Figure 2: A short text and its ATN representation

The ATN was used to determine the specific sequence according to which landmarks were described by each subject and to group them into classes of sequences. The software checked for matching between the subject's description and the configuration. Then, it attempted to match the sequence used with the sequences available in a repertory of typical descriptive sequences. In case of matching failure, the new sequence was incorporated into the repertory. Sequences were then grouped together as a function of a typology.

Classification of Descriptive Sequences

In this section, we analyze descriptive sequences underlying subjects' productions, as reflections of the way subjects coped with the linearization problem. As will be evident below, the majority of descriptions were systematic to some degree, in that landmarks were entered in an order which totally departed from random. These descriptions revealed the "projection" of a spatial framework onto the material to describe, or the application of systematic scanning over the configuration, or even reference to some preexisting Gestalt. In most cases, systematic descriptions involved a high degree of referential continuity. In our typology, any description that did not show at least one of the above features was considered to be non-systematic. The systematic descriptions were distributed among five classes, as summarized in Table 1.

Table 1: Distribution of descriptive strategies

SYSTEMATIC	**85**
Strict linear horizontal	18
Non-strict linear	33
Circular	15
Star-shaped	7
Other systematic	12
NON-SYSTEMATIC	**15**

The first class of descriptions were those reflecting a strict linear horizontal strategy. The subjects scanned the three "lines" of the configuration in turn, from top to bottom and from left to right, following the typical scanning mode of occidental reading. A total of 18 subjects (21 % of those who produced systematic descriptions) used this strategy, whose clearcut sequence is summarized in Figure 3, example (a).

The second class included descriptions which reflected the use of some form of linear scanning, but departed from strict linearity at some point. These strategies were therefore called non-strict linear strategies. They were adopted by 33 subjects (39 % of those who

Strict Linear Horizontal Strategy

(a)	1	2	3
	4	5	6
	7	8	9

Non-Strict Linear Strategies

(b)	1	2	3	(c)	1	3	2
	7	8	9		4	6	5
	4	5	6		7	9	8
(d)	1	6	7	(e)	1	4	5
	2	5	8		2	6	7
	3	4	9		3	8	9

"Circular" Strategies

(f)	3	4	5	(g)	1	2	3
	2	1	6		8	9	4
	9	8	7		7	6	5

"Star-Shaped" Strategies

(h)	1	5	2	(i)	6	2	7
	7	9	8		4	1	5
	3	6	4		8	3	9

Other Systematic Strategies

(j)	1	2	3	(k)	1	2	3
	5	4	9		9	8	4
	6	7	8		6	7	5

Figure 3: Examples of systematic descriptive sequences

produced systematic descriptions). This class is a composite one. For instance, instead of describing the first, second, and third line successively, each from left to right, one deviation from the strict horizontal strategy consisted of describing the three lines in some other sequence, for instance the upper, lower, and middle line, as is shown in example (b) of Figure 3. Example (c) preserved the top-down exploration of the three lines, but for each line, the two extreme landmarks were described before the middle one. This class included a few occurrences of other linear strategies, including the "boustrophedon" variant based on sinusoidal scanning (d) and combinations of vertical and horizontal linear strategies (e).

Fifteen subjects (18 % of those who produced systematic descriptions) made use of "circular" strategies. These descriptions had several variants, in particular clockwise or counterclockwise. Examples (f) and (g) of Figure 3 show two further variants on the clockwise strategy, starting from or arriving at the central landmark.

Seven subjects (8 % of those who produced systematic descriptions) adopted a strategy in which the set of landmarks was divided into two subsets, one with the four landmarks at the four cardinal points and the other with the four landmarks in the four angles of the island. Their descriptions consisted of combining these two subsets, resulting in a sort of an eight-branch star. These subjects mentioned the central landmark either at the beginning or at the end of the description. Two examples, (h) and (i), are shown in Figure 3. These modes of description were grouped together under the label of "star-shaped" strategies.

The last class contained 12 subjects (14 % of those who produced systematic descriptions) whose descriptions were considered systematic, without clearly belonging to one of the above classes. Although being somewhat less strict than their counterparts, these subjects nevertheless adopted a systematic approach, for example by segmenting the whole set of landmarks into two or three subsets or by using a mixed strategy. Such examples, (j) and (k), are shown in Figure 3. In this group, three subjects used an organization dependent on the semantics of landmarks, for instance mentioning first all natural landmarks, then man-made buildings.

The remaining 15 subjects produced descriptions which did not reflect any systematic approach. Cross-examination by two independent judges did not reveal any understandable organization in their descriptive sequences. No explicit or implicit framework could be uncovered even after careful analysis of descriptions. The poor quality of these descriptions apparently did not result from subjects' lack of application during the execution of the task. Subjects followed instructions attentively and their behavior was as serious as that of the other subjects during the experiment. Nevertheless, their descriptions gave the impression that they did not apply any sort of structured framework which would have accommodated the spatial characteristics of the configuration. Figure 4 shows an example of such a non-systematic description.

Thus, 85 % of the subjects used an identifiable descriptive strategy while completing the task. Their efforts to produce systematic descriptions were thought to reflect that they actually attempted to follow the instructions to produce a description which would help an addressee to build a representation of the island in spite of not having been shown it. There is no evidence that the remaining 15 % of subjects did not intend to communicate efficiently with their addressees. Nevertheless, they did not succeed in producing descriptions containing those features commonly recognized for their communicative value.

> *"The island is composed of different sites. We can see high mountains with eternal snow. They are located in the northeast of the island. To get to these mountains, you must first pass through a large forest which surrounds them except on the coastal side. On the eastern side of the island, there is a village through which runs a bridge linking the mountain to the farm. The farm is located farther south. Between the farm and the village there is a wide meadow, in the middle of which lies a small, quiet lake. At the extreme west of this island, the airport is by the sea and at the edge of the desert which occupies the southern quarter of the island."*

Figure 4: Example of a non-systematic description

The classes defined above were combined into larger classes according to a criterion expected to produce a useful hierarchy. Three levels were considered to concern both the systematic nature of descriptions and their expected communicative value. Level 1 resulted from the grouping of subjects who produced either strict linear horizontal, circular, or star-shaped strategies (N = 40). All these descriptions were very systematic and used a very strong framework for spatial organization. These descriptions used a preexisting spatial schema activated by the configuration, but which remained to some extent independent from it. Thus, the subjects' responses reflected their capacity to use a spatial framework available in long-term memory, with the implicit assumption that this framework should help an addressee to build a correct representation of the configuration. These preexisting frameworks strongly predicted the order in which landmarks would be described. Level 2 included all subjects who used non-strict linear or other systematic strategies (N = 45). This level corresponds to moderately systematic organizations, less constrained than Level 1 organizations. Many of the descriptions in Level 2 resulted in fact from combinations of two or more strategies. Lastly, Level 3 included subjects who produced non-systematic descriptions, all of which lacked any identifiable organizing framework (N = 15).

The remaining sections of this chapter use this three-level classification consistently to attempt to show some correlation with other features of interest, such as choice of perspective, presence of introductory statement, modes of landmark location, and spatial terminology.

Choice of Perspective

Descriptions of spatial configurations are usually produced from one of two perspectives, survey or route perspective, or sometimes a mixture of both (cf. Tversky, 1991). The survey perspective takes a bird's eye view of the described configuration. The route perspective, on the other hand, places the addressee virtually on the described territory and requires him or her to follow a route through this territory. The resulting descriptions are clearly differentiated. Survey descriptions usually take a fixed perspective, whereas route descriptions take the perspective of a moving observer. Figure 5 contrasts examples of descriptions made from each perspective.

The descriptions collected in the experiment were classified as reflecting either survey or route perspective. The great majority of subjects (89 %) produced descriptions mainly from a survey perspective, while only 11 % produced descriptions from a route perspective. This distribution was uncorrelated with the systematic or non-systematic nature of descriptive strategies, as is shown in Table 2, which summarizes the distribution of survey and route descriptions for Levels 1, 2, and 3.

Survey Perspective

"The island is square shaped. It is characterized by symmetrical distribution of its landmarks. To the northwest are located mountains, to the north the forest, to the northeast the village, to the west the meadow, at the center the viaduct, to the east the lake, to the southwest the farm, to the south the airport and lastly to the southeast the desert."

Route Perspective

"When arriving at the southwest of the island, you reach a farm. Going east, you encounter an airport, then a desert. Then, going north, you find a lake, then a village. Continuing to west, there is a forest, then a mountain. Lastly, while going south, you see a meadow and finally reach the farm from which you started. You have been continuously travelling around a viaduct."

Figure 5: Examples of descriptions from survey or route perspectives

Presence of Introductory Information

The care taken by subjects to provide their addressees with some preliminary information (i.e., before describing the exact location of each landmark) was considered to be an indicator likely to have some relationship with the systematic nature of descriptive strategies. Even a minimal introductory sentence prior to landmark description reflects the subjects' concern for standard writing rules and their intention to increase text understandability by placing their description in a structured context. Ideally, an introduction should provide the addressee with a framework including spatial cues which will serve to locate the landmarks later on in the description.

Almost half (42 %) of the subjects provided their addressees with some kind of preliminary information. There were four types of introductory sentences: (a) mention of

Table 2: Distribution of survey and route descriptions

	Survey	Route	Total
LEVEL 1	**34**	**6**	**40**
Strict linear horiz.	18	0	18
Circular	11	4	15
Star-shaped	5	2	7
LEVEL 2	**41**	**4**	**45**
Non-strict linear	30	3	33
Other systematic	11	1	12
LEVEL 3	**14**	**1**	**15**

number of landmarks (for instance, "The island contains nine landmarks"); (b) reference to the shape of the island (for instance, "The island is square-shaped"); (c) explanation on the method used for description (for instance, "I will start the description from northwest, then I will proceed clockwise"); (d) instructions to the addressee (for instance, "Imagine a grid made of three lines by three columns, thus forming nine squares"). Table 3 shows the distribution of these four types of introductory statements, including the cases where subjects combined two of them. However, the majority of subjects who produced such preliminary information limited it to a single sentence.

Table 3: Distribution of four types of introductory sentences and their combinations

	a	b	c	d	a+b	a+c	a+d	b+c	b+d	c+d	Total
LEVEL 1	**4**	**6**	**4**	**4**	**0**	**4**	**0**	**2**	**0**	**0**	**24**
Strict linear horiz.	2	2	2	2	0	0	0	1	0	0	9
Circular	0	3	2	2	0	3	0	0	0	0	10
Star-shaped	2	1	0	0	0	1	0	1	0	0	5
LEVEL 2	**6**	**3**	**3**	**1**	**2**	**1**	**0**	**0**	**0**	**0**	**16**
Non-strict linear	5	3	2	1	0	1	0	0	0	0	12
Other systematic	1	0	1	0	2	0	0	0	0	0	4
LEVEL 3	**0**	**0**	**0**	**0**	**2**	**0**	**0**	**0**	**0**	**0**	**2**

Table 4 summarizes the distribution of descriptions according to the presence or absence of an introductory statement, for Levels 1, 2, and 3. The data clearly indicate that subjects at Level 1 were most likely to include an introductory statement. Fewer Level 2 subjects used introductions and Level 3 subjects were unlikely to provide them. Chi square tests were carried out on the data. Differences in frequencies were attested by a significant value of $\chi 2$ (2) = 11.14, $p < .01$.

Table 4: Distribution of descriptions with or without an introductory sentence

	With	Without	Total
LEVEL 1	**24**	**16**	**40**
Strict linear horiz.	9	9	18
Circular	10	5	15
Star-shaped	5	2	7
LEVEL 2	**16**	**29**	**45**
Non-strict linear	12	21	33
Other systematic	4	8	12
LEVEL 3	**2**	**13**	**15**

The presence of an introductory statement, even a minimal one, is thus a reliable predictor of a highly structured descriptive sequence. It is also likely to reflect the good quality of the representational spatial schema used by the describer. A description that starts with an introductory sentence is likely to be a highly systematic one. On the other hand, the absence of any introduction cannot be considered as a definite predictor of a lack of organization. Among the 58 subjects who produced no introductory sentence, 45 of them (78 %) eventually produced a description having some systematic form (Levels 1 and 2). In other words, a text that started with an introductory sentence would probably use a highly systematic descriptive strategy. But the fact that it did not did not guarantee that the description would be poorly structured. Half of the subjects who used a strict linear horizontal strategy may have considered an introduction superfluous, if they expected their strategy to be part of their addressee's knowledge. Such postulates of shared knowledge or assumptions are frequently found in spatial descriptions. For instance, when mentioning some location at the first floor, nobody will specify that the "first floor" is located with reference to the ground, since for everybody, the implicit reference point for a building is the ground (cf. Garrod and Anderson, 1987).

Modes of Landmark Location

Descriptive strategies selected by people, as well as the perspective they take (and invite their addressees to take) on the configuration, are potent factors in the construction of the overall structure of a representation by the addressees. They may be considered as reflections of the structure of the representation that describers construct during the task of description. In addition to these macrostructural decisions, describers also have to deal with local problems. One of these problems is describing the location of each landmark using appropriate reference points.

Absolute Location
- of a single landmark :
"An X in the northwest."
"An X where the diagonals cross."
- of adjacent landmarks :
  - explicit :
  "An X in the top left corner ; a Y in the middle ; a Z on the right."
  " At the top, from left to right, an X, a Y and then a Z."
  - explicit/implicit :
  "At the top, from left to right, an X in the west, a Y in the middle, then a Z."
  "An X in the top left corner, then a Y and a Z."
  "You will find an X at the top, a Y and a Z."
- of all the landmarks :
"From left to right and from top to bottom, you will find an X, then a Y..."
"As you would read a page of text, according to occidental writing habits..."

Relative Location
- of a single landmark :
  - with precise positioning :
  "To the right of the X, a Y."
  "And below the X, a Y."
  - with imprecise positioning (ambiguity removed by the context) :
  "An X in the northeast ; on its left, a Y ; and after that, a Z."
  "Starting again from northwest, you will find an X on the right and, going straight on, a Y."
  "The X is beside the Y and the Z."
- of more than one landmark :
  - linear description :
    - continuous :
  "An X at the northwest ; starting from the X, going to east, there is a Y and a Z."
  "Then, after the Y which is on its right, there is a Z."
    - discontinuous :
  "In the first column, the X at the top is separated from the Z by the Y."
  "In the upper row, the X on the left is connected to the Z by the Y."
  - non-linear description :
    - deviating from linearity :
  "Above the X which is to its right, there is a Y."
  "Then, to the right of the X which is above, there is a Y."
    - star-shaped (one landmark used as pivot) :
  "Then, starting again from the W, you find an X on the left, a Y on the right and a Z below."

Figure 6: ATN-based classification of location modes

Location may be defined as absolute if the describer uses an organizational scheme defined by space coordinates. In this case, landmarks are situated independently of each other (for example, "There is a mountain in the northwest, a forest in the north, a desert in the southeast"). Location is relative when landmarks are situated relative to each other (for example, "To the east of the mountain, there is a forest"). In this case, the describer must have situated one landmark in advance as an absolute reference point in order to achieve consistency. Figure 6 shows some examples from the ATN-based classification of location modes.

We first identified those subjects who only used the absolute location mode (in contrast to those who used both absolute and relative modes). It is reasonable to expect that the use of absolute reference points exclusively should be more strongly associated with a preference for highly structured spatial frameworks (such as those based on cardinal points, for instance). Our assumption was thus that more subjects using absolute locations should also give the most systematic descriptions (Level 1). This level is characterized by the use of a strong preexisting framework, independent of the spatial configuration to be described. Consequently, highly systematic descriptive strategies should be accompanied by a preference for absolute modes of landmark location.

Table 5 shows the number of descriptions using exclusively absolute or mixed (absolute and relative) locations. Most of the subjects who opted for absolute locations were at Level 1, while the reverse was true for Level 2. None of the Level 3 subjects used exclusively absolute locations. The differential distribution of subjects who used either exclusively absolute or mixed location modes was significant, $\chi2$ (2) = 29.84, $p < .001$. This confirms our hypothesis of a link between the systematic approach and the mode of landmark location. The more systematic the descriptive strategies, the more likely was the use of exclusively absolute location.

Table 5: Distribution of descriptions using exclusively absolute or mixed (absolute + relative) modes for locating landmarks

	Absolute	Mixed	Total
LEVEL 1	**28**	**12**	**40**
Strict linear horiz.	16	2	18
Circular	8	7	15
Star-shaped	4	3	7
LEVEL 2	**11**	**34**	**45**
Non-strict linear	9	24	33
Other systematic	2	10	12
LEVEL 3	**0**	**15**	**15**

We also carried out an analysis which compared the average numbers of absolute and relative locations at Levels 1, 2, and 3. The individual statements produced by subjects to locate landmarks were used to compute the number of statements using absolute or relative locations. Table 6 shows such average values, from which it appears that Level 1 subjects consistently produced more statements using absolute than relative locations,

$t(39) = 6.90$, $p < .001$. There were no significant differences for either Level 2, $t(44) = 1.35$, or Level 3, $t(14) < 1$. Therefore, subjects giving the most systematic descriptions preferred absolute locations, whereas the other subjects used both location systems in about similar proportions.

Table 6: Mean number of statements using absolute or relative modes to locate landmarks.

	Absolute	Relative	Total
LEVEL 1	**7.90**	**1.98**	**9.88**
Strict linear horiz.	8.56	0.50	9.06
Circular	7.33	3.20	10.53
Star-shaped	7.43	3.14	10.57
LEVEL 2	**5.49**	**4.22**	**9.71**
Non-strict linear	5.45	3.88	9.33
Other systematic	5.58	5.17	10.75
LEVEL 3	**3.67**	**4.00**	**7.67**

Table 6 indicates that the sharp drop in the use of absolute locations by Level 3 subjects was not totally offset by the increase in the occurrence of relative locations. The overall number of location statements was greater for Level 1 than for Level 3 subjects. The more systematic the subjects were in their strategies, the more likely they were to opt for an absolute mode of landmark location.

We also examined the possibility of a relationship between the location modes used by subjects and their production of introductory statements. If descriptions based on absolute location modes reflect the use of highly structured frameworks of spatial reference, it is reasonable to expect some relationship between the location modes and the care taken by subjects to provide preliminary structuring information in the form of an introductory sentence.

Table 7 shows the distribution of Level 1 subjects into four cells resulting from crossing of the two indicators (exclusive vs. non-exclusive use of absolute locations and presence vs. absence of an introductory sentence). The presence of an introductory statement was closely associated with the absolute location mode in these subjects who produced the most systematic descriptive strategies. This effect was attested by a value of $\chi 2(1) = 8.76$, $p < .01$. Analyses of Level 2 and Level 3 subjects revealed no significant relationship between the two variables. Thus, the exclusive use of absolute locations and an introductory statement were two further reliable characteristics of subjects who generated the most systematic strategies.

Table 7: Relationship between location modes and the presence of an introductory statement in the Level 1 group

	Introductory Statement	
	Present	Absent
Location Mode		
Absolute	21	7
Mixed	3	9

We also examined landmark locations to detect errors in the descriptions. Surprisingly, given the very schematic material used in this experiment, many of the subjects made errors of location when describing the map, mostly by confusing "east" and "west". These errors were particularly remarkable given that the subjects had the map in front of them throughout the description.

Table 8 shows the distribution of descriptions containing at least one error or no errors. The proportions of error for Level 1 and Level 2 subjects were very similar. In both groups, the number of correct protocols significantly exceeded those which contained at least one error ($p < .001$, in both cases). On the other hand, relatively more Level 3 subjects made errors, so that the number of correct protocols was not significantly larger than the number of protocols with at least one error. It is not surprising that more errors were produced by subjects who showed the greatest difficulty in adjusting their description to a well-structured preexisting spatial framework.

Table 8: Distribution of descriptions containing at least one error or no errors

	At least one	None	Total
LEVEL 1	**6**	**34**	**40**
Strict linear horiz.	2	16	18
Circular	3	12	15
Star-shaped	1	6	7
LEVEL 2	**8**	**37**	**45**
Non-strict linear	5	28	33
Other systematic	3	9	12
LEVEL 3	**5**	**10**	**15**

Spatial Terminology

Space-related terminology is an extremely rich one, with many variants for expressing topological relations. In their descriptions, our subjects used canonical spatial terms referring to the four cardinal points ("north", "south", "east", "west"), as well as substitute terms less closely connected to an absolute frame of reference ("top", "bottom", "right", "left"). A number of prepositional expressions ("above", "below", etc.) and space-related verbs ("to skirt around", "to run along", etc.) were also used (cf. Aurnague and Vieu, 1993; Vandeloise, 1986).

Table 9 shows the number of descriptions using only the four cardinal points as references. The exclusive use of such terms typically reflects a subject's preference for an absolute location mode, which is itself correlated with a preference for highly systematic descriptive strategies. The data confirmed that Level 1 and Level 2 subjects consistently preferred the exclusive use of cardinal terminology (83 % of Level 1 and 73 % of Level 2 subjects opted for the exclusive use of canonical terms). By contrast, less systematic describers exhibited the opposite tendency, since they used less cardinal than mixed terminology (33 % versus 67 %). The overall effect was significant, $\chi 2$ (2) = 13.30, $p < .01$. More rigorous terminology (based on terms referring to cardinal points) was thus closely associated with more systematic descriptive strategies.

Table 9: Distribution of descriptions using exclusive or non-exclusive canonical spatial terminology (cardinal points)

	Exclusive	Non-exclusive	Total
LEVEL 1	**33**	**7**	**40**
Strict linear horiz.	16	2	18
Circular	11	4	15
Star-shaped	6	1	7
LEVEL 2	**33**	**12**	**45**
Non-strict linear	22	11	33
Other systematic	11	1	12
LEVEL 3	**5**	**10**	**15**

The last step of our investigation was a detailed analysis of predicative language. The variety of expressions used to locate landmarks was first approached by identifying five categories of expressions related to landmark description: (a) neutral descriptive expressions (e.g., "There is a mountain", "A mountain is situated"); (b) expressions related to the describer (e.g., "The lake I have just mentioned", "We can see an airport from here"); (c) expressions related to the addressee (e.g., "You have a mountain in front of you", "You will encounter a village"); (d) expressions related to the island (e.g., "The island contains a forest"); (e) intrinsic landmark descriptions (e.g., "The viaduct straddles", "A desert stretches away").

Each protocol was analyzed and the expression used to describe each landmark location was coded. The resulting figures are shown in Table 10. The average number of expressions per protocol was 9.00, 8.64, and 7.93 for subjects of Levels 1, 2, and 3, respectively. Neutral expressions were frequently used by subjects at all levels, but the frequency of occurrence steadily decreased from Level 1 through Level 3.

Table 10: Mean number of expressions used to locate landmarks

	Neutral Descriptive	Related to the Describer	Related to the Addressee	Related to the Island	Intrinsic to the Landmark	Total
LEVEL 1	**5.63**	**2.23**	**0.38**	**0.23**	**0.55**	**9.00**
Strict linear horiz.	5.17	2.39	0.72	0.44	0.33	9.06
Circular	6.27	2.20	0.13	0.00	0.33	8.93
Star-shaped	5.43	1.86	0.00	0.14	1.57	9.00
LEVEL 2	**5.13**	**1.89**	**0.18**	**0.60**	**0.84**	**8.64**
Non-strict linear	4.94	2.09	0.24	0.39	0.82	8.48
Other systematic	5.67	1.33	0.00	1.17	0.92	9.08
LEVEL 3	**4.00**	**0.53**	**0.33**	**0.80**	**2.27**	**7.93**

Expressions of categories (a), (b), and (c) were grouped together for further statistical analysis, since all used expressions with references external to the configuration. On the other hand, the expressions in categories (d) and (e) all directly referred to the configuration. The number of expressions belonging to the first and second supercategories were 8.23 and 0.78 for Level 1, 7.20 and 1.44 for Level 2, and 4.87 and 3.07 for Level 3. Differences were statistically significant for Level 1, $t(39) = 13.80$, $p < .001$, and Level 2, $t(44) = 8.85$, $p < .001$. The difference was not significant for Level 3, $t(14) < 1$. This analysis supported the claim that the more systematic a descriptive strategy, the more likely the subject was to use expressions based on external reference. Less systematic describers, who failed to use a highly structured preexisting framework, tended to use more expressions directly referring to the configuration than systematic describers.

Conclusions

A wide range of strategies is available to subjects who have to describe spatial configurations like maps, but the structural quality and communicative value of these strategies can vary greatly. In most cases, descriptive strategies reveal the describers' intention to facilitate the addressees' on-line integration of the constituent elements of the configuration. The most expert describers use procedures which allow them to provide their addressees with a global view of the configuration, and then to deliver detailed information in a systematic fashion.

The experiment reported in this chapter first showed that when subjects were asked to describe a spatial configuration, the great majority of them attempted to organize their descriptions systematically, indicating that they followed a structured plan. Subjects solved the "linearization problem" by selecting descriptive sequences expected to deliver information to addressees in an order which should help them to integrate ongoing information. The most systematic strategies were based on preexisting spatial schemata available in long-term memory. Only a minority of subjects produced descriptions of poor communicative value, using a sequence which would be difficult to apply to another configuration. Well-differentiated levels in the systematic organization of descriptive sequences were easily evidenced.

The second piece of information provided by this study was that subjects who produced highly systematic descriptive sequences exhibited several additional consistent features. First, they took great care to provide their addressees in advance with a spatial framework in the form of a relevant introductory statement. Secondly, they consistently preferred absolute modes of landmark location. Thirdly, they showed a clear preference for using canonical spatial terms referring to cardinal points, and expressions based on external reference. The modes of processing used at local levels (e.g., landmark location and spatial terminology) were closely correlated with the modes of processing used at the macrostructural level (definition of overall descriptive structure).

When subjects linearized their descriptions of the configuration, not only did most of them reveal their cognitive capacities of recoding information from a two-dimensional to a one-dimensional structure, but they also showed their willingness to take into account the cognitive capacities of their adressees. It is sensible for describers to use descriptive structures or schemata that the addressees are also likely to possess, and which may help them to integrate incoming information. Expectation is a central mechanism for addressees. More efficient processing will occur when any element provided to them fulfills their current expectations. Addressees expect pieces of information at critical points of the description, but they also expect the descriptive macrostructure to be easily identifiable. The experimental situation and data reported here provide clear indications that language and spatial cognition may interact efficiently in human communication in spite of the essential structural differences between them.

Acknowledgements: The research reported in this chapter was supported by grants from the French Ministry of Research and Technology (under contract 91 V 0808) and the National Center of Scientific Research (Cogniscience Program, National Thematic Project on "Representation of Space"). Thanks to André Bisseret for helpful comments on a first draft of this chapter. Requests for reprints should be sent to Dr. Marie-Paule Daniel, Groupe Cognition Humaine, LIMSI-CNRS, Université de Paris-Sud, BP 133, 91403 Orsay Cedex, France.

References

Aurnague, M. and Vieu, L. (1993). A three-level approach to the semantics of space, In *The semantics of prepositions: From mental processing to natural language processing* (C. Zelinsky-Wibbelt, ed.), pp. 393-439. Berlin: Mouton de Gruyter.

Bisseret, A. and Montarnal, C. (1993). *Stratégies de linéarisation lors de descriptions textuelles de configurations spatiales*, Rapport de recherche No. 1927, INRIA, Grenoble.

Brambring, M. (1982). Language and geographic orientation for the blind, In *Speech, place, and action* (R.J. Jarvella and W. Klein, eds.), pp. 203-218. Chichester: Wiley.

Conklin, E.J. and McDonald, D.D. (1982). Salience: The key to the selection problem in natural language generation, In *Proceedings of the 20th Annual Meeting of the Association for Computational Linguistics*, pp. 129-135, Toronto, 16-18 June 1982.

Denis, M. (1994). *La description d'itinéraires: Des repères pour des actions. Notes et Documents du LIMSI*, No. 94-14, Juillet 1994.

Denis, M. and Denhière, G. (1990). Comprehension and recall of spatial descriptions. *European Bulletin of Cognitive Psychology* 10, 115-143.

Denis, M., Robin, F., Zock, M. and Laroui, A. (1994). Identifying and simulating cognitive strategies for the description of spatial networks, In *Comprehension of graphics* (W. Schnotz and R.W. Kulhavy, eds.), pp. 77-94. Amsterdam: North-Holland.

Ehrich, V. and Koste,r C. (1983). Discourse organization and sentence form: The structure of room descriptions in Dutch. *Discourse Processes* 6, 169-195.

Ehrlich, K. and Johnson-Laird, P.N. (1982). Spatial descriptions and referential continuity. *Journal of Verbal Learning and Verbal Behavior* 21, 296-306.

Foos, P.W. (1980). Constructing cognitive maps from sentences. *Journal of Experimental Psychology: Human Learning and Memory* 6, 25-38.

Garrod, S. and Anderson, A. (1987). Saying what you mean in dialogue: A study in conceptual and semantic co-ordination. *Cognition* 27, 181-218.

Klein, W. (1983). Deixis and spatial orientation in route directions, In *Spatial orientation: Theory, research, and application* (H. L. Pick Jr. and L.P. Acredolo, eds.), pp. 283-311. New York: Plenum.

Levelt, W.J.M. (1982). Linearization in describing spatial networks, In *Processes, beliefs, and questions* (S. Peters and E. Saarinen, eds.), pp. 199-220. Dordrecht, The Netherlands: Reidel.

Levelt, W.J.M. (1989). Speaking: *From intention to articulation*, Cambridge, MA: The MIT Press.

Linde, C. and Labov, W. (1975). Spatial networks as a site for the study of language and thought. *Language* 51, 924-939.

Maass, W. (1993). A cognitive model for the process of multimodal, incremental route descriptions, In *Spatial information theory: A theoretical basis for GIS* (A.U. Frank and I. Campari, eds.), pp. 1-13. Berlin: Springer-Verlag.

Mani, K. and Johnson-Laird, P.N. (1982). The mental representation of spatial descriptions. *Memory and Cognition* 10, 181-187.

Shanon, B. (1984). Room descriptions. *Discourse Processes* 7, 225-255.

Taylor, H.A. and Tversky, B. (1992). Descriptions and depictions of environments. *Memory and Cognition* 20, 483-496.

Tversky, B. (1991). Spatial mental models, In *The psychology of learning and motivation: Advances in research and theory* (G.H. Bower, ed.), Vol. 27, pp. 109-145. New York: Academic Press.

Vandeloise, C. (1986). *L'espace en français: Sémantique des prépositions spatiales*, Paris: Editions du Seuil.

Marie-Paule Daniel, Luc Carité and Michel Denis
Groupe Cognition Humaine, LIMSI-CNRS
Université de Paris-Sud, Orsay, France

Part Three: Specific Themes

MODELING DIRECTIONAL KNOWLEDGE AND REASONING IN ENVIRONMENTAL SPACE: TESTING QUALITATIVE METRICS

Daniel R. Montello and Andrew U. Frank

Abstract: Researchers from a variety of disciplines have proposed models of human spatial knowledge and reasoning in order to explain spatial behavior in environmental spaces, such as buildings, neighborhoods, and cities. A common component of these models is a set of hypotheses about the geometry of spatial knowledge, particularly with respect to the roles of topological and metric knowledge. Recently, mathematicians and computer scientists interested in formally modeling everyday intelligent spatial behavior have developed models incorporating "qualitative" spatial reasoning ("naive" spatial reasoning). One branch of this effort has been the development of so-called "qualitative metric" models to solve problems such as wayfinding. A qualitative metric employs more sophisticated geometry than just topology but at a relatively imprecise or coarse-grained level. Such models essentially reason with a small finite number of quantitative categories for direction and/or distance. In this chapter, we evaluate the abilities of qualitative metric models to account for human knowledge of directions by comparing simulations derived from qualitative metrics to empirical data and theorizing derived from human-subjects testing.

Introduction

Researchers have proposed models of spatial knowledge and reasoning in order to explain human spatial behavior in environmental spaces (the relatively large-scale spaces of buildings, neighborhoods, and cities). Although varying in comprehensiveness, these models have typically included ideas about processes of knowledge acquisition, the form of stored knowledge, and its retrieval and manipulation in working-memory. Such models have been provided by researchers from a wide variety of disciplines: geographers (Couclelis et al., 1987), psychologists (Piaget and Inhelder, 1948/1967; Siegel and White, 1975), and computer scientists (Kuipers and Levitt, 1988; McDermott and Davis, 1984), among others.

A common aspect of these models is a set of hypotheses about the geometric sophistication of spatial knowledge acquired from direct locomotor experience in the environment (cf. Golledge and Hubert, 1982; Kuipers and Levitt, 1988; Landau et al.,

J. Portugali (ed.), The Construction of Cognitive Maps, 321–344.

1984; Mandler, 1988; McDermott and Davis, 1984; McNamara, 1992; Montello, 1992, in press). Some models proposed for human spatial knowledge of the environment have suggested that it is best described as metric. A metric geometry describes spaces that have properties such as symmetry and the triangle inequality, properties that define quantitative measurement on spatial dimensions (see Montello, 1992; Shepard, 1964). Others have suggested that spatial knowledge is characterized by a less sophisticated geometry than a metric geometry. Geometries that are less than metric (e.g., topologies) do not define such quantitative properties, but include qualitative properties such as connectivity and containment. Still others have proposed compromise models that include multiple knowledge stores, one or more that is metric and one or more that is nonmetric.

Qualitative Metrics

Within the past decade, there has been considerable work within the AI (artificial intelligence) community to develop a formal model of spatial reasoning that can reason well without the necessity of very precise metric knowledge or elaborate decision algorithms. This work has taken place within the larger context of *qualitative reasoning*, a term describing models that reason fairly effectively about a variety of problems without sophisticated and precise calculation abilities. Work in *naive* or *qualitative physics*, for example, has attempted to predict the motion of pulley systems without the use of precise physical data and rules of calculus (e.g., Forbus et al., 1991). In turn, qualitative reasoning models have derived much of their inspiration from the now robust topic of *fuzzy logic* (Dutta, 1990; McDermott and Davis, 1984; Zadeh, 1975).

For instance, Dutta (1988) provides a fuzzy model of spatial knowledge in which a statement about distance and direction is modeled as two fuzzy categories, each category consisting of a center value, and left and right intervals of spread. The statement "object A is about 5 miles away", for example, is modeled as having a center of 5 miles and 1 mile spreads around 5 miles. The statement essentially says that the distance is between 4 and 6 miles. The statement "object A is in about a north-easterly direction" is modeled as having a center at 45° and 10° spreads around 45°. The statement essentially says that the direction is between 35° and 55°. In both cases, the correct value is modeled as having some nonzero probability of falling within the category spreads. As we will discuss below, however, modelers such as Dutta provide no a priori reasoning or empirical evidence as to deciding how large this spread should be.

Such imprecise and inelaborate models of reasoning about spatial quantities have been dubbed *qualitative metrics*. They hold promise as models of human environmental spatial knowledge. Presumably, the models could account for human abilities and limitations at skills such as navigation and communication about space. Qualitative modelers have noted several difficulties with information processing in the real world, including perceptual imprecision, temporal and memory limitations, the availability of only approximate or incomplete knowledge, and the need for rapid decision-making (Dutta, 1988, 1990). One of the attractive properties of such approaches is possibly providing a way to incorporate both the metric skills and metric limitations of human spatial behavior without positing separate metric and topological knowledge structures.

Most of the work on qualitative metrics has focused on knowledge of directions in the environment necessary for navigation and spatial communication. Although the details of these proposals vary, they agree in positing a model of directions which consists of a small number of coarse angular categories, commonly four 90° categories (front, back, left, right) or eight 45° categories (front, back, left, right, and the four intermediate). Frank (1991a, 1991b) provides good examples of such approaches. His models consist of either 4 or 8 "cones" or "half-planes" of direction. Values along the category boundaries are considered "too close to call" and result in no decision about direction. He also provides a set of operators for manipulating these values. Other writers provide similar models of directional knowledge (Freksa, 1992; Hernández, 1991; Ligozat, 1993; Zimmermann, 1993).

It must be noted that AI researchers in general, and qualitative spatial modelers in particular, are not motivated exclusively or even primarily by a desire to simulate human knowledge and behavior accurately. In many cases, they may simply wish to design an intelligent system that works. Such an approach may only implicitly or incidentally produce a model of human spatial thought, if at all. But it should also be stressed that qualitative metric modelers have definitely taken inspiration from what they consider a realistic approach to human reasoning about space (and time):

> "It is a truism that much of human reasoning is approximate in nature. Spatial reasoning is an area where humans consistently reason approximately with demon-strably good results." (Dutta, 1988: 126).

> "Spatial reasoning is ubiquitous in human problem solving.
> Significantly, many aspects of it appear to be qualitative" (Forbus et al., 1991: 417)

> "Much of the knowledge about time and space is qualitative in nature. Specifically, this is true for visual knowledge about space." (Freksa, 1991: 365)

> "Our goal is to establish qualitative spatial relations between objects in a cognitively plausible way." (Hernández, 1991: 374)

> "a new approach is presented...to combine knowledge about distances and positions in a qualitative way. It is based on perceptual and cognitive considerations about the capabilities of humans navigating within their environments." (Zimmermann, 1993, 69).

In spite of all of this apparent wisdom about human spatial reasoning, these modelers have nearly completely avoided citing any behavioral research to support such conclusions. Our purpose in the work reported below is to make an initial attempt to evaluate qualitative metric models against empirical data from human subjects. The empirical data consists of estimates by humans of the sizes of turns of pathways they have walked.

Empirical Data and Simulation Approach

A two-stage approach was taken in order to evaluate qualitative metric models of human directional knowledge empirically. In Stage 1, existing models from the qualitative reasoning literature were used to develop testable simulations that were maximally faithful to those models. The results of Stage 1 were used to design improved simulations in Stage 2. These Stage 2 simulations went beyond the existing qualitative metric models, but nevertheless attempted to retain the fundamental insight of a qualitative metric. That insight is the proposal that humans employ a small, finite number of quantitative categories to organize spatial knowledge. In both stages, Monte Carlo simulations were carried out in which estimates of turns were generated by randomly sampling within the discrete categories suggested by the model.

Some writers have proposed models in which entire single categories constitute responses (e.g., a forward response is a cone with a range of 90°); the precision of the response is not greater than the entire category (see Freksa, 1992; Zimmermann, 1993). Such a model is incompatible with the requirement of many behavioral studies, including the study we compare to our simulations in the present research, for subjects to estimate at much higher levels of precision. Of course, the fact that people readily provide estimates at the level of precision of one or a few degrees does not ensure that their knowledge is stored or is accurate at that level of precision, but it does suggest that similar precision should be designed into the simulations in order to ensure comparability. Therefore, random responding at the level of the single degree within discrete categories was considered the most realistic way to model the qualitative metrics in the simulations below.

The data used to evaluate qualitative metric models of directional knowledge came from the results of some research by Sadalla and Montello (1989). This research was an investigation of subjects' knowledge of turn sizes after walking pathways containing a single turn. Vision-restricted subjects who could see only the floor down around their feet walked an 8.3 m pathway marked on the floor containing two straight segments and one turn. There was a .5 m gap between the end of the first segment and the beginning of the second segment (thus not providing a completed visible angle). On different trials, the size of the turn from straight ahead was varied. All subjects walked and estimated 11 different turns in different random orders, ranging in size from 15° to 165° from straight ahead and separated by 15° increments (Figure 1). Thus, the least extreme turn from straight ahead is labeled *15°* and the most extreme turn is labeled *165°*. Half of the subjects walked turns to the right, the other half to the left. This variable was unrelated to estimation performance and was not considered further by Sadalla and Montello, nor is it considered below.

After walking to the end of the path, subjects used a circular pointer to provide three separate measures of their knowledge of the angular size of the pathway turn. Two of these measures are used below to evaluate the simulations. Measure 1, henceforth called

Figure 1

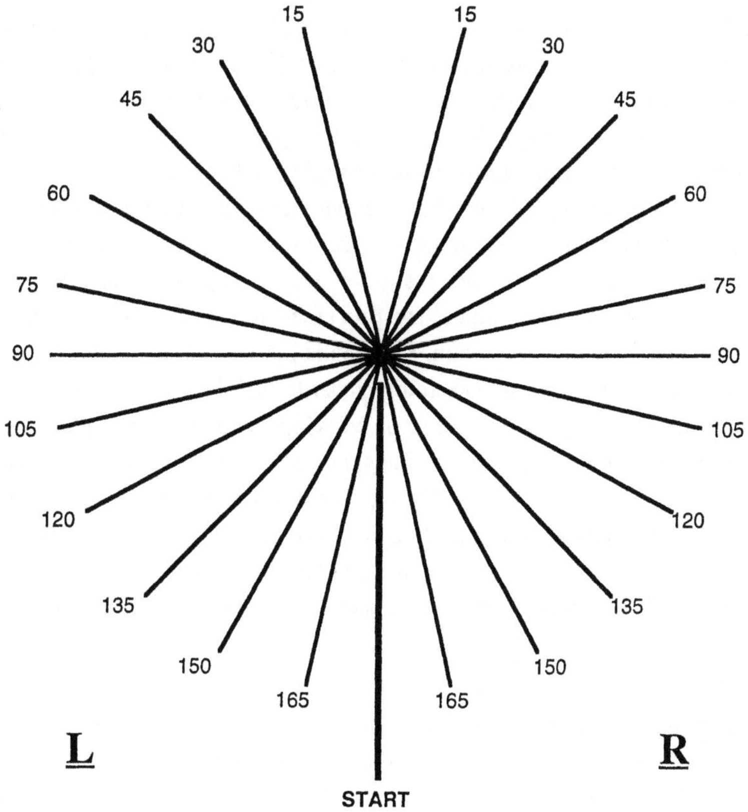

Figure 1: Angular pathways walked by subjects in Sadalla and Montello (1989).

Turn Size, required subjects to "reproduce the size of the turn"; measure 2, called *Original Direction*, required subjects to "point in the original direction of travel". Thus, correct answers to the two measures would be mirror images for any given turn. The latter measure is expected to result in poorer performance, however, because it compounds error from two sources: estimation of the angle of the turn size and estimation of the angle of the original direction from the heading direction at the end of the path. Measure 3 from Sadalla and Montello required subjects to "point back to the start location" and is not used here to evaluate the simulations because it involves both distance and directional knowledge.

Our thinking about human knowledge of directions in the environment is guided by a simple psychological process model that describes the acquisition, storage, and use of that knowledge (e.g., Simon, 1979). The model consists of five stages: (1) perception, (2) encoding, (3) long-term memory storage, (4) retrieval and recoding in working memory, and (5) behavioral output. Information about directions is perceived from the environment or from body movement. This information is encoded and stored in long-term memory. When the information is needed (e.g., for wayfinding decisions or researchers' requests), it is retrieved from long-term memory and placed in working memory (short-term memory). Various recoding processes (e.g., scale construction, image manipulation, verbalization) are brought to bear on the working-memory representations in order to produce behavioral outputs such as turn reproductions.

At various stages in the model, processes occur which produce error in the directional knowledge. These error processes result in both inaccuracies and imprecision in knowledge, and in behavioral output. The processes are of three types: systematic bias, categorization, and random fluctuation. Systematic biases lead to reliable inaccuracies, as when a turn is repeatedly recalled as being closer to 90° than it actually was. Biases are thought to operate during the encoding or working-memory recoding stages, or both. Categorization leads to the imprecision of angular knowledge that characterizes qualitative reasoning. Categorization is essentially a process whereby continuous information is coded into discrete intervals. It is also thought to operate during the encoding or working-memory stages, or both. Finally, random fluctuations lead to inaccuracies that are not, however, reliable or repeatable. They are expected to operate at all stages of the model.

Simulation Stage 1

Perhaps the most critical issue in designing the simulation models concerns the number of quantitative categories of knowledge to incorporate, essentially the level of precision expressed by the model. As discussed further below, existing proposals have not decided this issue in any empirically principled way. Rather, they have generally attempted to produce adaptive reasoning with as few categories as possible. Two levels of precision of directional knowledge were tested in the Stage 1 simulations, a 4-cone (cones of 90° each) and an 8-cone model (cones of 45° each) (Figure 2). Because the idea of a qualitative metric is that the reasoner has no reliable information more precise than at the level of the spatial category, the directional cones were modeled as uniform random distributions during Stage 1 (as opposed to normal distributions, for instance).

Figure 2

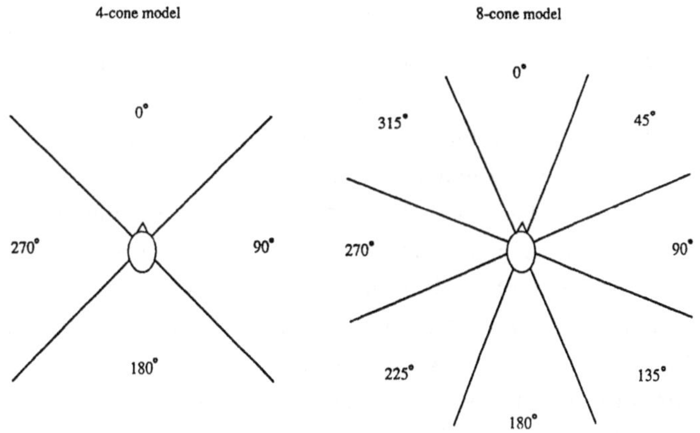

Figure 2: Homogeneous 4-cone and 8-cone models tested in Stage 1.

A second important issue in designing the simulation models concerns the accuracy of knowledge. Unfortunately, existing models do not address the accuracy issue; they assume perfect accuracy within their limits of precision. That is, existing models do not incorporate any systematic biases in knowledge. All estimates for any turn falling within a particular cone will be sampled from that cone. This approach to accuracy is modeled in Stage 1 in two ways. First, all estimates for a given pathway turn are consistently sampled from within a single correct cone, referred to as *single-cone* sampling (turns falling on cone boundaries were considered ambiguous, however, and were sampled equally from the two neighboring cones). Single-cone sampling will result in constant errors that vary across turns (imperfect accuracy). Sampling during Stage 1 was done in a second way by proportionately sampling from two neighboring cones so that no average inaccuracy resulted for any turns, referred to as *proportional* sampling. Proportional sampling results in patterns of no constant error, perfect accuracy across turns (within limits of sampling error). Each simulation was run enough times to generate 200 estimates for each turn.

Both the single-cone and proportional sampling approaches to the question of accuracy have implications for variability of performance as well. Of course, such perfect accuracy, even if imprecise, is almost certainly a poor model of human spatial knowledge. This is addressed directly in the Stage 2 simulations.

Results of Stage 1

In order to evaluate the responses of the simulations, patterns and magnitudes of both constant error and variability from the simulations are compared to those from the empirical data set. Circular statistics (Batschelet, 1981) is used to calculate mean directions and mean angular deviations. These statistical techniques are appropriate for use with variables that consist of directional responses in 360°, called *circular variables* (or any periodic variable that shows cyclical trends). The techniques allow calculation of a mean angle or direction. Subtracting the mean direction from the correct answer provides a measure of *constant error* (systematic bias in one direction or the other). *Mean angular deviation* is also calculated as a measure of between-case variability (sometimes called variable error) in performance. It is the angular analogue to standard deviation. Mean angular deviation equals 0 when all directional estimates are exactly the same, and it reaches a maximum at just over 80° when directional estimates are maximally distributed around 360° (i.e., no agreement between subjects).

Constant errors for the empirical data (right and left turns collapsed) are depicted in Figure 3, both for Turn Size and Original Direction. Figure 4 depicts constant error for all four Stage 1 simulations: 4-single, 8-single, 4-proportional, and 8-proportional sampling. In all graphs of constant error, distortions toward 90° are graphed as positive errors, those away from 90° as negative errors.

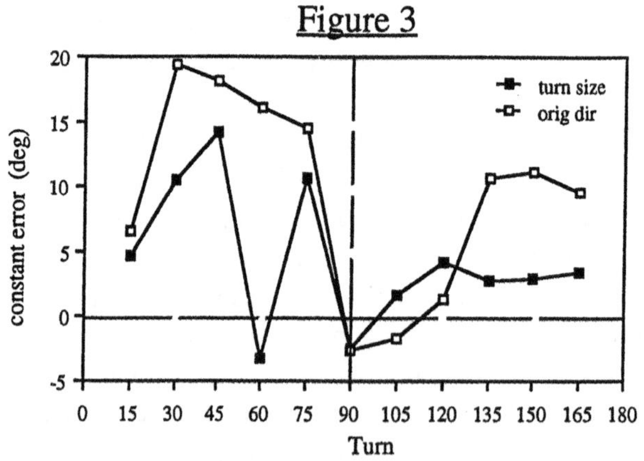

Figure 3: Constant error for empirical data from Sadalla and Montello (1989). Positive errors are distorted towards 90°.

Figure 4

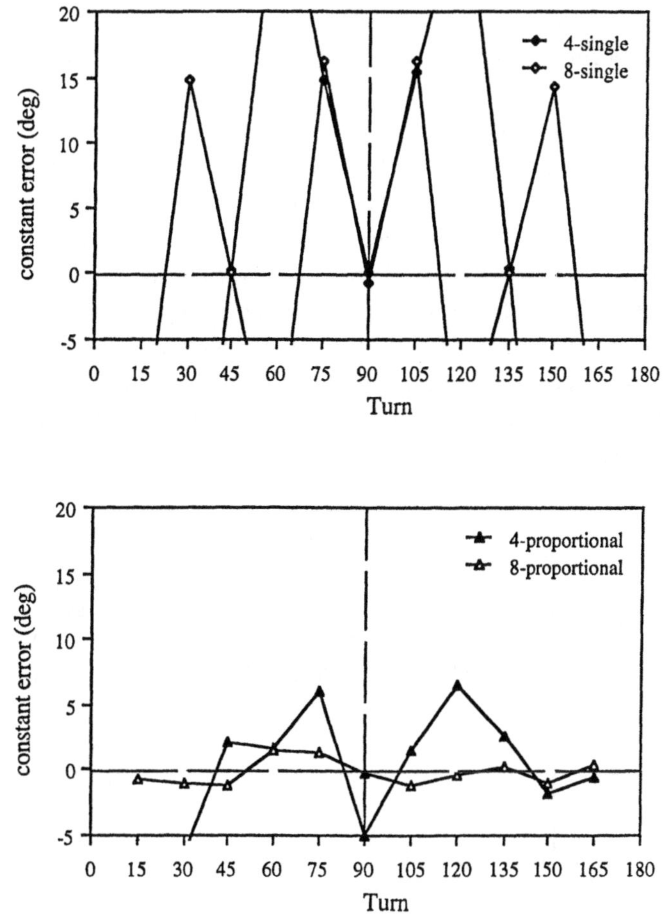

Figure 4: Constant error for Stage 1 simulation data. Positive errors are distorted towards 90° .

Constant error in the empirical data shows a distinctive pattern. There is a clear tendency for subjects to distort their turn estimates toward 90° (positive errors), with the error for Turn Size at 120° a clear exception that is likely due to a mistake in the original data collection. The range of distortion across turns is about 15°, with more distortion for acute than for obtuse turns (*acute* turns are 15-75°, *obtuse* turns are 105- 165°). The distortion is greater for Original Direction than for Turn Size. As discussed above, this is expected insofar as the former measure compounds the angular processing required for the latter measure.

None of the simulations mimic this pattern of accuracy well; most notably, none show a consistent bias toward right angles. It is true that the range of distortion is fairly well reproduced by the 4-proportional model, but this results from chance sampling error only. The 8-proportional model is nearly flat, showing very little distortion at all. Both single-cone models show a range of distortion across turns that is far too extreme, about 60° for the 4-cone model and 30° for the 8-cone model.

Figure 5 depicts mean angular deviations (variability) for both empirical measures. The corresponding results for the four simulations are depicted in Figure 6. Variability in the empirical data shows a distinctive pattern that was in fact the focus of the original analysis by Sadalla and Montello (1989). There is low variability for turns at or near the orthogonal axes (0°, 90°, 180°), with gradually increasing variability towards oblique turns (45°, 135°). Average variability is about 30°, with a range across turns of about 20°. As with constant error, there is more variability for acute than for obtuse turns. Also like the constant error, there is greater variability for Original Direction than for Turn Size, at least in the acute quadrant.

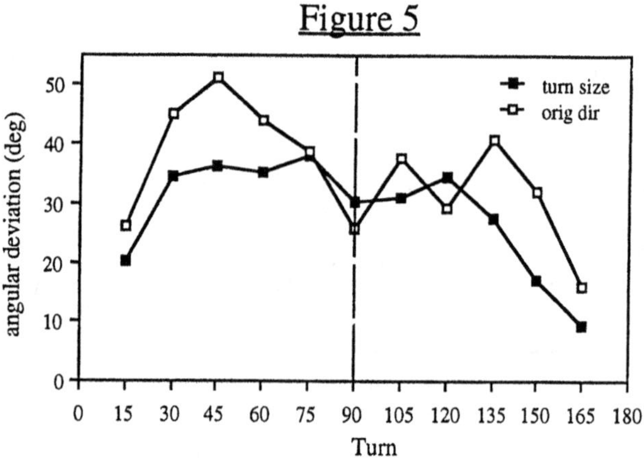

Figure 5: Mean angular deviation (variability) for empirical data from Sadalla and Montello (1989).

Interestingly, this pattern is reproduced to some extent by the 4-cone models: both proportional and single-cone sampling show the high agreement near orthogonal axes and the low agreement at oblique turns. The 4-proportional sampling results in a gradual increase to the obliques, while the 4-single sampling is flat for turns not exactly at 45° or 135°. But both 4-cone models produce overall variability that is too high (45-55°) and a range across turns that is too severe (30°). The 8- single model, on the other hand, results in a magnitude of variability (about 20°) that is too small in comparison to the empirical data and no change in variability across turns at all. Only the 8-proportional model results

Figure 6

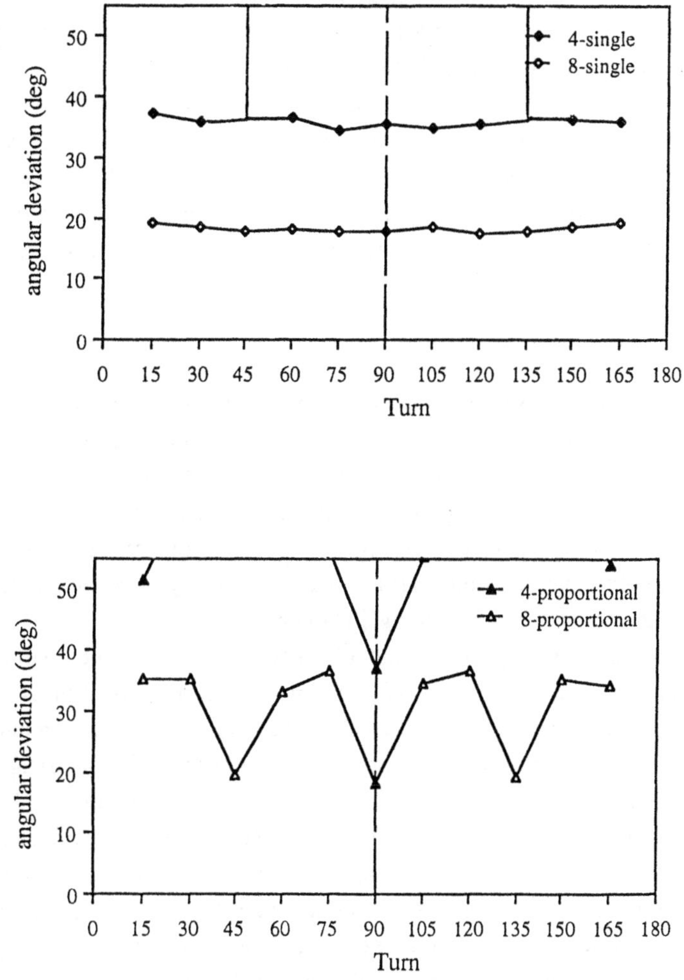

Figure 6: Mean angular deviation (variability) for Stage 1 simulation data.

in a magnitude of variability that matches the empirical data very well. This model results in an overall variability of about 25-30°, almost the same as the empirical data, and a range across turns of about 15°. However, neither 8-cone model reproduces the distinctively shaped pattern of the empirical variability. The 8-single model is flat, as mentioned above, showing no change in variability across turns at all. And the 8-proportional model results in lower variability at both 90° and the obliques, 45° and 135°. This drop at the obliques is the opposite of what is found in the empirical data. In addition, this model does not produce a decline in variability near 0° and 180°.

As a final approach to evaluating the performance of the simulations, circular correlations (Jammalamadaka and Sarma, 1988) are calculated between the correct turn values and both the empirical estimates and the simulated estimates. Circular correlation provides a measure of relationship between two circular variables. These are calculated within-case and averaged across cases using Fisher's r-to-z transformation to calculate mean correlations. The empirical estimates correlate very highly with the correct values, .94 for Turn Size and .89 for Original Direction. Both 8-cone models are similarly highly correlated with the correct values, .95 for single and .94 for proportional sampling. The 4-cone models also correlate strongly with the correct values but less so, .81 for single and .71 for proportional sampling.

Discussion of Stage 1

Our Stage 1 attempt to evaluate qualitative metric models empirically suggests their possible viability as models of human directional knowledge. A qualitative metric model consisting of 8 45°-cones sampled proportionally reproduces the magnitudes of variability quite well. And both 8-cone models produce estimates that correlate with the actual turn sizes almost exactly as strongly as do the empirical estimates. The 4-cone models, on the other hand, produce far too much variability in performance and did not correlate with the actual turn sizes strongly enough.

Unfortunately, the 8-proportional model failed to reproduce the distinctive and oft replicated empirical pattern of minimal variability near orthogonal turns and maximal variability at oblique turns. Nor did it reproduce the empirical pattern of minimal constant error near orthogonal turns and distortion of turns toward 90° (empirical evidence of these patterns is cited and presented in Franklin et al., under review; Loftus, 1978; Sadalla and Montello, 1989; Tversky, 1992). These failures probably stem to a large extent from the lack of a reasonable approach to knowledge accuracy in existing qualitative metric models (we discuss this further at the end of the chapter). In our Stage 2 simulations, therefore, we attempt to improve the fit of the simulation results primarily by incorporating into the models some empirical and theoretical ideas about knowledge accuracy.

Simulation Stage 2

The Stage 1 simulations were attempts to test qualitative metric models as they are currently described in the literature. Our approach in the Stage 2 simulations is to design more promising qualitative metric models of human spatial knowledge using the Stage 1 results as a guide. Because the 8-proportional model from Stage 1 did a good job of replicating the empirical magnitudes of variability and the within-case correlations with the actual turn sizes, it was decided to use 8-proportional approaches as a basis for our Stage 2 simulations.

Our discussion of the Stage 1 results above suggests several ways to modify and improve our initial simulations. Our first modification is to use heterogeneous cone sizes in an attempt to produce maximal variability near oblique turns and minimal variability near orthogonal turns. This is done in two ways: orthogonal cones of 30° and oblique cones of 60°, or orthogonal cones of 20° and oblique cones of 70°. In addition, we tried to decrease the variability for turns near 0° and 180° even more by splitting these two cones in half, producing cone sizes of 15°- 60°-30°-60°-15° or 10°-70°-20°-70°-10° (these are actually 10- cone models). These models are depicted in Figure 7. Decreasing the sizes of cones directly in front and behind the body by splitting them in half is consistent with the theoretical primacy of the front-back over the left-right axis of egocentric space (Franklin et al., under review; Shepard and Hurwitz, 1984) and the resulting maximal acuity of directional judgments near 0° and 180°. Franklin and Tversky (1990), for example, in a discussion of their *spatial framework*, found that subjects responded faster to locational queries about objects located in front or in back than about objects located to the left or right. The primacy is seen in our empirical data by the fact that variability for the 90° turn is generally greater than that for the 15°.

Figure 7

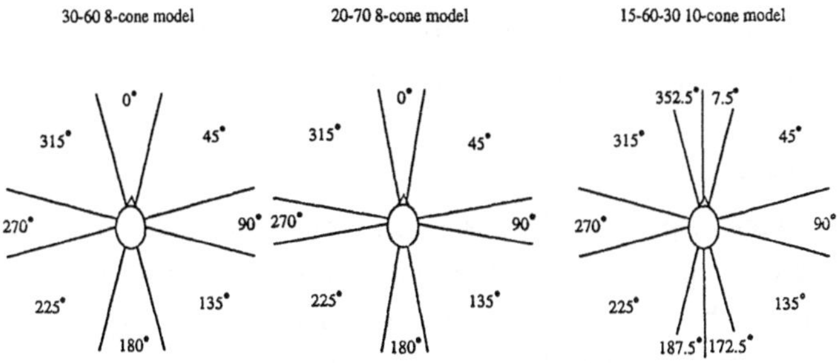

Figure 7: Heterogeneous 8 and 10-cone models tested in Stage 2.

Our second modification is designed to increase variability, particularly for turns near and at 45° and 135°. This "variability adjustment" is done by modifying the sampling of the 45°, 135°, and 90° turns. For all three turns, 10 cases are shifted from the cone containing the actual turn value to each of the two neighboring cones on either side. This adjustment method includes the 90° turn because the Stage 1 simulations had suggested that sampling entirely within a 20° or 30° cone would not produce enough variability for the 90° turn. Such an adjustment is also consonant with the lesser acuity for the left-right vs. front-back dimensions of space mentioned above. Each simulation was again run enough times to generate 200 cases.

Our third modification is to build in a *right-angle heuristic*. Any turn that deviates from straight ahead or straight behind is judged to be more nearly a right-angle turn than is actually the case. This is done by oversampling cones towards 90° for all turns other than the 90° turn, and is over and above the variability adjustments just described. Such a heuristic should produce the characteristic empirical pattern of distortion towards 90°. The heuristic is implemented in two ways: a fixed number of estimates are shifted one cone towards the 90° cone, or a percentage of estimates are shifted to the 90° cone. In either case, this adds sampling of the 90° cone for those turns that did not already sample it (such as the 15° turn).

In addition, the right-angle heuristic is made asymmetric by oversampling to a greater extent for acute than for obtuse turns. In the fixed number method, 10 cases are shifted toward 90° for the 15° and 165° turns, 40 cases are shifted for all other acute turns, and 20 cases are shifted for all other obtuse turns. In the percentage method, 20% of cases are shifted to the 90° cone for acute turns, and 10% are shifted for the obtuse turns. Either method should produce greater distortion for acute turns than for obtuse turns, and may also produce greater variability for estimates of acute turns.

Finally, we also examined the effect of using a different distribution to sample within each cone. All cones were sampled according to uniform distributions in the Stage 1 simulations. For the Stage 2 simulations, we also try sampling according to a normal, or Gaussian, distribution. These normal distributions were designed so that ±2 standard deviations covered the ranges of the corresponding cone (e.g., 30° or 60°).

Results of Stage 2

As was done with the Stage 1 simulations, patterns and magnitudes of both constant error and variability are compared to those from the empirical data set. All of the simulation variations described above were systematically varied across all possible combinations, but only the results from four simulations are considered here in detail. *Uni30fix* samples uniformly from 30° and 60° cones and biases according to a fixed number of cases. *Uni20fix* does the same but for 20° and 70° cones. *Uni15%* samples uniformly from 15°, 30°, and 60° cones (10-cone model), and biases according to a percentage of cases. *Norm15%* does the same but samples normally.

Constant errors for Uni30fix and Uni20fix are depicted in Figure 8, along with the empirical results for Turn Size and Original Direction. Constant errors for Uni15% and Norm15% are similarly graphed against the empirical results in Figure 9. Distortions toward 90° are again graphed as positive errors, those away from 90° as negative errors.

Figure 8

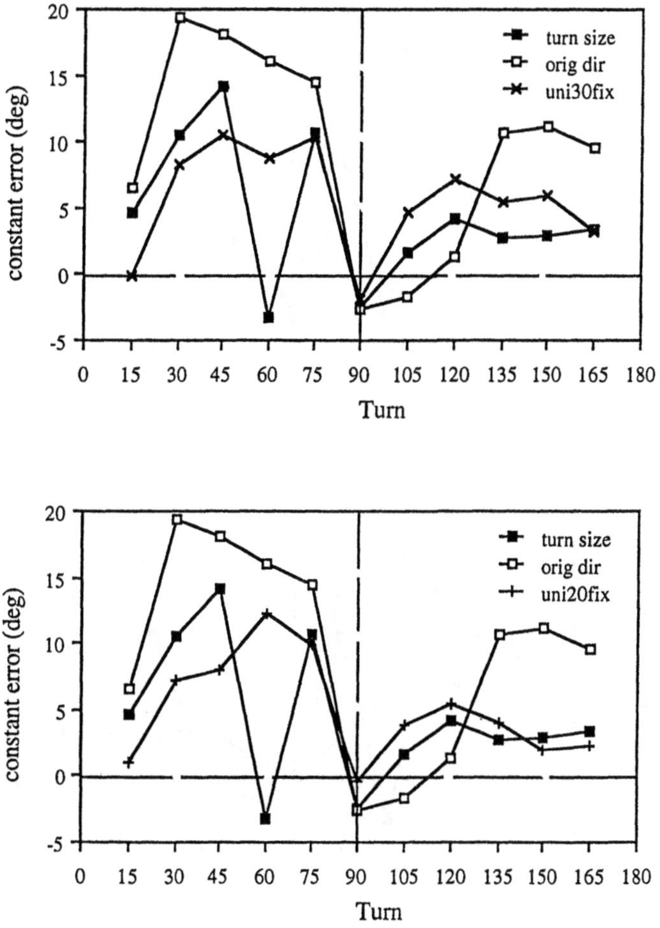

Figure 8: Constant error for Stage 2 simulation data (Uni30fix, Uni20fix) and empirical data. Positive errors are distorted towards 90°.

All four simulations recreate the empirical pattern of constant error quite well: minimal distortion near orthogonal turns, bias toward 90°, and greater bias for acute turns. The magnitudes of distortion match those for Turn Size very closely, with the exception of too little bias for the acute turns of 15-45°. Of note is the fact that the two alternatives for cone size are nearly equivalent; whether 30° and 60° or 20° and 70° cones are used makes very little difference (Figure 8). Similarly, whether uniform or normal distributions are used to model the cones makes very little difference (Figure 9). These conclusions were supported throughout all of the conducted simulations.

Figure 9

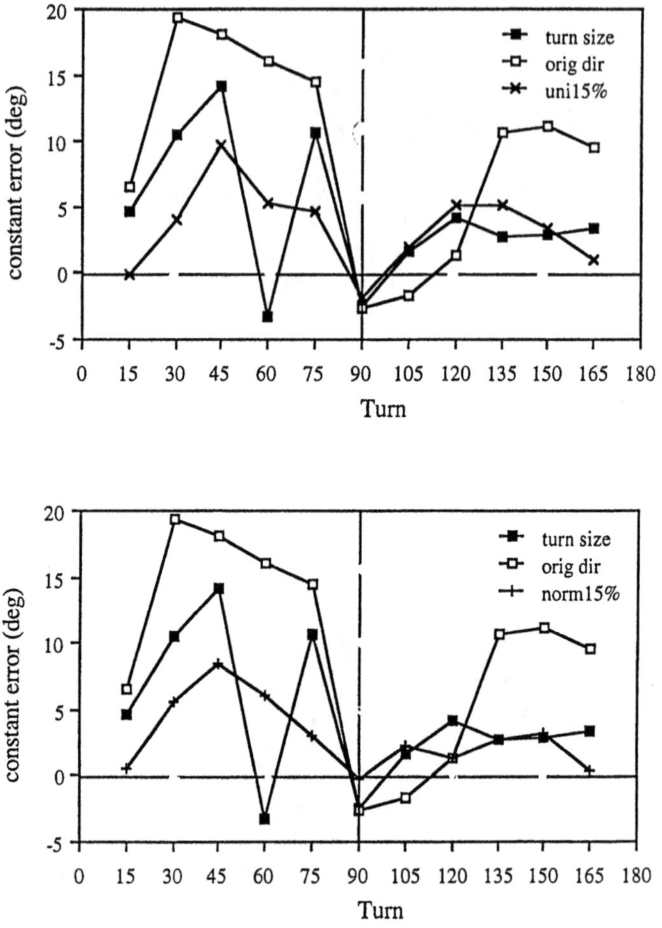

Figure 9: Constant error for Stage 2 simulation data (Uni15%, Norm15%) and empirical data. Positive errors are distorted towards 90°.

Similarly close fits are seen from an examination of variability in performance. Figure 10 depicts angular deviation for Uni30fix and Uni20fix, along with the empirical results for Turn Size and Original Direction. Figure 11 shows angular deviation for Uni15% and Norm15%. Again, all four simulations recreate the empirical pattern quite well: minimal variability near orthogonal turns, maximal variability for oblique turns, and greater variability for acute turns. The magnitudes of variability match those for Turn Size very closely. It can again be seen that the choice of cone sizes and sampling distributions is of little consequence, a result that was also echoed throughout all of the conducted

Figure 10

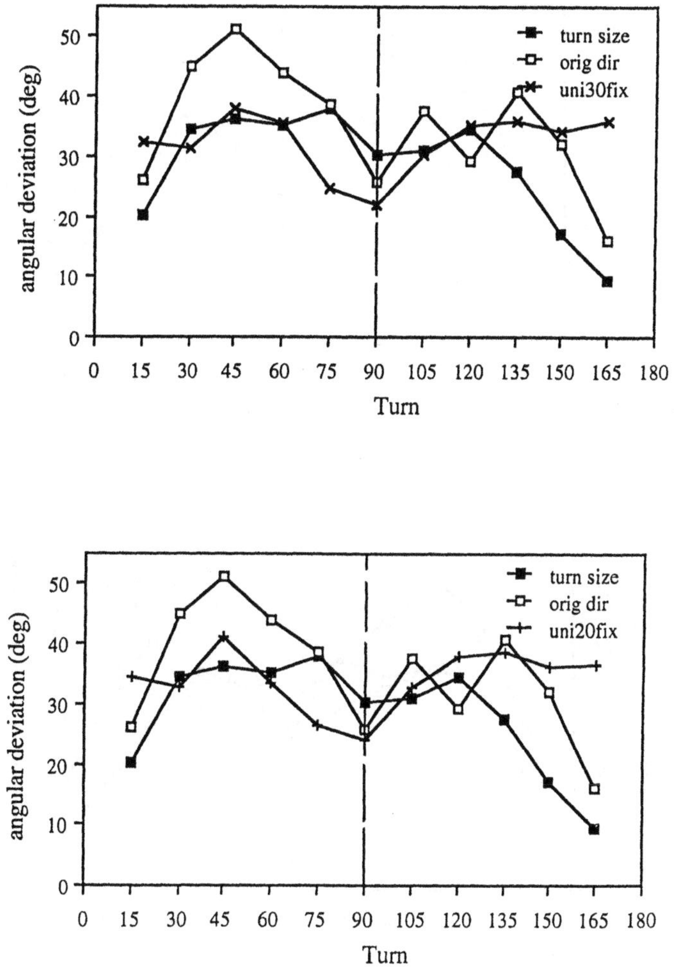

Figure 10: Mean angular deviation (variability) for Stage 2 simulation data (Uni30fix, Uni20fix) and empirical data.

simulations. One modification that does make a difference was splitting the 0° and 180° cones in half in the Uni15% and Norm15% simulations. This reduces the excessive variability for turns of 15° and 165° seen in the first two simulations, though still not to the level of the empirical data.

Finally, circular correlations between the correct turn values and the four sets of simulated estimates are calculated as in Stage 1. The mean correlations for Uni30fix, Uni20fix, and Uni15% are all .90, and that for Norm15% is .91. These compare

Figure 11

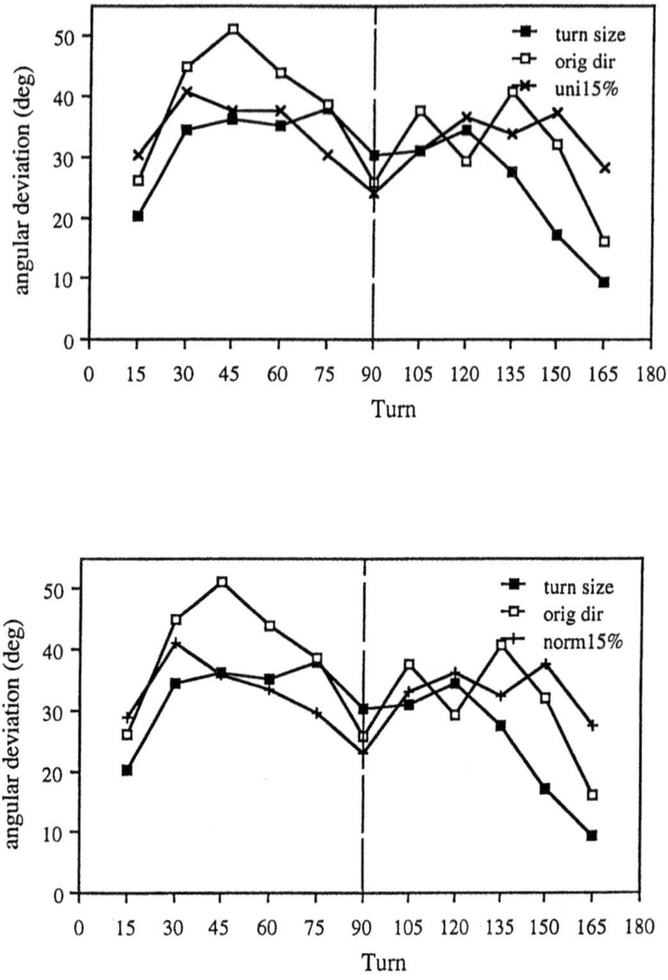

Figure 11: Mean angular deviation (variability) for Stage 2 simulation data (Uni15%, Norm15%) and empirical data.

favorably with the correlations of the empirical estimates with the correct values presented above, .94 for Turn Size and .89 for Original Direction.

Discussion of Stage 2

The Stage 2 simulations incorporated insights from the Stage 1 simulations and from other behavioral literature on directional knowledge. Their performance matched the patterns of constant error and variability in the empirical data quite well, though not perfectly. This occurred without completely forgoing the essential insight of the qualitative metric literature. Each estimate was drawn randomly from one of a small number of cones, providing support for the idea that directional knowledge is imprecise and consists of a small number of quantitative categories of directions. Undoubtedly, other specific approaches to designing qualitative simulations could have been taken, though the relative insensitivity of the Stage 2 simulations to factors such as the exact cone size and nature of the sampling distributions suggests that additional variations in the specifics of the simulations would produce only modest improvements in their fit to the empirical data.

Summary and Conclusions

The results of our simulations based on qualitative metric models provide some support for their promise as models of human spatial knowledge, but also suggest some of their serious weaknesses. Their major insight, that human knowledge is metric but only in a very imprecise way, receives support from these evaluations. In particular, some variety of proportional sampling within 8 or 10 directional cones produces magnitudes and patterns of constant error and variability that coincide closely with empirical patterns obtained from human-subjects testing. The exact sizes of the cones is apparently not critical, as long as orthogonal cones are smaller than oblique cones. It is also important to include in the model in some way a heuristic that oversamples cones towards 90°.

These simulations point to what is probably the most notable weakness of existing qualitative metric models, namely their lack of a reasonable approach to modeling knowledge accuracy. This conclusion derives from the empirical fact that none of the Stage 1 simulations, which were designed to be maximally faithful to existing models, produce anything like the pattern of bias towards right angles found in the empirical data used here or found in other research. Restricting sampling to single cones leads to unrealistically inaccurate performance; sampling proportionally from neighboring cones without implementing any biasing heuristic leads to unrealistically accurate performance. Also telling, however, is the difficulty and arbitrariness we experienced in trying to interpret existing proposals to guide how qualitative categories should be constructed and sampled. There was a definite ad hoc character to the decisions we made about how many quantitative categories to use, how large their respective ranges should be, and how they should be sampled. In its current state, therefore, the qualitative reasoning literature does not provide suitable models of human directional knowledge because it lacks an a priori principled approach to some of the central issues of such models.

As far as the Stage 2 simulations indicate, it is not important whether the cones are modeled as uniform or normal distributions. Models in which cones are described as consisting of central prototype values and normal variability around the prototype have been proposed in the behavioral literature (Huttenlocher et al., 1991). It is in fact a common approach to modeling fuzzy categories in the qualitative reasoning literature (e.g., Dutta, 1988, 1990; see also the discussion by McDermott and Davis, 1984). The distinction between uniform cones without internal structure and cones consisting of variability around a central value cannot be decided by these simulations.

The work in this chapter leads to ideas very similar to those found in Franklin and Tversky's (1990) discussion of their *spatial framework*. Notably, both research programs stress the idea of a Cartesian frame for organizing egocentric spatial knowledge. Two apparent contradictions with their work call for comment, however. One is the empirical finding by Sadalla and Montello (1989) that performance on acute turns (to the front) is more distorted and less precise than on obtuse turns (to the back). Franklin and Tversky not only found that subjects responded faster to front-back queries than to left-right queries, but also that responses to front queries were faster than to back queries. Franklin et al. (under review) replicated this and also found that directional pointing to objects in the front was more precise than to objects in the back across repeated trials (however, they also found that the range of directions which subjects would consider "to the front" was greater than to other directions). On the face of it, these results seem to contradict those by Sadalla and Montello. After all, trials with obtuse turns required subjects to turn around and walk "back", and to point in a backwards direction for both measures. And yet these trials resulted in lower variability than did performance after walking acute turns.

However, the tasks in the two sets of studies were different in some ways that may explain this apparent contradiction. Sadalla and Montello did not actually require subjects to point "to" anything behind their bodies; rather, these subjects had to move a pointer to show certain angular relationships on a circular pointer. The pointer had a movable wire and a radius drawn on its surface, thus providing a visible angle to subjects. Obtuse turns required the production of an acute angle on the pointing circle; acute angles are reproduced more precisely because of the stability provided by the context of the neighboring wire and radius line (cf. Pratt, 1926). Further, these subjects were responding with respect to a turn they had actually walked; subjects in Franklin et al. pointed to an object that they perceived visually. These considerations suggest that the quadrant asymmetries found by Sadalla and Montello may not necessarily be generally characteristic of egocentric directional knowledge. The issue calls for further clarification.

The second apparent contradiction with the *spatial framework* involves the superiority of at least an 8-cone model over a 4- cone model in the present simulations. The *spatial framework* essentially posits a 4-cone model, though not framed in that way (see especially, Franklin et al., under review). The empirical work by Franklin and her colleagues has not involved testing of qualitative metrics, however, and has not attempted to establish the number or sizes of cones necessary to produce appropriate estimation variability from random responding. And much of their work has concerned referents for natural language terms for direction ("front", "right", etc.) rather than nonlinguistic knowledge of directions. The two need not be synonymous.

The distinction between linguistic and nonlinguistic spatial knowledge is probably an important one to make. Several of the qualitative modelers in fact take inspiration from spatial information as expressed in natural language (Dutta, 1988, Fisher and Orf, 1991; Frank, 1991a, 1991b; Hernández, 1991; Zadeh, 1975; see also, Mark and Frank, 1991). Either they see language as equivalent to knowledge in general, or they have a specific interest in trying to model linguistic knowledge, for instance to improve natural language queries in Spatial Information Systems. However, there is a growing consensus that linguistic and nonlinguistic knowledge are not the same. The two involve at least partially separate systems, psychologically and physiologically. Furthermore, although linguistic knowledge of space is relatively limited in the precision of information it typically expresses (great metric precision is usually unnecessary), nonlinguistic knowledge may be much more metrically precise (Jackendoff and Landau, 1991; Landau and Jackendoff, 1993; McNamara, 1992; Rybash and Hoyer, 1992). Haber et al. (1993) found evidence for this difference directly relevant to our concern with directional knowledge. In their research, blind subjects indicated directions via several verbal and nonverbal methods; the verbal methods produced noticeably less precision than the nonverbal methods.

Linguistic issues aside, empirical studies could be conducted that would provide a more direct approach to designing and testing qualitative metric models than that provided by Sadalla and Montello (1989). The estimation variabilities from their research were actually obtained from between-subject performance. Within-subject variability is more to the point, however, because humans may organize knowledge with somewhat idiosyncratic quantitative categories. In an improved study, subjects would repeatedly walk each of a series of turns several times, providing multiple estimates of each turn. Furthermore, one could directly determine category ranges and boundaries by psychophysically establishing relative thresholds for angles and distances in a design that varies turn angles gradually over many repeated trials. A straightforward approach would require subjects simply to state which of two walked turns is greater or less than the other, or whether they are in fact equal.

Finally, it would be valuable to investigate qualitative models of distance knowledge to complement work on directional knowledge. Much less work on distance models has been reported in the qualitative literature. Frank (1991b) briefly discussed qualitatively modeling distance as consisting of two (near, far) or three (near, intermediate, far) categories; category size would depend on context. Fisher and Orf (1991) discuss a fuzzy set model of "near" and "close". Zimmermann (1993) exploits the information provided by two sets of half-planes connected by a directional vector. A triangulation between the two centers of the half-planes and a third point to be estimated results in some coarse knowledge about the distance of the third point relative to the two centers. In any case, it is likely that a successful qualitative model of distance would also incorporate heterogeneous category sizes (the smallest distance categories would have shorter ranges) and a biasing heuristic to produce overestimation of short distances relative to long distances (Montello, 1991).

In conclusion, this work leads to fruitful ideas about how to model human knowledge in a way that respects its metric qualities without imparting it with unrealistically excessive precision and accuracy. An interesting but as yet untested possibility is that the precision of human spatial knowledge decays over time (during long-term memory storage?). Spatial knowledge may exhibit rather precise metric qualities during perception (see Attneave and Pierce, 1978) and in memory after brief delays, delays characteristic of most empirical research. Over time, perhaps after delays on the order of months or years, there may be a continual degradation of spatial knowledge so that distortion and imprecision both increase. As one of us has suggested in the past (Montello, in press), such considerations support the need for rigorous very long-term longitudinal studies of environmental spatial knowledge.

References

Attneave, F., and Pierce, C. R. (1978). Accuracy of extrapolating a pointer into perceived and imagined space, *American Journal of Psychology 91*, 371-387.

Batschelet, E. (1981). *Circular statistics in biology*, London: Academic.

Couclelis, H., Golledge, R.G., Gale, N. and Tobler, W. (1987), Exploring the anchor-point hypothesis of spatial cognition, *Journal of Environmental Psychology 7*, 99-122.

Dutta, S. (1988). Approximate spatial reasoning, In *First International Conference on Industrial and Engineering Applications of Artificial Intelligence and Expert Systems*, pp. 126-140, Tullahoma, TN: ACM Press.

Dutta, S. (1990). Qualitative spatial reasoning: A semi-quantitative approach using fuzzy logic. In *Design and implementation of large spatial databases* (A. Buchmann, O. Gnther T. R. Smith, and Y. F. Wang, eds.), pp. 345-364. New York: Springer-Verlag.

Fisher, P.F. and Orf, T.M. (1991). An investigation of the meaning of near and close on a university campus. *Environment and Urban Systems 15*, 23-35.

Forbus, K. D., Nielsen, P. and Faltings, B. (1991). Qualitative spatial reasoning: The CLOCK project. *Artificial Intelligence 51*, 417-471.

Frank, A.U. (1991a). Qualitative spatial reasoning with cardinal directions. In *Proceedings of the Seventh Austrian Conference on Artificial Intelligence*, pp. 157-167. Springer-Verlag: Berlin.

Frank, A. U. (1991b). Qualitative spatial reasoning about cardinal directions. In *Autocarto 10* (D. Mark and D. White, eds.), pp. 148-167.

Franklin, N., Henkel, L. A. and Zangas, T. (under review). Parsing surrounding space into regions.

Franklin, N. and Tversky, B. (1990). Searching imagined environments. *Journal of Experimental Psychology: General 119*, 63-76.

Freksa, C. (1991). Qualitative spatial reasoning. In *Cognitive and linguistic aspects of geographic space* (D.M. Mark and A.U. Frank, eds.), pp. 361-372. Dordrecht, The Netherlands: Kluwer Academic Publishers.

Freksa, C. (1992). Using orientation information for qualitative spatial reasoning. In *Theories and methods of spatio-temporal reasoning in geographic space* (A. U. Frank I. Campari, and U. Formentini, eds.), pp. 162-178. Berlin: Springer-Verlag.

Golledge, R.G., and Hubert, L. J. (1982). Some comments on non- Euclidean mental maps. *Environment and Planning A 14*, 107-118.

Habe,r L., Haber, R. N., Penningroth, S. Novak, K. and Radgowski H. (1993). Comparison of nine methods of indicating the direction to objects: Data from blind adults. *Perception 22*, 35-47.

Hernández, D. (1991). Relative representaion of spatial knowledge: the Z-D case. In *Cognitive and linguistic aspects of geographic space* (D.M. Mark and A.U. Frank, eds.), pp. 373-386. Dordrecht, The Netherlands: Kluwer Academic Publishers.

Huttenloche, J., Hedges, L.V. and Duncan S. (1991). Categories and particulars: Prototype effects in estimating spatial location. *Psychological Review 98*, 352-376.

Jackendoff, R., and Landau, B. (1991). Spatial language and spatial cognition. In *Bridges between psychology and linguistics: A Swarthmore Festschrift for Lila Gleitman* (D.J. Napoli and J.A. Kegl, eds.), pp. 145-169. Hillsdale, NJ: Lawrence Erlbaum.

Jammalamadaka, S.R. and Sarma, Y.R. (1988). A correlation coefficient for angular variables. In *Statistical theory and data analysis II* (K. Matusita, ed.), pp. 349-364. North- Holland: Elsevier Science Publishers B.V.

Kuiper,s B.J. and Levitt, T.S. (1988). Navigation and mapping in large-scale space. *AI Magazine, summer*, 25-43.

Landau B., and Jackendoff R. (1993). "What" and "where" in spatial language and spatial cognition. *Behavioral and Brain Sciences 16*, 217-265.

Landau, B., Spelke, E., and Gleitman, H. (1984). Spatial knowledge in a young blind child. *Cognition 16*, 225-260.

Ligozat, G.F. (1993). Qualitative triangulation for spatial reasoning. In *Spatial information theory: A theoretical basis for GIS* (A. U. Frank and I. Campari, eds.), pp. 54-68. Berlin: Springer-Verlag.

Loftus, G.R. (1978). Comprehending compass directions. *Memory & Cognition 6*, 416-422.

Mandler J.M. (1988). The development of spatial cognition: On topological and Euclidean representation. In *Spatial cognition: Brain bases and development* (J. Stiles-Davis, M. Kritchevsky, and U. Bellugi, eds.), pp. 423-432. Hillsdale, NJ: Lawrence Erlbaum Associates.

Mark, D.M. and Frank A. U. (eds.). (1991). *Cognitive and linguistic aspects of geographic space.* Dordrecht, The Netherlands: Kluwer Academic Publishers.

McDermott, D. and Davis, E. (1984). Planning routes through uncertain territory. *Artificial Intelligence 22*, 107-156.

McNamara, T.P. (1992). Spatial representation. *Geoforum 23*, 139- 150.

Montello D. R. (1991). The measurement of cognitive distance: Methods and construct validity. *Journal of Environmental Psychology 11*, 101-122.

Montello, D.R. (1992). The geometry of environmental knowledge. In *Theories and methods of spatio-temporal reasoning in geographic space* (A. U. Frank, I. Campari, and U. Formentini, eds.), pp. 136-152. Berlin: Springer-Verlag.

Montello, D.R. (in press). A new framework for understanding the acquisition of knowledge about large-scale environments. In *Sptial and Temporal Reasoning in Geographic Information Systems* (R.G. Golledge and M. Egenhofer, eds.).

Piage, J. and Inhelder, B. (1948/1967). *The child's conception of space.* New York: Norton.

Pratt, M.B. (1926). The visual estimation of angles. *Journal of Experimental Psychology 9,* 132-140.

Rybash, J.M. and Hoyer W. J. (1992). Hemispheric specialization for categorical and coordinate spatial representations: A reappraisal. *Memory & Cognition 20,* 271- 276.

Sadalla, E.K., and Montello, D.R. (1989). Remembering changes in direction. *Environment and Behavior 21*, 346-363.

Shepard, R.N. (1964). Attention and the metric structure of the stimulus space. *Journal of Mathematical Psychology 1*, 54-87.

Shepard, R.N. and Hurwitz, S. (1984). Upward direction, mental rotation, and discrimination of left and right turns in maps. *Cognition 18*, 161-193.

Siegel, A.W. and White, S.H. (1975). The development of spatial representations of large-scale environments. In *Advances in child development and behavior* (H. W. Reese, ed.), pp. 9-55. New York: Academic.

Simon, H.A. (1979). *Models of thought.* New Haven, CT: Yale University Press.

Tversky, B. (1992). Distortions in cognitive maps. *Geoforum 23,* 131-138.

Zadeh, L.A. (1975). Fuzzy logic and approximate reasoning. *Synthese 30*, 407-428.

Zimmermann, K. (1993). Enhancing qualitative spatial reasoning -- combining orientation and distance. In *Spatial information theory: A theoretical basis for GIS* (A. U. Frank and I. Campari, eds.), pp. 69-76. Berlin: Springer-Verlag.

Daniel R. Montello
Department of Geography
University of California, Santa Barbara

Andrew U. Frank
Department of Geo-Information
Technical University of Vienna

MAPPING AS A CULTURAL UNIVERSAL

David Stea, James M. Blaut and Jennifer Stephens

Abstract: This chapter discusses the hypothesis that mapping behavior, the making of map-like models, is a cultural universal, an important component of ecological behavior. The essay presents a theoretical framework for the hypothesis and discusses three categories of evidence – developmental, prehistoric, and cross-cultural – which support the hypothesis. Humans must visualize, analyze, describe, and communicate the nature of large environments perceived atomistically, and therefore they create material representations depicting environments as if seen as a whole, from overhead. The result is an organized sign system with certain linguistic properties, including two syntactic transformations (rotation/projection and scale reduction), and the semantic representation of landscape features as iconic or abstract signs. This concept of the map yields useful criteria for the identification and study of maps in culture, history, and behavior. Many examples of prehistoric imagery, mostly parietal, extending to periods earlier than the Neolithic (of both geographical hemispheres), appear map-like, giving evidence of rotated, scale-reduced, and abstracted depiction of the environment and suggesting that mapping may have represented a form of adaptive behavior for modern humans. In a few cases, which are discussed, the representation depicts a real local landscape. Ethnographic studies, while in general not concerned with mapping, have provided evidence that mapping activity occurs in many contemporary cultures. Studies of the behavior of very young children, finally, indicate that mapping abilities appear much earlier than generally supposed, and seem to play an important role in early development.[1]

Introduction

It is suggested in this chapter that mapping of the macroenvironment is a behavioral or cultural universal; specifically, that (1) all humans from early in life – as soon as they begin to acquire some competence in manipulating the material world of objects and surfaces – engage in mapping behavior and make maps, (2) maps have been made since very early times, at least since the Upper Paleolithic, and (3) all cultures, everywhere – even those with limited material culture – make maps in one form or another. We will explain what we mean by "map" and "mapping" in the following section, but it should be made clear at the outset that a map, as discussed here, is something broader than the conventional cartographic product. Something will be called a map if it is a 2- or 3-dimensional model of a macroenvironment (landscape, place), reduced in scale, rotated to an overhead perspective, and semantically marked with meaningful iconic or noniconic

J. Portugali (ed.), The Construction of Cognitive Maps, 345–360.

signs. Thus we focus our attention on material maps, not merely cognitive maps, and on maps as spatio-temporal products, not spatial abstractions.

The emphasis of this chapter is upon mapping theory, and with this the chapter begins and ends: a theoretical overview at the start and a theoretical critique at the conclusion. In between, we will attempt to summarize, briefly, a body of relevant research accomplished, and of evidence uncovered, by the present authors and others.

The multidisciplinary nature of the research underlying the theoretical framework to be presented derives from the fact that highly varied lines of evidence are brought together. These include: ontogenetic or developmental data concerning the origins and early development of maplike behavior in very young children; phylogenetic – that is, historic and prehistoric – data, about the earliest appearance and early evolution of maplike representations in cultural evolution; ethnographic or cross-cultural data about mapmaking and map use in the ethnographic present; some insights obtained from other workers in the cognitive, neurophysiological, and brain sciences which may provide us with evidence on the evolution of mapping ability within the central nervous system; and finally, theoretical judgments about the possible relationship between natural language acquisition and the acquisition of mapping ability.

Theoretical Overview

The theoretical framework to be presented is fundamentally cultural-ecological. What will be called macroenvironmental behavior (roughly synonymous with "geographical behavior") forms the starting point for theorizing about mapping behavior. Macroenvironmental behavior is markedly though not entirely different from other categories of human behavior and calls for mapping as a specific adaptive mechanism and, crucially, a specific kind of sign behavior.

The human individual interacts with an environment that poses three dissimilar situations of action. This is not simply an arbitrary classification of parts of a single continuum; the three situations are relatively distinct and the human response pattern to each of the situations is also distinct. To begin with there is a marked difference between situations which present themselves immediately as essentially social, as involving interaction with other human beings, and those which present themselves as essentially material or, broadly speaking, as environmental. Although it is true that the categories "social" and "non-social" are not fully separable, and that there is an element of value in all interaction with inert things as there is a material element in all interaction with other people, it is also true that humans deploy different behavior, and confront different problems, when they act with other humans rather than with material things, large or small. (See Tolman, 1951; Blaut and Stea, 1969; Ittelson, 1973; Bruner, 1986;

Proshansky and Fabian, 1987.) No less fundamental is the difference between situations of action which are macroenvironmental and situations which are microenvironmental. The difference is partly a matter of scale, of relative size of the person and the situation which he or she deals with (See Mead, 1934, on "distant experience"; Brunswik, 1955, and Barker, 1968, on "molar behavior"). Macroenvironments thus are generally larger than people; microenvironments are generally smaller. The prototype of a macroenvironment is a "place;" the prototype of a microenvironment is an "object." Macroenvironmental behavior and learning can thus be called "place behavior" and "place learning," and microenvironmental behavior and learning can be called "object behavior" and "object learning" (Blaut and Stea, 1969; Proshansky and Fabian, 1987; Blaut, 1991). The specificity of place behavior is very clear. For instance, in place behavior, sensory modalities are deployed in a specific way: e.g., exteroceptors play a larger role, proprioceptors a lesser role, than in object behavior (Stea and Blaut, 1973a). But most importantly, in place behavior, the person ordinarily cannot perceive the entire place (landscape, macroenvironment) as a whole from a single earthbound perspective, and can interact with only a small part of it at any one time and location.

In human macroenvironmental behavior, it is necessary to visualize, analyze, describe and communicate the complex characteristics of the vast environment within which all humans live. Since this environment is too large and complex to be perceived from any single vantage point on the ground, in cognizing it or learning about it or describing it, humans need to map it. Cognitive mapping is thus ubiquitous. But humans must also, in many situations, produce a material map, a map-like model of the landscape (or a gestural model, which may be elaborated into map-like dance, games, drama, etc.) To produce such a map in such a way that the product conveys meaningful ecological information to others, a person performs three very specific mapping operations. The first two operations are syntactic: the environment must be reduced in scale and it must be rotated or projected in perspective to a virtual position more or less overhead. The third operation is the semantic representation of landscape features and forms as signs, either iconic or non- iconic. The map is thus an organized sign system, a language specialized for use in ecological behavior. This conception of (macroenvironmental) map and mapping derives from our understanding of the essential needs of humans to operate ecologically, and yields rather precise criteria for the identification of maps in adult and child behavior, in ethnography, and in cultural history and prehistory. A map is a two dimensional depiction or a three dimensional model of a large environment, in which the environment is seen rotated, reduced, and semantically abstracted. One example with which we are all familiar is a standard road map. Another less obvious example is the product of children's play on the floor or ground – individually or cooperatively – with landscape-like toys. In this essay we will use the term "map" as short-hand for "map-like model," while agreeing that a "real" or cartographic map has a more specific character, and also that the maps which are products of contemporary cartographic science take many other forms. In simple terms, map-like models are the ecological and probably evolutionary source of cartographic maps.

Thinking about maps and mapping in this way, emphasizing the role of mapping in ecological behavior, we find that the following hypothesis is very attractive: In all cultures, people make maps. In all cultures, situations must surely arise in which map-making is called for. There is no a priori reason to believe that people of some cultures today – in the ethnographic present – lack the psychological abilities needed to make, read, and use maps (the idea of the "primitive mind" no longer has many adherents and we reject it; see Blaut, 1993). By extension, we can assume that people have possessed this mental equipment for many millennia. Therefore, in advance of any search for evidence, a good theoretical case can be made for the assertion that mapping is universal in human culture and has been so for a very long time. But of course we must have empirical evidence. To this matter we now turn.

Empirical Findings

Earlier research by the present authors has shown that children as young as five can read vertical aerial photographs at appropriate scale without prior instruction, perceive such a photograph to be a macroenvironment – a landscape – and successfully (veridically) carry out, in play, make-believe (simulated) navigation games on the photo – i.e., use the photo as a map (Blaut, 1969, 1991; Blaut, McCleary, and Blaut, 1970; Blaut and Stea, 1969, 1971, 1974; Stea, 1969, 1976, 1982; Stea and Blaut, 1973a,b). These black-and- white air photos, usually about 1:2000 to 1:5000 in scale, clearly show individual landscape features with which a child is familiar, but show these from a vertical, overhead perspective, a view which most of the children studied had not previously encountered. The fact that children can interpret and use such representations without any prior instruction is strong evidence that this maplike behavior is somehow very fundamental in human development. Additional research was accomplished with children aged three through six observing children's macroenvironmental toy play (Blaut and Stea, 1969, 1974; Stea and Taphanel, 1974; Blaut, 1991). Children's toy play provides a particularly good indicator, because it is a natural activity engaged in, presumably, all cultures and relates clearly to a child's learning about the larger environment. The research was carried out in the U.S., Puerto Rico, and St. Vincent, with a few observations also in Mexico and Venezuela and among New Zealand Maaori. The main findings have been confirmed, directly and indirectly, and extended, by many investigators (see, e.g., Blades and Spencer, 1987; Bluestein and Acredolo, 1979; Hart, 1979; Matthews, 1984a; McGee,

1982; Spencer, et al., 1980; Spencer, Blades, and Morsley, 1989; Walker, 1980. Indirect support is to be found in, e.g., Atkins, 1981; Blades and Spencer, 1986, 1994; Conning and Byrne, 1984; Landau, 1986; Landau and Spelke, 1985; Matthews, 1984b, 1985; S. Muir, 1985; Ottosson, 1988; Presson, 1982; Siegel and Schadler, 1977).[2] Mention should be made of a series of studies by M. Blades and C. Spencer (summarized in Blades and Spencer, forthcoming) which show that children aged four can use maps to find real-world places (e.g., hidden objects, pathways through a maze).

At present, another study with three-, four-, and five-years-olds, undertaken by M. Blades, S. Sowden, C. Spencer, and ourselves, is underway and partial results have been obtained. This research proposes to study children in several cultures, including (initially) England and Mexico, to determine, cross-culturally, the ability of young children to read an arial photograph without training and perform a navigation task on the photo: to show how one would go from one position on the photo to another in a real-world veridical manner, e.g., not crossing over housetops. If the navigation task is performed successfully, we have evidence that a child fully reads the photo as a map of the landscape and that the child can use this map in a make-believe wayfinding exercise. Thus far we have tested 21 children, aged four years to four years and six months, in Sheffield, England. Sixteen of the 21 children succeeded fully at the task; as to the remainder, it is not yet clear whether lack of success was due to lack of ability or unwillingness to respond. These data, with back-up transcripts of the children's comments, provide excellent evidence that four-years-olds are capable of a very simple, but very real, form of mapping behavor.

The findings obtained with children are interpreted by us in terms of a three-element model: (1) early mapping is carried out in toy play and related activities; (2) mapping becomes explicit, important, and visible as children acquire mobility and thus begin seriously to navigate in, learn about, and cope with the macroenvironment; (3) the mapping ability, however, seems to emerge in rudimentary form much earlier, and appears to represent an elaboration of the infant's rapidly-developed ability to structure its external environment. Having seen that such mapping ability is present in very young children, in several cultures, it has seemed reasonable to hypothesize that it may be a cultural universal, and that it may reflect a basic and fundamental human ability, perhaps as universal and ancient as complex linguistic ability and artistic ability.

Work on language acquisition points to the possibility that grammatical transformations appear very early, perhaps contemporaneously with the early mapping ability of mobile children (Bremner and Bryant, 1985; Pinker, 1990). If grammatical structuring in natural language and in mapping (scale, projection or rotation, and perhaps other elements of syntax [see Head, 1984; Schlichtmann, 1985; Lyutyy, 1986; Blaut, 1991]) are related, suggestions (e.g., by Chomsky, 1988) that basal grammatical structure may have roots or precursors that are innate lead to speculation that mapping may also have innate precursors, perhaps with identifiable neural correlates. If so, this may provide a good argument for the universality of mapping, analogous to the arguments about linguistic universals. In any case, the parallel (or, as we suspect, homology) between the construction of linguistic message-sending actions, oral or gestural, and the construction of maps (including map-like models, such as toy assemblages), may indicate common roots. The former is spatio-temporal, not temporal (i.e., not merely sequential production of sounds: see recent work on sign language, on brain processes, etc.). It is well-known that primitive cognitive mapping is observed in infrahuman species: see Thinus-Blanc, 1988). But whether or not there is a predisposition of some sort to engage in mapping behavior, it is clear that the ability emerges very early. Workers in cognitive and developmental psychology have in recent years shown that perceptual and cognitive development proceeds more rapidly than was previously believed, consistent with findings cited above that mapping skills develop early in life, much earlier than some developmental theorists and researchers (e.g., Piaget and Inhelder, 1956; Downs, Liben, and Daggs, 1988) have postulated.

Existing information on mapping across the ethnographic spectrum is more limited. It is well-known from the contemporary ethnographic record that indigenous mapping and maps are found in a number of highly distinctive non-Western cultures, including Inuit, Marshall Islanders, Native Australians, and other groups in both geographical hemispheres (see, e.g., Blakemore, 1981; Blaut, 1991; Delano Smith, 1987; Davenport, 1960 [Marshall Islands]; Gould, 1970; Heth and Cornell, 1985; Mountford and Walsh, 1943; Munn, 1973 [Native Australians]; Spink and Moodie, 1972 [Inuit]; Stea, personal observation; Ritchie, 1977, and personal communication [Maaori]; Bassett, forthcoming [Africa]; Denniston, 1994 [Cuna]). We believe that ethnographers have not, in general, sought this kind of data, although it may be anticipated that a systematic search of existing ethnographic reports will disclose information not previously recognized as relevant to mapping, and future ethnocartographic field research (including a study which we plan to carry out in Mexico) will provide additional information. A particularly important subject for future research is the cross-cultural search for map-like toy-play and toys (and surfaces) representing landscape elements.

In the historical and prehistoric dimension, there is a wide array of extant examples. The best-known is the Catal Hüyük map from Anatolian Turkey, dated about 6,200 BC, thousands of years earlier than the oldest known writing system. This is in essence a plan of the town itself (Mellaart, 1967; Delano Smith, 1987). A three-dimensional model of part of the Mayan city of Tikal, dated to the Early Classic period, also appears to be a partial town plan. Doolittle (1986) describes a Pre-Columbian engraving on a large boulder in the Valley of Sonora, Northwest Mexico, which clearly is a map of the nearby agricultural landscape. A parietal representation from Picacho Point, Arizona, relatively dated to 950-1150 AD, almost certainly depicts the nearby terrain (Wallace and Holmlund, 1986). The foregoing are examples of prehistoric representations which, we argue, confirm themselves to be maps (or map-like models) because of the observed association of map and the local macroenvironment. However, there are many other examples which give strong evidence of being map-like representations, depicting macroenvironments both real and symbolic, e.g., map-like rock paintings from Mesolithic Central India (Neumayer and Tewari in Lorbranchet, 1992); a rock painting from Thailand's Bronze Age (perhaps 2,000 B.C.) which nearby villagers describe as a map of rice fields (Bullen in Lorbranchet, 1992); map-like engravings, probably Neolithic, on boulders in the Black Desert region of Jordan (Betts and Helms, 1984); and a map-like rock-face painting in Satan Canyon, Texas, which is at least 2,000 years old and probably much older (Grieder, 1966). Early contact-era maps have also been found in Mexico: these appear to be cartographically indigenous, and not artifacts of diffusion from Europe. Even 16th century maps annotated by the Spanish are said to "illustrate a wide range of...indigenous cartographic techniques" (Butzer and Williams, 1992, p. 536). Maps dated to just a decade after the conquest of Central Mexico have been analyzed by Yoneda (1981) in terms of specified sets of indigenous cartographic symbols. Delano Smith (1985) has described rock painted maps from Italian proto-history, at Bedolina. Marshack (1977, 1979) has described numerous engravings from the Eastern European Upper Paleolithic as displaying water-related motifs (conceivably river meanders). In addition, an engraved mammoth tusk from central Europe, and presumed by B.Klima to be a map (Keeley, personal communication), may be more than 20,000 years old. A number of other, mainly parietal, examples of what appear to be map-like prehistoric representations have been discussed in the literature, and our search will doubtless turn up further examples of prehistoric maps.

Let us summarize this evidence, then. Our research, and that of many others, has shown that mapping is carried out by very young children in several cultures. We cannot say that this holds true for all cultures, but it seems very probable that this is the case. (The first place to look is at toy- play, constructing fun landscapes on the floor or ground.) As to other cross- cultural information, although the evidence of mapping is scattered and incomplete, it does come from a great diversity of cultures, including at least one land-based, highly mobile society with limited material culture (Native Australians), and another whose economy and culture revolve about ocean navigation (Insular Polynesians). Given such information, and other data not described, it seems very possible that maps are indeed made by adults across the entire range of contemporary cultures. As to the historical or phylogenetic dimension, enough evidence exists from enough places to confirm in principle the hypothesis that humanity was making maps prior to the invention of writing and prior even to the Agricultural Revolution, with some evidence also suggesting origins in the Upper Paleolithic. It is not inconceivable that mapping, art, and grammatically complex language all emerged in the same epoch.

Theoretical Discussion

For many years, and until quite recently, the realms of theory, research, and application in early human development had been dominated by the theoretical legacy of Jean Piaget (see, e.g., Piaget and Inhelder, 1956). Piaget's contention that early development proceeds through a series of immutable stages has since assumed the level of dogma in many circles. Orthodox Piagetian theory was for a time very widely accepted in scholarship, including geography, and, following Piaget, it was argued that young children, below the age of six or seven, have not yet reached the stage at which map-thinking and mapping behavior is possible because (Euclidean) spatial concepts have not yet been acquired. Orthodox Piagetian ideas have been subjected to thoughtful critical review over the past twenty-five years (some of the criticisms are summarized in Berk, 1994). During this period, many workers have demonstrated that complex spatial behavior and mapping appear much earlier than had been deduced from orthodox Piagetian theory. It is evident that the argument of this chapter (like earlier work by the present authors) falls within this latter paradigm: we argue that mapping appears very early in development and we connect this proposition to a broader theory of human mapping. In recent years, however, some Piagetians, notably Liben and Downs (see, e.g. Downs, Liben, and Daggs, 1988) have urged a return to the Piagetian view, and have questioned, mainly on grounds of deduction from Piagetian theory, the findings of such workers as C. Spencer, M. Blades, and ourselves. This controversy deserves a comment here because, if the Piagetian view, unmodified, is correct, mapping is or was not carried out by young children, by so-called primitives – see below – and by our stone-age ancestors.

Actually, two areas of psychology which all too rarely interact are involved in this controversy: development and learning. So antithetical is the psychology of learning to the most orthodox Piagetians that they eschew the term "learning," employing such words as "experience" instead. To evaluate the current state of affairs, and the findings on the development of human spatial cognition presented in this chapter in the context of that state of affairs, we must look briefly at development and learning, two major contributors to human maturation, together.

From the structuralist perspective of Piaget, the stages of human development are pre-programmed; that is, they are part, in effect, of the genetic endowment of the neonate, and function as something akin to conceptual "gates" which are "unlocked" at certain critical points in the maturation of the child. The driving forces are thus largely (or basically) internal. The driving forces in the forms of learning studied by psychologists involved in learning research (who define learning as "changes in behavior due to practice") may be largely external, or an interaction of external and internal factors.

From the orthodox Piagetian stance, then, children cannot exhibit the behavior characteristic of more advanced stages until they have "grasped" the "concepts" essential to those stages. The relation among stages is mathematically transitive: if stage "A" precedes "B" and "B" precedes "C," then "A" must precede "C." Further, the separation between stages is never zero; they can never be made to coincide. The "time-distance" between one stage and another is always greater than zero and at least roughly specifiable (the range of end-points define allowable maxima and minima): the set consisting of developmental stages and allowable relations among them thus approximates a weak metric space.

In this reductionist view, concepts must precede behavior; and the inability of the child at stage X-1 of development to "grasp" concepts "achieved" only in stage X inhibits the effectiveness of experience – or, in other words, prevents further learning. In this sense, the Piagetian developmental view is deductive. By contrast, an essentially inductive learning perspective sees experience-based changes in behavior as prerequisite to development. Put otherwise, one view sees lack of development as a barrier to learning, while the other sees learning as a facilitator of development (e.g. Downs and Stea, 1977). In the mapping realm, Piagetians would view the achievement of mapping concepts as prerequisite to mapping behavior, while learning theorists would see mapping behavior as prerequisite to the achievement of mapping concepts.

A major Piagetian issue revolves about the relation between stages and ages. Piaget himself, and nearly all Piagetians, have argued that the two cannot be rigidly equated, that the transition between two stages can occur over a range of ages. Unfortunately, this point has largely been lost in the application of Piagetian ideas to child-rearing and education. Piagetian child-rearing manuals, read by parents much more familiar with their children's ages than their developmental stages, may make Piagetian theory into a self-fulfilling prophecy when children reared by Piagetian manuals are tested to see whether they conform to Piagetian stages of development. That is, if independently-reared youngsters are confined to the same set of experiences at the same points in time, it is impossible to distinguish developmental similarities from absence of learning opportunities. There is further confusion between developmental criteria and adult learning. Thus, concepts that adults have learned, such as Euclidean geometry and Newtonian physics, are taken as developmental standards for children – in speaking of the "achievement" of Euclidean spatial concepts, for example. Similarly, primary schooling is indexed by ages and grades, not stages, and educators, understandably immersed in their system of ordering, have translated Piagetian stages into their own indices: thus it is common to hear or to read that children cannot understand map concepts before age 11 and that map education prior to late elementary school (or even later) is therefore contra-indicated (see, e.g., Towler and Nelson, 1968; Satterley, 1973; Naish, 1982).

It was this latter contention that the two senior authors set out to test in the late 1960s, and further refuted in the decades that followed – as summarized in this chapter. What has been described above may be termed the "strong" Piagetian view of stately progression through immutable stages. An interactionist alternative, a "weaker" position much more in accord with our own findings and those of many others (e.g. Blades and Spencer, 1986, forthcoming; Spencer, Blades, and Morsley, 1989; Spencer, Harrison, and Darvizeh, 1980), preserves stages – though not conceived as qualitatively distinct or separated by leaps of maturation – but sees these interacting with learning in such a way that learning (experience or teaching) may, while not reordering development, accelerate it (e.g., Downs and Stea, 1977, pp. 203-205). So-called gifted children are often products of "enriched" environments, for example. In terms of educational applications, the "strong" and "weak" developmental perspectives (vis-a-vis Piaget) have quite opposite implications: from the "strong" view, enhanced experience may be seen as relatively pointless, while from the "weak" perspective wherein development and learning are interactive, enhanced experience is pivotal.

As late as 1971, Piaget drew upon the earlier work of the anthropologist Lévy-Bruhl to suggest that cognitive development among "primitive" cultures may stop at the concrete operational stage; that they may never achieve the more advanced stage of cognitive development exhibited by Europeans:

> [What] is the adult operatory level in tribal organization in respect to technical intelligence, the solution of elementary logico-mathematical problems, and so forth?...[It] is quite possible, and this is the impression we have from known ethnographical work, that in many societies adult thought does not go beyond the level of concrete operations and therefore does not reach that of propositional operations which develop between the ages of twelve and fifteen in our milieus (Piaget, 1971, p. 61)

In fact, Piaget then called for the cross-cultural evaluation of his findings. The response has been thin and, unfortunately, the uncritical view of "primitive mind," though roundly trounced in recent years, still has some adherents.

There is a logical problem, here, for Piaget. Let us suppose (for a moment only) that there really is "primitive mind": logically, it must then be biologically or culturally based. If, on the one hand, the Piagetian progression through developmental stages is in fact part of the organism's structural endowment and a psychological universal, this ontogenetic progression ought to have been present, and manifested, since the most recent changes in human brain structure. If, on the other hand, the terminal stages vary with biology – phenotype and gender (the latter suggested by M. McGhee, 1979) – then Piaget's lawful progression loses its universality. Further, if these progressions and their terminal stages vary with culture and the levels of technological sophistication that presumably divide so-called "primitive" from "advanced" societies, then they are not immutable: i.e. they are subject to alteration according to circumstances. Accepting "primitive mind," then, implies that Piagetian stages are either non-universal or non-immutable, or both. Our own theoretical framework, drawn from data that strongly refute "primitive mind," directs us, while accepting both the basic concept of stages and the possible existence of a structural engram for mapping, to reject both uniform time periods between stages and stage immutability, and to search for ways in which structural development and experience/learning interact.

While there are extant theories concerning the early development of spatial cognition and of mapping behavior in young children, our theoretical critique reveals a paucity of theory concerning mapping in the ethnographic present and an almost complete absence of theoretical perspectives on evidences of prehistoric mapping behavior. One evident problem both in constructing theory and in seeking evidence of mapping behavior diachronically and cross-culturally is that maps done by non-Westerners, or prehistoric maps, may not bear a close resemblance to the products of cartography that Westerners have come to recognize and to accept. For they have often been constructed with specific purposes in mind which may differ greatly from those of the makers of, for example, an automobile club road map. Basically, all maps, however executed (graphically, in performance, etc.), are meant to communicate from one person or group to herself or themselves (information storage) or from one person or group to another (information transmission). To communicate, maps maximize relevant information, or presentation of spatial information in the most relevant form. To do this, mapmakers "distort." Distortion, as such, is not confined just to children and primitives, nor to the cognitive maps studied by environmental psychologists, but is practiced by contemporary Western mapmakers who systematically distort spatial information for propaganda, advertising, or other purposes (Monmonier, 1991). Apparent distortions from accepted veridicality may also be the result of attempts to increase the map's informational value: thus, the medieval map depicting Marco Polo's travels which graced the cover of an issue of *Science* two decades ago showed what, had it been drawn by a child, some developmental psychologists might have interpreted as "failure to coordinate perspectives": that is, horses and men were shown in elevation, rather than plan. Intuitively, however, this "uncoordinated perspective" is in fact more informative. Is that what children have been trying to tell us all along? In any event, these and other "distortions" may make it more difficult for those who hold some conventional views of what a map is to recognize prehistoric maps; or maps produced by some non-Western cultures, or by children.

Conclusions

Is a "map" more or less than a "model?" Modelling the environment is an important part of culture and one outcome of such environmental modelling is, generically speaking, the map. We argue that an important source of scientific understanding can be derived from an ecological view of mapping as one formal outcome of environmental modelling, and that such understanding has significance for a number of disciplines and bodies of theory concerned with culture, cultural ecology, and human development and evolution. Indeed, map- like modelling in toy play may be one of the first – perhaps even the first – kind of "model thinking" by the human mind.

Rather than starting with maps as artifacts, as independent variables with related characteristics of human communities as dependent variables, we propose deriving maps and mapping from the universal needs of human communities. These needs, which appear in early childhood and increase with the process of maturation, are extremely basic: where food and water, safe and dangerous places, holy places, etc., are to be found. More than just a part of religious ritual – although that as well – early maps are likely to have been the very substance of survival.

The evidence presented here suggests that mapping is characteristic of a wide range of contemporary cultures – we suspect all cultures – and extends back to a time before the Neolithic – we suspect long before the Neolithic, well into the Upper Paleolithic. Mapping, then, is not just an historical phenomenon, nor simply the product of literate societies. This chapter suggests that, while modes of mapping are themselves learned, mapping has occurred across time and cultures, and that the ability to map, to express in material form the cognition of large-scale environments from wholly or partially aerial perspectives, is indeed a cultural universal.

Acknowledgment: Research supported in part by the U.S. National Science Foundation (Grant SBR-9423865) and in part by the Vice-Chancellor for Research, University of Illinois at Chicago.

Notes

1. Some material in this chapter was contained in a symposium presentation at the 1994 Annual Meeting, Association of American Geographers, San Francisco. The authors acknowledge helpful comments by their symposium colleagues, and those of participants in the Faculty Seminar in Environmental Studies at Mount Holyoke College in November, 1994. We also appreciate the suggestions (some of which we had the wisdom to accept) made to us by Bill Doolittle, Paul Hockings, Larry Keeley, Malcolm Lewis, Alexander Marshack, Jack Prost, Catherine Delano Smith, and Chris Spencer.
2. Some of these extensions have transcended the realm of children, to include the interactive effects of gender and social class on demonstrated environmental knowledge (Stea and Taphanel, 1974), and the application of environmental modeling to the process of public participation in various aspects of physical planning (e.g. Stea, 1984, 1987; Wisner, Stea, and Kruks, 1991).

References

Atkins, C. (1981). Introducing basic map and globe concepts to young children. *Journal of Geography* 80, 228-233.

Barker, R. (1968). *Ecological psychology*. Stanford: Stanford University Press.

Bassett, T. (forthcoming, 1995). African maps and mapmaking. In The *Encyclopedia of the History of Science, Technology, and Medicine in Non-Western Cultures*. New York: Garland.

Berk, L. E. (1994). Why children talk to themselves. *Scientific American* 271(5), 78-83.

Betts, A., and Helms, S (1986). Rock art in eastern Jordan: "Kite" carvings? *Paleorient* 12(1), 67-72.

Blades, M. and Spencer, C. (1987). The use of maps by 4-to 6-year-old children in a large-scale maze. *British Journal of Developmental Psychology* 5, 19-24.

Blades, M. and Spencer, C. (1986). Map use by young children. *Geography* 71, 47-52.

Blades, M., and Spencer, C. (1994). The development of children's ability to use spatial representations. in *Advances in Child Development and Behavior*.

Blades, M. and Spencer, C. (forthcoming). Young children's use of spatial relationships in tasks with maps and models. University of Sheffield, *Cartographica.*

Blakemore, M. (1981). From way-finding to map-making: the spatial information fields of aboriginal peoples. *Progress in Human Geography* 5, 1-24.

Blaut, J. (1969). Studies in developmental geography. *Place Perception Research Report No. 1.* Worcester, Massachusetts: Clark University.

Blaut, J. (1987). Notes toward a theory of mapping behavior. *Children's Environments Quarterly* 4(4), 27-34.

Blaut, J. (1991). Natural mapping. *Transactions, Institute of British Geographers* 16, n.s., 55-74.

Blaut, J. (1993). *The Colonizer's Model of the World: Geographical Diffusionism and Eurocentric History*. New York and London: Guilford Press.

Blaut, J., McCleary, G., and Blaut, A. (1970). Environmental mapping in young children. *Environment and Behavior* 2(3) 335-49.

Blaut, J. and Stea, D. (1969). Place learning. *Place Perception Research Report No. 4*. Worcester, Massachusetts: Clark University.

Blaut, J. and Stea, D. (1971). Studies of geographic learning. *Annals of the Association of American Geographers* 61, 387-393.

Blaut, J. and Stea, D. (1974). Mapping at the age of three. *Journal of Geography* 73, 5-9.

Bluestein, N. and Acredolo, L. (1979). Developmental changes in map-reading skills. *Child Development* 50(3), 691-97.

Bremner, J. and Bryant, P. (1985). Active movement and development of spatial abilities in infancy. In *Children's searching: the development of search skills and spatial representation* (H. Wellman, ed.), pp. 53-72.

Bruner, J. (1986). *Actual minds, possible worlds*. Cambridge, Massachusetts: Harvard University Press.

Brunswik, E. (1955). The conceptual framework of psychology. In *International Encyclopedia of Unified Science,* Part 2 (O. Neurath, R. Carnap, and C. Morris, eds.). Chicago: University of Chicago Press.

Butzer, K. and Williams, B. (1992). Addendum: Three indigenous maps from New Spain dated ca. 1580. *Annals of the Association of American Geographers* 82, 536-542.

Chomsky, N. (1988). *Language and Problems of Knowledge*. Cambridge, Massachusetts: MIT Press.

Conning, A. and Byrne, R. (1984). Pointing to pre-school children's spatial competence: a study in natural setting. *Journal of Environmental Psychology* 4, 165-175.

Davenport, W. (1960). Marshall Islands navigational charts. *Imago Mundi* 15, 19-26.

Delano Smith, C. (1985). The origins of cartography, an archeological problem: Maps in prehistoric rock art. *Papers in Italian Archeology* 4, pt. 2, C. Malone and S. Stoddart, eds. B.A.R. International Series 244.

Delano Smith, C. (1987). Cartography in the prehistoric period in the Old World: Europe, the Middle East, and North Africa. In *The History of Cartography, Vol. 1: Cartography in Prehistoric, Ancient, and Medieval Europe and the Mediterranean*, (J. Harley and D. Woodward, eds), pp. 54-99. Chicago: University of Chicago Press.

Denniston, D. (1994). Defending the land with maps. *World Watch* 7(1) 27-31.

Doolittle, W. (1988). Pre-Hispanic occupance in the Valley of Sonora, Mexico: Archeological confirmation of early Spanish reports. Anthropological Papers of the University of Arizona, No. 48. Tucson: University of Arizona Press.

Downs, R. and Stea, D., eds. (1973). *Image and Environment: Cognitive Mapping and Spatial Behavior.* Chicago: Aldine.

Downs, R. and Stea, D. (1977). *Maps in Minds.* New York: Harper and Row.

Gould, R. (1970). Spears and spear-throwers of the Western Desert Aborigines of Australia. American Museum Novitates 2403, 1-42.

Grieder, T. (1966). Periods in Pecos Style pictographs. *American Antiquity* (5), 710-720.

Hart, R. (1979). *Children's Experience of Place.* New York: Irvington.

Head, G. (1984). The map as natural language: a paradigm for understanding. In *New Insights in Cartographic Communication. Cartographica Monograph* 31 (C. Board, ed.), pp. 1-32.

Heth, D., and Cornell, E. (1985). A comparative description of representation and processing during search. In *Children's searching: the development of search skills and spatial representation* (H. Wellman, ed.). Hillsdale, New Jersey: Erlbaum.

Ittelson, W. (1973). Environment perception and contemporary perceptual theory. In *Environment and Cognition* (W. Ittelson, ed.), pp. 1-19. New York: Seminar Press.

Landau, B.(1986). Early map use as an unlearned ability. *Cognition* 22, 201-223.

Landau, B. and Spelke, E. (1985). Spatial knowledge and its manifestations. In *Children's searching: the development of search skills and spatial representation* (H. Wellman, ed.). Hillsdale: Erlbaum.

Liben, L. S., A. H. Patterson, and N. Newcombe, eds. (1981). *Spatial Representation and Behavior Across the Life Span: Theory and Applications.* New York: Academic Press.

Lloyd, R. (1989). Cognitive maps: encoding and decoding information. *Annals of the Association of American Geographers* 79, 101-124.

Lorbranchet, M., ed. (1992). *Rock Art in the Old World.* New Delhi: Indira Gandhi National Center for the Arts.

Lyutyy, A. (1986). On the essence of the language of maps. *Mapping Sciences and Remote Sensing* 23, 127-139.

Marshack, A. (1977). The meander as a system: the analysis and recognition of iconographic units in Upper Paleolithic compositions In *Form In Indigenous Art: Schematisation In the Art of Aboriginal Australia and Prehistoric Europe* (P. Ucko, ed.). London: Duckworth.

Marshack, A. (1979). Upper Paleolithic symbol systems of the Russian plain: cognitive and comparative analysis. *Current Anthropology* 20, 271-295, 303-311.

Matthews, M. (1984a). Cognitive maps: a comparison of graphic and iconic techniques. *Area* 16, 33-40.

Matthews, M. (1984b). Environmental cognition of young children: images of journey to school and home area. *Transactions of the Institute of British Geographers* 9, 89-105.

Matthews, M. (1985). Young children's representation of the environment: A comparison of techniques. *Journal of Environmental Psychology* 5, 261-78.

McGee, C. (1982). Children's perception of symbols on maps and aerial photographs. *Geographical Education* 4, 51-59.

McGee, Mark G. (1979). *Human Spatial Abilities: Sources of Sex Differences.* New York: Praeger.
McGhee, P. E. (1979). *Humor: Its Origins and Development.* New York: W. H. Freeman.
McGhee, P. E., ed., (1989). *Humor and Children's Development.* New York: The Haworth Press.
Mead, G.H. (1934). *Mind, Self, and Society.* Chicago: University of Chicago Press.
Mellaart, J. (1967). *Çatal Hüyük: A Neolithic Town in Anatolia.* New York: Thames and Hudson.
Monmonier, M. (1991). *How to Lie With Maps.* Chicago: University of Chicago Press.
Mountford, C. and Walsh, G. (1943). A stone tjuringa of unusual form from the Aranda tribe of Central Australia. *Mankind* 3, 113-115.
Muir, M. and Blaut, J. (1969). The use of aerial photographs in teaching mapping to children in the first grade: an experimental study. *The Minnesota Geographer* 22(4), 1-19.
Muir, S. (1985). Understanding and improving students' map reading skills. *Elementary School Journal* 86, 207-216.
Munn, N. (1973). *Walbiri Iconography: Graphic Representation and Cultural Symbolism In a Central Australian Society.* Ithaca: Cornell University Press.
Naish, M. (1982). Mental development and the learning of geography. In *New UNESCO Sourcebook for Geography Teaching* (N. Graves, ed.). Paris: UNESCO.
Neumayer, E. (1992). Rock art of India. In *Rock Art in the Old World.* (M. Lorbranchet, ed.). New Delhi: Indira Gandhi National Center for the Arts.
Ottosson, T. (1988). What does it take to read a map? *Cartographica* 25, 28-35.
Pericot-Garci, L., Galloway, J., and Lommel, A. (1967). *Prehistoric and Primitive Art.* New York: Harry N. Abrams.
Piaget, J. (1971). *Psychology and epistemology.* New York: Grossman Publishers.
Piaget, J., and Inhelder, B. (1956). *The Child's Conception of Space.* New York: Humanities Press.
Pinker, S. (1990). Language acquisition. In *Foundations of Cognitive Science* (M. Posner, ed.). Cambridge, MA: MIT Press.
Presson, C. (1982). The development of map-reading skills. *Child Development* 53, 196-9.
Proshansky, H. and Fabian, A. (1987). The development of place identity in children. In *Spaces for children: the built environment and child development* (C. Weinstein and T. David, eds.), 21-40.
Ritchie, J. E. (1977). Cognition of place: The island mind. *Ethos* 5(2), 187-194.
Satterley, D. (1973). Skills and concepts involved in map drawing and map interpretation. In *Perspectives in Geographical Education* (J. Bale, N. Graves, and R. Walford, eds). Edinburgh: Oliver and Boyd.
Schlichtmann, H. (1985). Characteristic traits of the semiotic system "map symbolism". *Cartographic Journal* 22, 23-30.
Siegel, A. and Schadler, M. (1977). Young children's cognitive maps of their classrooms. *Child Development* 48, 388-94.
Spelke, E. (1988). Origins of visual knowledge. In *Visual Cognition and Action* (D. Osherson, S. Kosslyn, and J. Hollerbach, eds.). Cambridge, MA: MIT Press.
Spencer, C., Harrison, N. and Darvizeh, Z. (1980). The development of iconic mapping ability in young children. *International Journal of Early Childhood* 12, 57-64.
Spencer, C., Blades, M. and Morsley, K. (1989). *The Child in the Physical Environment: The Development of Spatial Knowledge and Cognition.* New York: Wiley.
Spink, J. and Moodie, D. (1972). Eskimo maps from the Canadian eastern arctic. *Cartographica Monograph 5.*
Stea, D. (1976). *Environmental Mapping.* Milton Keynes: Open University.
Stea, D. (1982). Cross-cultural environmental modelling. In *Mind, Child, Architecture* (A. Lutkus and J. Baird, eds.). Hanover, New Hampshire: University Press of New England.
Stea, D. (1984). Participatory planning and design for the Third World. In *Architectural Values and World Issues* (W. Gilland and D. Woodcock, eds.). Silver Springs, MD: International Dynamics.

Stea, D. (1987). Participatory planning and design in intercultural and international practice. In *Ethnoscapes* (D. Canter, M. Krampen, and D. Stea, eds.). Aldershot (U.K.): Avebury.

Stea, D., ed. (1969). Working papers in place. perception. *Place Perception Research Report No. 2.* Worcester, Massachusetts: Clark University.

Stea, D. and Blaut, J. (1973a). Notes toward a developmental theory of spatial learning. In *Image and Environment: Cognitive Mapping and Spatial Behavior* (R. Downs and D. Stea, eds.). Chicago: Aldine.

Stea, D. and Blaut, J. (1973b). Some preliminary observations on spatial learning in Puerto Rican school children. In *Image and Environment: Cognitive Mapping and Spatial Behavior* (R. Downs and D. Stea, eds.). Chicago: Aldine.

Stea, D. and Taphanel, S. (1974). Theory and experiment on the relation between environmental modelling ("toy play") and environmental cognition. In *Psychology and the Built Environment* (D. Canter and T. Lee, eds.). New York: Wiley.

Stea, D., and Turan, M. (1993). *Placemaking.* Aldershot (U.K.): Avebury.

Thinus-Blanc, C. (1988). Animal spatial cognition. In *Thought Without Language* (L. Weiskrantz, ed.). Oxford: Clarendon Press.

Towler, J. and Nelson, L. (1968). The elementary school child's concept of scale. *Journal of Geography* 67, 24-28.

Walker, R. (1980). Map using abilities of five to nine year old children. *Geographical Education* 3, 545-54.

Wallace, H., and Holmlund, J. (1986) *Petroglyphs of the Picacho Mountains: south central Arizona.* Institute for American Research, Anthropological Paper No. 6.

Wallace, R. (1989). Cognitive mapping and the origin of language and mind. *Current Anthropology* 30, 518-526.

Wellman, H., ed (1985). *Children's Searching: The Development of Search Skills and Spatial Representation.* Hillsdale, NJ: Erlbaum.

Wisner, B., Stea, D., and Kruks, S. (1991). Participatory and Action Research Methods. In *Advances in Environment, Behavior, and Design*, Vol. 3 (E. H. Zube and G. T. Moore, eds.). New York: Plenum.

Wolfenstein, M. (1978). *Children's Humor: A Psychological Analysis.* Bloomington: Indiana University Press.

Yoneda, K. (1981). *Los mapas de Cuautinchan y la historia cartografica pre-hispanica.* Mexico, D.F.: Archivo General de la Nación.

David Stea
Universidad Internacional de México
Centro Internacional para la Cultura y el Ambiente
and Mount Holyoke College

James M. Blaut
University of Illinois at Chicago

Jennifer Stephens
University of Illinois at Chicago

SUBJECT INDEX

The GeoJournal Library

19. C.J. Campbell: *The Golden Century of Oil 1950–2050.* The Depletion of a Resource. 1991 ISBN 0-7923-1442-5
20. F.M. Dieleman and S. Musterd (eds.): *The Randstad: A Research and Policy Laboratory.* 1992 ISBN 0-7923-1649-5
21. V.I. Ilyichev and V.V. Anikiev (eds.): *Oceanic and Anthropogenic Controls of Life in the Pacific Ocean.* 1992 ISBN 0-7923-1854-4
22. A.K. Dutt and F.J. Costa (eds.): *Perspectives on Planning and Urban Development in Belgium.* 1992 ISBN 0-7923-1885-4
23. J. Portugali: *Implicate Relations.* Society and Space in the Israeli-Palestinian Conflict. 1993 ISBN 0-7923-1886-2
24. M.J.C. de Lepper, H.J. Scholten and R.M. Stern (eds.): *The Added Value of Geographical Information Systems in Public and Environmental Health.* 1995 ISBN 0-7923-1887-0
25. J.P. Dorian, P.A. Minakir and V.T. Borisovich (eds.): *CIS Energy and Minerals Development.* Prospects, Problems and Opportunities for International Cooperation. 1993 ISBN 0-7923-2323-8
26. P.P. Wong (ed.): *Tourism vs Environment: The Case for Coastal Areas.* 1993 ISBN 0-7923-2404-8
27. G.B. Benko and U. Strohmayer (eds.): *Geography, History and Social Sciences.* 1995 ISBN 0-7923-2543-5
28. A. Faludi and A. der Valk: *Rule and Order. Dutch Planning Doctrine in the Twentieth Century.* 1994 ISBN 0-7923-2619-9
29. B.C. Hewitson and R.G. Crane (eds.): *Neural Nets: Applications in Geography.* 1994 ISBN 0-7923-2746-2
30. A.K. Dutt, F.J. Costa, S. Aggarwal and A.G. Noble (eds.): *The Asian City: Processes of Development, Characteristics and Planning.* 1994 ISBN 0-7923-3135-4
31. R. Laulajainen and H.A. Stafford: *Corporate Geography.* Business Location Principles and Cases. 1995 ISBN 0-7923-3326-8
32. J. Portugali (ed.): *The Construction of Cognitive Maps.* 1996 ISBN 0-7923-3949-5

KLUWER ACADEMIC PUBLISHERS – DORDRECHT / BOSTON / LONDON

9 780792 339496

GPSR Compliance

The European Union's (EU) General Product Safety Regulation (GPSR) is a set of rules that requires consumer products to be safe and our obligations to ensure this.

If you have any concerns about our products, you can contact us on ProductSafety@springernature.com

In case Publisher is established outside the EU, the EU authorized representative is:

Springer Nature Customer Service Center GmbH
Europaplatz 3
69115 Heidelberg, Germany

Batch number: 10584510

Printed by Printforce, the Netherlands